Human Social Evolution

Human Social Evolution

THE FOUNDATIONAL WORKS OF RICHARD D. ALEXANDER

Edited by Kyle Summers
and
Bernard Crespi

OXFORD
UNIVERSITY PRESS

Oxford University Press is a department of the University of Oxford.
It furthers the University's objective of excellence in research, scholarship,
and education by publishing worldwide.

Oxford   New York
Auckland   Cape Town   Dar es Salaam   Hong Kong   Karachi
Kuala Lumpur   Madrid   Melbourne   Mexico City   Nairobi
New Delhi   Shanghai   Taipei   Toronto

With offices in
Argentina   Austria   Brazil   Chile   Czech Republic   France   Greece
Guatemala   Hungary   Italy   Japan   Poland   Portugal   Singapore
South Korea   Switzerland   Thailand   Turkey   Ukraine   Vietnam

Oxford is a registered trademark of Oxford University Press
in the UK and certain other countries.

Published in the United States of America by
Oxford University Press
198 Madison Avenue, New York, NY 10016

Library of Congress Cataloging-in-Publication Data

Human social evolution : the foundational works of Richard D. Alexander / edited by Kyle
Summers and Bernard Crespi.
pages cm
Includes bibliographical references.
ISBN 978-0-19-979175-0 (alk. paper)
1. Alexander, Richard D.—Influence.   2. Alexander, Richard D. —Criticism and
interpretation.   3. Human evolution.   4. Social evolution.   5. Natural selection.
I. Summers, Kyle.
GN281.H8495 2013
599.93'8—dc23
2013017153

9 8 7 6 5 4 3 2 1
Printed in the United States of America
on acid-free paper

CONTENTS

CONTRIBUTORS

Laura Betzig
The Adaptationist Program
Ann Arbor, MI, 48109 USA
lbetzig@gmail.com

Stan Braude
Department of Biology
Washington University in St. Louis
Campus Box 1137, One Brookings Drive
St. Louis, MO 63130-4899 USA
braude@wustl.edu

Bernard J. Crespi
Department of Biological Sciences
Simon Fraser University
8888 University Drive
Burnaby, B.C. Canada V5A 1S6
crespi@sfu.ca

Robin I.M. Dunbar
Department of Psychology
Magdalen College
University of Oxford
Oxford, OX1 2JD, United Kingdom
robin.dunbar@psy.ox.ac.uk

Mark Flinn
Department of Anthropology
University of Missouri
Columbia, Missouri 65211-1440 USA
FlinnM@missouri.edu

Steven A. Frank
Department of Ecology and
Evolutionary Biology
University of California
Irvine, CA 92697-2525 USA
safrank@uci.edu

William Irons
Department of Anthropology
Northwestern University
1810 Hinman Avenue
Evanston, IL 60208-1310 USA
w-irons@northwestern.edu

David C. Lahti
Department of Biology
Queens College, City University of
New York
65-30 Kissena Boulevard
Flushing, NY 11367 USA
david.lahti@qc.cuny.edu

Bobbi Low
School of Natural Resources and
Environment
University of Michigan
Ann Arbor, MI 48109 USA
bobbilow@umich.edu

David Queller
Department of Biology
Washington University in St. Louis
Campus Box 1137, One Brookings Drive
St. Louis, MO 63130-4899 USA
queller@wustl.edu

Paul W. Sherman
Department of Neurobiology and
Behavior
W307 Seeley G. Mudd Hall
Cornell University
Ithaca, NY, 14853 USA
pws6@cornell.edu

Karl Sigmund
Faculty for Mathematics
University of Vienna
Nordbergstrasse 15
A-1090, Vienna, Austria
Karl.Sigmund@univie.ac.at

Beverly I. Strassmann
Department of Anthropology
University of Michigan
419 West Hall,
1085 S. University Ave.
Ann Arbor, MI 48109-1107 USA
bis@umich.edu

Kyle Summers
Department of Biology
East Carolina University
Greenville, NC 27858 USA
summersk@ecu.edu

Paul Turke
Department of Pediatrics and
Communicable Diseases
University of Michigan
Ann Arbor, MI 48109 USA
paulturke@gmail.com

Mary Jane West-Eberhard
Smithsonian Tropical Research
Institute
Apartado Postal 0843-03092
Panamá, República de Panamá
mjwe@sent.com

PREFACE

This book is a tribute to one of the great minds in evolutionary biology, Richard D. Alexander. His help and encouragement during our graduate careers at the University of Michigan was invaluable to both of us, and we miss the penetrating discussions of complex topics in human and animal behavior and evolution that he loved to engage in. His contributions to science, and the humanities, should become standard reading for generations to come, and we hope this volume will help to make that goal a reality. Dr. Alexander provided unstinting help with various facets of the process of developing and producing this volume, and we thank him for his efforts.

We also would like to take this opportunity to thank all of the people who contributed to this volume—their contributions have served to highlight Dr. Alexander's work, and illuminate the many contributions he has made to our understanding of human social evolution. These contributors also illustrate how Dr. Alexander's legacy is being passed on through the scientists that he trained and influenced during the course of his career. He taught so many of us how to think about evolution, and humanity, and how to turn these thoughts into productive science.

We also thank our families, who have tolerated our absent-mindedness, and absences during the long hours and late nights required to complete this volume.

K.S. and B.C.

Human Social Evolution

Introduction

Kyle Summers and Bernard J. Crespi

After decades rife with science strife
It seems appropriate to join
The slice of life that plies the knife
Along the flip side of the coin

R. D. ALEXANDER, 2011

Richard D. Alexander is a farmer and rancher, horse trainer, poet, story teller, folk singer, song writer, musician, author, and a philosopher, as well as a husband (to Lorrie Alexander), a parent, and a grandparent. He and his wife have run a large farm in Manchester, Michigan for more than thirty-five years. Alexander grew up in rural Illinois, the child of two school teachers turned livestock farmers. His childhood passed without many of the conveniences of modern life, such as electricity and indoor plumbing. His mother cooked on a wood stove, and light after dark came from kerosene lamps. His family raised cows, pigs, and chickens on feed they grew themselves, selling meat, eggs, and cream. Alexander grew up doing "chores" that most people would consider hard labor, such as working his own threshing team of draft horses on different farms across the county. He went to school in a one room country schoolhouse with a single teacher for all grades. In 1946, Alexander attended Blackburn College, where he was consigned to a single dormitory with a mix of new high school graduates and veterans of World War II who were returning to school. In high school, Alexander had no thought of attending college, and when he first went to college he had no thought of a career in academia. From these rural origins sprang an intellect that has transformed our understanding of human social behavior and evolution and, we propose, ourselves.

Alexander's intellectual curiosity about human evolution probably sprang from his early experiences in church, where he found himself fascinated by the questions raised concerning human nature, yet dissatisfied by the answers proferred. Early in his college career, during his time at Blackburn College, he realized that in academia he could pursue any questions he thought were of interest. Although he pursued coursework in philosophy at Illinois Normal University after transferring from Blackburn, Alexander was struck by the lack of a model of human nature, and ultimately turned to biology to pursue his interests. Before entering graduate school, he did not have the opportunity to take a course in evolutionary biology, and even after deciding to pursue graduate study in biology at the University of Ohio, where he studied entomology (see ch. 1), there were few courses in evolutionary biology available. Nevertheless, over time his interests in evolution crystallized and motivated Alexander to pursue three of the most difficult and important questions in biology: how the diversity of life came into being through the process of speciation, how natural selection has shaped the complexity of life (including our own minds), and the meaning of the extreme social attributes of humans (e.g., art, music, dance, religion) from an evolutionary perspective. Alexander was able to pursue groundbreaking research on the first two questions as a graduate student working on singing insects, but it was only after he had become a faculty member at the University of Michigan that he began to focus on sociality, and particularly on human social behavior.

Over a long career at the University of Michigan, Alexander became, we would argue, the world's leading thinker on human social behavior from an evolutionary perspective. His publications on this topic trace back to a remarkable review in 1968 (with Donald Tinkle) of two books (*On Aggression* by Konrad Lorenz and *The Territorial Imperative* by Robert Ardrey), where he laid out a hypothesis concerning the influence of intergroup competition on human social behavior that continues to be influential today. It was obvious from this review that Alexander had been thinking about these issues for a long time, and from these beginnings sprang a long series of publications on human social evolution that have continued throughout his career at the University of Michigan, and beyond.

A key watershed occurred with the publication of Alexander's first book on human social evolution: *Darwinism and Human Affairs*, in 1979 (The University of Washington Press). In developing his ideas for this book, Alexander was greatly influenced by the profound insights of three contemporaries: George Williams, who taught biologists how to think about selection at the level of individual rather than species benefit, William Hamilton, whose inclusive fitness theory, and evolutionary stable strategy reasoning, taught us how to apply selection thinking at the levels of genes, families, and evolving traits, and Robert Trivers, who first explained how kin cooperation and kin conflict are necessarily enmixed, and how reciprocity can evolve, especially in species with powerful cognitive abilities, such as humans. This was a time of Darwinian revolution for the study of behavior, when the conceptual tools for understanding behavioral phenotypes, especially conflicts and confluences of interest, first came together. In *Darwinism and Human Affairs*, Alexander first applied many of the key theoretical approaches that these three researchers had developed—and many of his own—to human social behavior in a comprehensive and systematic way, providing the most complete and rigorous overview of the entire scope of human social behavior from an evolutionary perspective achieved to that point.

After the publication of *Darwinism and Human Affairs*, a veritable flood of research on human social behavior was initiated by evolution-minded scientists, not only in biology but in many other fields. The major themes that Alexander laid out in his first book continue to be the focus of intense interest and debate in the study of human social behavior today, including kinship and nepotism, direct and indirect reciprocity, ontogeny, life history and senescence, the evolution of culture, deceit and self-deception, innate and learned behavior, morality, law and justice, and the evolution of artistic expression, among others.

Many years have passed since *Darwinism and Human Affairs*, and in this time Alexander, his students and colleagues, and many others, have been constructing a new evolutionary synthesis upon these themes, a synthesis that has grown to encompass anthropology, psychology, psychiatry, economics, sociology, the arts, humanities, and religion. This volume is a celebration of Richard Alexander's work, and his diverse, enduring influences in the study of virtually all aspects of humanity. For each chapter we have chosen an excerpt from a key paper or chapter that represents a particular theme that Alexander wrote about over the course of his career. Each chapter is introduced by an expert in the field, most of whom are former students or colleagues of Dr. Alexander, who provides perspective on how his ideas have advanced scientific thought.

We believe this structure for the book is appropriate because Alexander constantly strove to inspire his students and colleagues to think carefully about evolution and human behavior, and to develop and test novel, integrative hypotheses. In his classic "Evolution and Human Behavior" course, taught for decades at the University of Michigan, Alexander would challenge any and all students (undergraduate and graduate) to try and find errors in his arguments, and to develop their own alternative hypotheses. Alexander would carefully read hundreds of essays, and provide firm yet helpful comments to all the students in this very popular (and hence very large) class. That some of his students took him up on this challenge is apparent from this volume. Beverly Strassmann, for instance, wrote a paper in this class as an undergraduate that has itself become a classic in the field of human social evolution (Strassmann 1981). But this is just a start—all of the students and colleagues of Alexander who have written introductory essays for this volume have gone on to develop their own research programs in social evolution, and those authors who did not directly interact with Alexander were nonetheless inspired by his published work as they developed their own theories. Even the essays themselves reveal the profound influence of Alexander's philosophy. For example, Paul Turke, in his essay, proposes a novel hypothesis connecting altriciality and neoteny to the delayed senescence that characterizes the human species relative to other primates. Even in the short space of an introduction, Alexander's associates cannot help but explore novel connections and new ideas. This is a major part of Alexander's legacy—a cadre of evolutionary thinkers who, inspired by his example, have spent their lives developing and testing hypotheses concerning social evolution, and especially the social evolution of that most complicated of species, ourselves.

One

Chapter 1 by one of Alexander's first graduate students, Mary Jane West-Eberhard, provides an overview of Alexander's early work on communication and mating behavior. West-Eberhard is a staff scientist at the Smithsonian Tropical Research Institute. She is one of the world's leading experts on the behavior and evolution of social wasps, and has also published on sexual and social selection in general, and in relation to speciation, as well as a groundbreaking body of theory connecting developmental mechanisms to evolutionary phenomena, including morphological and behavioral change under selection, and population divergence and speciation. The chapter provides a brief synopsis of some of Alexander's early work on communication, mating behavior, and speciation in insects, and some of his early thoughts on human evolution. West-Eberhard shows how one of his papers was a bridge between his earlier writings and the later ones on human evolution. She argues that Alexander's exceptional abilities to illuminate the evolutionary basis of human social behavior stemmed from his strong background in the systematics and evolutionary biology of the singing insects, and his pioneering work on Darwinian approaches to behavior in nonhuman animals.

Two

In chapter 2, Steven Frank explores a new view of the evolution of cooperation that was developed by Alexander, as illustrated by an excerpt from his second book on human social evolution, *The Biology of Moral Systems* (1987). Frank, a professor of evolutionary biology at the University of California at Irvine, is one of the world's leading evolutionary theoreticians. His work has transformed our understanding of inclusive fitness, multilevel selection, sex ratio evolution, parasite-host coevolution, and genetic conflict, among many other topics. Frank began his pursuit of a career in evolutionary biology after taking Alexander's class in animal behavior as an undergraduate. He was a graduate student of the late William Hamilton, but was also advised by Alexander. He points out that the extensive cooperative networks that are part and parcel of the vast nation-states that characterize human society are not easily explained by the twin pillars of cooperation developed in evolutionary biology: kinship and reciprocity. He argues that Alexander's work led to a new view of how these cooperative networks could evolve and remain stable: group suppression of conflict. Alexander's ideas in this vein followed from two parallel themes that he pursued over the course of his career: the evolution of morality and justice, and the evolution of individuals at different levels in the hierarchy of life over the course of evolution. These ideas were stimulated by the work of other great thinkers (such as John Rawls in the case of morality, and Egbert Leigh in the case of conflict suppression in the evolution of the hierarchy of life), yet he developed a unique synthesis that made the generality of the concepts clear. The theoretical underpinnings of this mechanism have now been developed and refined (by Steven Frank and others), and it has become another general principle upon which our understanding of the evolution of cooperation rests. The idea has far-reaching implications. For example, as Alexander stresses in the *Biology of Moral Systems*, the reproductive opportunity-leveling characteristic of large democratic nation-states may have led to their notable success at warfare and territorial expansion at the expense of more despotic regimes. In fact, the collapse of despotism (which was the rule rather than the exception during the long course of human history following the development of agriculture) may well have been driven by this dynamic. After all, the prospect of fighting and risking one's life for king and country is less appealing when the royal elite have monopolized most of the women.

Three

In chapter three, Paul Sherman introduces an excerpt from a classic paper on the evolution of eusociality (by Alexander, Noonan, and Crespi), published in 1991 in an edited volume entitled *The Biology of the Naked Mole-Rat*. Sherman, a professor of biology at Cornell University, is a world-renowned researcher in the field of animal behavior, having carried out groundbreaking studies on the social behavior of ground squirrels, naked mole rats, and wood ducks, among other organisms. He is also a leader in the field of evolutionary medicine, publishing innovative studies of spices as antimicrobial agents, allergies as anticancer mechanisms, and morning sickness as a toxin-avoidance mechanism, among many other topics. Sherman was a graduate student of Alexander, and they later worked together to establish the first naked mole-rat colonies in the United States, at the University of Michigan and at Cornell University. Sherman relates the story of how Alexander conceived of the key characteristics of a eusocial vertebrate as a thought experiment and prediction before anyone was aware of the existence of such an animal. Remarkably, Alexander's description almost perfectly described the naked mole-rat, which was the subject of research by the biologist Jennifer Jarvis in South Africa. The discovery of a eusocial vertebrate allowed profound insights into the ecological and evolutionary mechanisms that drove the evolution of eusociality, and in turn this led to a flood of research and publications, culminating in *The Biology of the Naked Mole-Rat*. Sherman points out that Alexander had been thinking about the evolution of eusociality for a long time before writing the chapter, beginning with a strong interest in the phenomenon as an entomologist studying social behavior. He notes that Alexander et al. (1991) made several key points with respect to the evolution of eusociality: First, the haplo-diploid system that characterizes the major eusocial insect groups (ants, bees, and wasps) is not a sufficient explanation for the evolution of eusociality; second, eusociality is a much more general phenomenon than initially appreciated, evolving convergently across vast spans of the tree of life; and third, both intrinsic (genetic and developmental) and extrinsic (ecological) factors must have been crucial for the evolution of eusociality. Alexander et al. (1991) argued that both "ecological constraints" (environmentally imposed constraints on the ability of individuals to breed on their own) in the form of the need for nest site protection from predation (an extrinsic factor), and levels of genetic relatedness (kinship, an intrinsic factor), are crucial for the evolution of eusociality. Sherman emphasizes that the arguments presented in Alexander et al. (1991) have become widely accepted, and the chapter has become an indispensible guide to understanding the evolution of eusociality.

Four

Chapter 4, by David Queller, introduces an article that represents Alexander's (1974) first exposition of general theory for the evolution of social behavior across all animals, including humans. Queller was a doctoral student with Alexander, and has since gone on to become one of the foremost researchers working on social evolution, in organisms from plants to wasps to slime molds to humans. He describes how Alexander's (1974) paper, integrating nepotism, reciprocity, and his new idea—parental manipulation—served as a nexus for future theory and research, which included extensions, inspiration, presages of much-later developments such as skew theory, and ultimately constructive assaults. Alexander (1974) also marks his transition from mainly studying crickets, to mainly focusing on the evolution of sociality, and his effective combining of studies of specific taxa and big questions in social evolution is mirrored in the work of many of his students.

Five

The fifth chapter, by Mark Flinn, expands on a paper elucidating Alexander's conception of how culture, long a bastion defended against biology, evolves in the contexts of the human psyche, human conflicts and confluences of interest, and human beliefs concerning how best to respond to particular social and material contingencies. As an evolutionary anthropologist who has also survived, and prospered, in the biology-versus-culture debate that was catalyzed by the changes in evolutionary theory that Alexander, Hamilton, Trivers, Williams, and others developed, Flinn is uniquely suited to describe how Alexander provides a novel behaviorally based perspective on human culture. Alexander's analysis opened the door for future researchers to study culture from the perspective of evolutionary biology. Flinn was one of the first through this door. He was influenced by

Alexander beginning in high school, then as an undergraduate, Master's student, and Postdoctoral Fellow at the University of Michigan. They have co-authored two papers on cultural evolution (Flinn & Alexander 1982, 2007). Flinn has developed a research program that integrates cultural with biological anthropology—helping to generate the recent, now-maturing field of evolutionary anthropology. He is currently president–elect of an organization that Alexander helped create, the Human Behavior & Evolution Society.

Six

In chapter 6, Bobbi Low introduces a 1968 review, written by Alexander and Tinkle, for the journal *Bioscience,* of two books (*On Aggression* by Konrad Lorenz and *The Territorial Imperative* by Robert Ardrey). Low is a professor of resource ecology in the School of Natural Resources and the Environment, and a faculty associate in population studies at the Institute for Social Research, both at the University of Michigan. As a long-time colleague of Alexander, she was strongly influenced by his work. Although she began her career studying life history evolution in nonhuman animals, Low later developed a strong interest in human social evolution, and has contributed seminal research to this field. For example, her work on the evolutionary ecology of human mating strategies and family relationships is widely cited, comprising a classic body of work. She was also a pioneer in the use of long-term, extensive historical demographic datasets from Scandanavian parish records to test predictions from evolutionary theory relevant to human behavior and life history (Low quips that her research concerns the "sex lives of dead Swedes"). As Low points out, Alexander and Tinkle (1968) was no run-of-the-mill book review, but rather presented a novel case for the importance of intergroup competition in driving the evolution of human cognition and behavior. She provides crucial context for the review, which was presented long before the current surge of interest in the relationship between warfare and human social evolution. Low also notes that Alexander and Tinkle developed the hypothesis that intergroup competition was focused on access to resources, and in particular, reproductive resources, long before behavioral ecologists began applying this sort of cost/benefit thinking to human behavior in any systematic way. Low also highlights the sad fact that public attitudes toward evolutionary explanations of human behavior (or anything else for that matter) have barely changed in the long span of time that has passed since the review was published. Low's introduction is indeed a potent reminder of the importance of bringing these issues to the attention of the public, even (or especially) now.

Seven

Chapter 7 reprints a classic paper on parental care and concealed ovulation by Alexander and Noonan, published in 1979. This paper is introduced by Beverly

Strassmann, a professor in the Department of Anthropology at the University of Michigan. Strassmann was a graduate student under Alexander's supervision, but was inspired by his ideas even before entering graduate school. In fact, while taking Alexander's "Evolution and Human Behavior" course at the University of Michigan as an undergraduate, she wrote a term paper extending Alexander and Noonan's 1979 argument, emphasizing the benefits of concealed ovulation for subordinate males. This paper (Strassmann 1981) has itself gone on to become a classic reference in the field. Strassmann notes that Alexander and Noonan's emphasis on the intimate connections between increased levels of intergroup competition, the complexity of social interactions within groups, and the importance of increased paternal care was novel and prescient, leading to a flood of new insights concerning human social evolution. She points out that connecting the evolution of concealed ovulation to the evolution of increased paternal care was an inspired and pivotal contribution, clarifying causal connections that before then had been completely obscure. Strassmann's paper further clarified these connections by showing that it was particularly subordinate males that were most likely to benefit from concealed ovulation, which promoted a male reproductive strategy that emphasized paternal effort over mating effort.

As Strassmann notes, both the hypothesis developed by Alexander and Noonan (1979) and her extension of that hypothesis have stood the test of time. There have been alternative hypotheses presented, but Alexander himself was able to discredit these hypotheses with a set of closely argued rebuttals in his own review of the topic (Alexander 1990).

Strassmann devotes the rest of her introductory essay to examining the evidence that ovulation is, in fact, concealed in humans. This claim has been the focus of considerable attention in the literature, and a number of researchers have disputed it. Strassmann provides an extensive and masterful overview of these studies, and shows that they do not provide convincing evidence against concealed ovulation, when the phenomenon is properly understood. She brings to bear evidence from the scientific literature, from her own ethnographic, behavioral and physiological studies of the Dogon people of Mali, Africa (work of unparalleled depth and detail), and from comparative studies across primates. This review supports the original claim of Alexander and Noonan (1979) that ovulation is indeed concealed in humans, and that claims to the contrary are based on weak evidence. She also critiques the assumption that any behavioral changes occurring over the menstrual cycle are necessarily adaptive.

Eight

In chapter 8, Paul Turke introduces work by Alexander on a novel feature of human primates—early-childhood physical altriciality coupled with psychological precociality— that has long withstood our understanding in terms of natural selection along the human lineage (Alexander 1990). This paper addresses the life-history component of how modern humans have evolved, centering on how tradeoffs of early survival with optimal trajectories of development to success as an adult have been alleviated in humans by extensive parental care. Such shifts in life-history selection (from just becoming a live adult, to becoming a *better* adult) thus freed human infants to become physically helpless grubs that are neurologically highly advanced over our ancestors. As for many other aspects of human evolution, a complex Alexandrian mix of social cooperation with competition drives the life-history transition to a long childhood, during which the social brain develops to full, advanced maturity. Turke, an evolutionary anthropologist immersed in traditional cultures and their patterns of child-rearing, was a Postdoctoral Fellow at Michigan with Alexander. He is now a (apparently, the world's first) Darwinian pediatrician, developing and applying evolutionary theory to help optimize child development and health.

Nine

Chapter 9, introduced by Karl Sigmund, addresses a question that has puzzled humans since the first written records of civilization: how and why humans exhibit morality. Following in his mission to lay foundations for all characteristics of humans, no matter how rarified or apparently problematic for evolutionary theory, Alexander (1987) expounds his concept of indirect reciprocity, which extends from roots in nepotism and direct reciprocity to encompass interactions between any and all pairs of humans, in any circumstance. As Sigmund, a leader in the mathematical study of human social interaction, describes, this early work by Alexander on the biological bases of indirect reciprocity, and its manifestations as morality, has helped to inspire a now wide and deep field of game-theoretical modeling that focuses on how humans cooperate and compete. This work, dovetailed with psychological and primatological studies, is uncovering the biological underpinnings of human prosociality, and its dark sides of ingroup-outgroup antipathy.

Ten

Robin Dunbar, one of the world's leading thinkers on the biological and psychological bases for human sociality, introduces Alexander's (1989) article on the evolution of the human psyche. He sets this prescient paper into its theoretical and historical context, as one of the first studies that comprehensively seeks to explain the "why" question of human consciousness and other core facets of the human mind. As Dunbar describes, Alexander proposes the stunningly new hypothesis that human mental capacities evolved in the context of selection for ability to "play" in the mind—to build and permute, consciously and unconsciously, alternative social strategies and tactics in a world where evolutionary success is deeply contingent on skills for social navigation. Language, brain size, the "social" components of the brain, complex cognition and affect, social-network structures, the neurological "default mode" system of functional brain activation—all follow naturally from Alexander's model for the evolution of the human psyche. Dunbar's work, like Alexander's, is deeply rooted within primatological and anthropological-archaeological contexts. This framework allows evaluation of how and why the modern human lineage has so spectacularly diverged from all others, and what this legacy may hold in store for, and suggest for improving, our runaway-technology future.

Eleven

Chapter 11 reprints a classic paper by Alexander (1982) on the evolution of morality. This subject represents the nexus of Alexander's approach to human social evolution, tying together a number of core themes that run through his work, including family relationships, kinship networks, direct and indirect reciprocity, ethics, law, justice, intergroup competition, warfare, and religious belief, with a consistent focus on conflicts of interest. Alexander focuses on the conflict between egoism and utilitarianism as a central paradox in moral philosophy, along with the inherent duality of human nature.

David Lahti, a professor of biology at Queens College, City University of New York, introduces the chapter. Lahti is eminently qualified to comment on the relationship between evolution and morality, as he holds doctorates in both evolutionary biology and philosophy and has published extensively on morality from an evolutionary perspective (e.g., Lahti 2003). He begins his introduction by highlighting a long-standing conflict within the biology department at the University of Michigan, in which Alexander and his students were seen as holding the view that society is "built on lies." This conflict traces to the extreme discomfort with which many people (including many scientists) react to the concept of "human interests" at the core of morality that Alexander's evolutionary approach reveals. Alexander begins the article by pointing out that previous attempts to analyze morality have suffered from the lack of a proper understanding of human interests, which ultimately center on efforts to promote the survival of the genes. Lahti notes that this revelation comes as an unpleasant shock to many (himself included). He then moves on to a brief but bracing review of the main objections of philosophers to incorporating an evolutionary perspective into any analysis of morality. He swiftly eviscerates some of the major objections to the evolutionary approach, and yet makes a reasoned appeal for an important place for philosophy as we develop an evolutionary approach to morality. Lahti then moves on to consider other ways in which learning of Alexander's detailed arguments relating evolution to morality might cause people to experience "nausea." Alexander's arguments imply that moral behavior evolved in the context of egoism (actually inclusive fitness), relative reproductive success, and intergroup competition (see ch. 6). Lahti points out that each of these historical foundations shocks the typical moral philosopher. But this is only the beginning of the discomfiture that Alexander's world view can engender. Alexander argues that an evolutionary view of morality explains why justice is incomplete, why moral systems are not ideal, why happiness is elusive, and why ethical dilemmas persist. Lahti notes that these views are anathema to many; they find them depressing and nihilistic. From an evolutionary perspective, every member of society has an interest in promoting moral codes that benefit society as a whole (at an individual cost), even though they may not, in fact, follow the code themselves. This kind of hypocrisy is (judging from virtually all sources of news and gossip) extremely common, yet we continue to aspire to high moral standards. Self-deception (in the service of deceiving others about our true intentions) is rampant in this view, and may well be the norm rather than the exception. Hence, Lahti argues, it is not surprising that many, and probably most, have responded to Alexander's arguments with some mix of outrage, denial, accusations of bigotry (or worse), and general aversion. What then, are we to do, given this window into our true natures (rather like the plastic windows inserted into the sides of cows at the Cornell Veterinary School to reveal their intestinal workings)? Alexander himself does not shy away from moral advice, even as he lays bare the evolutionary roots of our moral codes in excruciating detail. Lahti argues that several considerations may allow us to rescue morality from the primordial morass of its evolutionary origins, although in every case there is no guarantee that our evolutionary history will not, in fact, continue to influence our thinking and behavior. Thinking about how to behave after coming to grips with Alexander's evolutionary understanding of morality is a bit like peering between two mirrors facing each other. Nevertheless, Lahti elaborates three reasons that morality may well survive the evolutionary onslaught on its perfection. He concludes by arguing in favor of Alexander's proposition that we are better off as a society and a species if we understand the true evolutionary origins of our moralizing, rather than pretending that our moral codes are unsullied by the shadow of our evolutionary past.

Twelve

In chapter 12, Laura Betzig reviews the influence of the Alexander, Noonan, and Crespi (1991) chapter on the evolution of eusociality (see also ch. 3), particularly with regard to the importance of ecological constraints for human social evolution. Betzig, an evolutionary anthropologist, is particularly well qualified to comment on this issue, as she has carried out extensive cross-cultural analyses of the evolution of despotism and its relationship to differential reproduction in human societies. She has written a book on the topic (Betzig 1986), has published many more papers on the subject since, and is currently completing a book on the history of the West that also focuses on this issue. Betzig points out that the evolution of sterility that characterizes classical eusocial species such as ants and naked mole-rats, has also applied to human societies (or to portions of them). Alexander et al. (1991) emphasized the critical importance of ecological constraints in driving the evolution of cooperative breeding, and ultimately (in the extreme) the sterile castes of eusociality. This line of thinking can be traced to much-earlier papers by Alexander, such as his classic 1974 paper on the evolution of social behavior (see ch. 4). In turn, Alexander's arguments in this vein led to the development of an entire branch of theory within behavioral ecology (Reproductive Skew Theory; see Vehrencamp 1983), as well as influencing the work of Betzig and others working specifically on human behavioral ecology. As Betzig illustrates in her introductory essay, the historical record strongly confirms the role of ecological constraints in enabling despots to impose sterility on (some of) their subjects, across societies and across the course of human history.

Thirteen

Stan Braude introduces chapter 13, an article by Alexander (1986) on the social-evolutionary significance of one of the most curious facets of human social life and language, humor. Braude, who was a doctoral student of Alexander's studying naked mole-rats, explains how Alexander's attack on this topic epitomizes his belief that evolutionary reasoning and analytic hypothesis-testing methods can and should be applied to all human phenomena, even the most seemingly arbitrary and far-removed from survival and reproductive functions. Alexander proposes and evaluates a suite of ideas concerning the significance of humor in human social interactions—displays of social-linguistic prowess, ingroup-outgroup manipulations, tests of social-intellectual skill, and strong links to mental scenario-building—all of which revolve around his central thesis of shifting balances of social competition and cooperation, at different levels, as prime movers of human evolution. Alexander's analysis of humor, unlike most of his other pioneering works, has yet to be built much upon by others. Perhaps this is because he has reached so far in this case from biology into everyday life, or perhaps it is because empirical tests of hypotheses for humor would appear so challenging. We hope that the article will inspire such studies, to better draw social biology, art, and language together, and understand a fundamental feature of what it means to laugh—and in doing so be human.

Fourteen

Chapter 14 presents an essay on the evolution of religion entitled "Religion, Evolution and the Quest for Global Harmony," which is published in this volume for the first time. This chapter is introduced by William Irons, a professor of anthropology at Northwestern University, and a longtime friend and associate of Dr. Alexander. His accomplishments in the field of evolutionary anthropology are too numerous to recite here, but his contributions rank among the best in the field. Irons has made substantive contributions to the study of religion from an evolutionary perspective himself, having developed the hypothesis that deep religiosity can serve as a signal of commitment to the group, one that is difficult or impossible to counterfeit (Irons 2001). Irons points out that there have been two major themes in the recent literature on the evolution of religion. A number of researchers have focused on religion as a byproduct of psychological mechanisms that predispose us to believe in unseen but imagined entities, such as gods. Research in this area focuses on identifying aspects of our "environment of evolutionary adaptedness" that might have selected for such tendencies, and on testing for specific psychological mechanisms in this context. Another group of related theories focus on the idea that religious beliefs function to enhance cooperation among individuals within groups. Irons places Alexander's essay within the context of recent theory in this area, and argues that his essay provides a novel approach that focuses on the central values upon which religious belief and the "concept of god" are based. Alexander develops the idea that the concept of god serves as a metaphor for the community at large, consisting of the "kindred" (an amorphous composite of all an individual's relatives, that shifts in composition based on individual perspective), and other individuals tied together through the bonds of reciprocity (direct and indirect). In this sense, we have indeed evolved to "serve god," as we serve our circles of kin and friends, and this concept provides an evolutionary basis for the origin and evolution of religious beliefs that is firmly based in inclusive fitness theory, a foundation of evolutionary biology. Irons stresses a point also emphasized by Alexander: It is not productive to denigrate other individuals for their religious faith, and this conception of god and the meaning of religion may ultimately reconcile scientific and religious views of the world.

Final

We introduce the final chapter, within which Alexander (2008) presents and evaluates his hypotheses for how natural selection and human evolution have led to the appearance, in modern humans, of art. Both of us received our doctoral degrees at Michigan with Alexander as an adviser—and we both apparently imprinted intellectually on him in following career paths that led from studies of behavior and evolution of specific taxa (poison dart frogs and gall insects), to studies that more or less directly address questions of human social evolution (e.g., despotism, social-brain disorders, and others). We refer to the arts as the "capstone" of Alexander's evolutionary edifice for explaining and analyzing human traits, because in developing his theory for the arts he has, taken together with previous work, conceptually unified virtually all human endeavors, including, arguably, the most challenging and apparently far-removed from the machinations of natural selection. Art is a reflection of culture. But, according to Alexander, it is also a product of absorption and transmission of cultural entities by perceived inclusive-fitness maximizing behavior, and a direct result of human scenario-building in forged connections between minds. As such, art is beneficial, from an evolution-ary perspective, to both artists and those who appreciate their creations. This bio-logical basis for the arts should consummate the Darwinian syntheses of biology with culture, science with the arts, and inclusive fitness theory with religion, com-munity, and human well-being. Whether or not it truly does so will depend on how Alexander's ideas, here and in the previous chapters, are received and built upon further. This will be up to you.

We conclude by arguing that many of the themes that Alexander championed during his life's work have become major subjects of interest and debate, and many of his arguments have been amply supported by later work. For example, building on Darwin's (1871) argument that human cooperative tendencies developed in the context of intergroup competition, Alexander developed the idea that intergroup competition was a runaway process that ultimately led to the evolution of human intellect and the primacy of social strategies in the evolution of intelligence (see ch. 6 and its introduction by Low). The subject of intergroup competition and warfare in relation to within-group cooperation in humans (and chimpanzees) has recently become a hot topic, attracting both theoretical and empirical atten-tion (e.g., Bowles, 2006, 2009; Puurtinen & Mappes 2009; Ginges & Atran 2011; Matthew & Boyd 2011). Alexander's focus on the importance of monogamy and reproductive opportunity leveling in hominid evolution (see ch. 7 and its intro-duction by Strassmann) is supported by recent results from the analysis of family structure across hunter-gather societies (e.g., Hill et al. 2011).

It is now quite clear that, just as Alexander (1987) argued, reputation in the context of indirect reciprocity has a profound effect on an individual's treatment by other individuals within their society (e.g., Milinski et al. 2002; Semmann et al. 2004; Nelissen 2008). Indirect reciprocity is a potent and viable mechanism for maintaining cooperation, even in large societies (e.g., Nowak & Sigmund 1998; Panchanathan & Boyd 2004; Rockenbach & Milinski 2006: see also ch. 9 and its introduction by Sigmund). Key aspects of reputation building (or losing) such as honor and heroism (and shame) have strong influences on human cooperative behavior, just as Alexander predicted (e.g., Jaquet et al. 2011). Punishment, shun-ning, and shaming are ubiquitous and potent components of human social inter-action, providing powerful incentives to cooperate (e.g., Brandt et al. 2003; Price 2005; Barclay 2006; Rockenbach & Milinski 2006; dos Santos et al. 2011; Marlowe et al. 2011). There is even evidence that individuals prefer to work in institutions with substantial penalties for violating group norms, in the interest of ultimately reaping higher rewards from participation (e.g., Gurek et al. 2006).

Hence, it is quite likely that the fitness of individuals within social groups has been closely linked to cooperativeness and willingness to contribute to group goals over the long course of human history. This is not to say that intergroup competition has been unimportant in human evolution (quite the reverse), but there is every indication that effective mechanisms of opportunity leveling (see ch. 2 and Frank's introduction thereto) including negative (shame, punishment, coercion) and positive (reputation-based rewards, honors) mechanisms, have been highly effective in aligning the interests of the individual with that of the group in human societies.

In conclusion, there seems to be a consensus emerging that human cooperative behaviors, and associated moral codes, have evolved through a delicate dance of incentives (e.g., honor, recognition, and reward) and disincentives (e.g., punishment, shaming, and shunning) among individuals who are balanced on a knife edge of competition for hierarchy negotiation within groups and competition and warfare between groups. This consensus is precisely what Alexander predicted in his foundational body of work.

References

Alexander, R.D. 1974. The evolution of social behavior. *Ann. Rev. Ecol. Syst.* 5: 352–383.

Alexander, R.D. 1979. Darwinism and Human Affairs, Seattle: University of Washington Press.

Alexander, R.D. 1982. Biology and the moral paradoxes. *J. Social Biol. Struct.* 5: 389–395.

Alexander, R.D. 1986. Ostracism and Indirect Reciprocity: The Reproductive Significance of Humor. *Ethol. Sociobiol.* 7:253–270.

Alexander, R.D. 1987. *The Biology of Moral Systems.* Hawthorne, NY: Aldine de Gruyter.

Alexander, R.D. 1989. The evolution of the human psyche. Pp. 455–513 *In* C. Stringer and P. Mellars (eds), *The Human Revolution.* Edinburgh, Univ. of Edinburgh Press.

Alexander, R.D. 1990. How Did Humans Evolve? Reflections on the Uniquely Unique Species. *Univ. Mich. Zool. Spec. Pub.* 1: 1–38.

Alexander, R.D. 2008. Evolutionary Selection and the Nature of Humanity, Pp. 301–348 In V. Hosle and C. Illies (eds), *Darwinism & Philosophy.* Notre Dame, IN: University of Notre Dame Press.

Alexander, R.D. and Tinkle, D.W. 1968. A comparative review (of *On Aggression* and *The Territorial Imperative*). *BioScience* 18:245–248.

Alexander, R.D. and Noonan, K. M. 1979. Concealment of ovulation, parental care and human social evolution. Pp. 436–453 In: N. A. Chagnon and W. G. Irons (eds.) *Evolutionary Biology and Human Social Behavior: An Anthropological Perspective.* North Scituate, MA: Duxbury Press.

Alexander, R.D., Noonan, K.M., and Crespi, B. J. 1991. The evolution of eusociality. In: Sherman, P., Jarvis, J.U.M, and Alexander, R.D. (eds.), *The Biology of the Naked Mole-Rat* Princeton, NJ: Princeton University Press, pp. 3–44.

Barclay, P. 2006. Reputational benefits for altruistic punishment. *Evol. Hum. Behav.* 27:344–360.

Betzig, L.L. 1986. *Despotism and Differential Reproduction: A Darwinian View of History*. New York, NY: Aldine-de Gruyter.

Bowles, S. 2006. Group competition, reproductive leveling, and the evolution of human altruism. *Science* 314:1569–1572.

Bowles, S. 2009. Did warfare among ancestral hunter-gatherers affect the evolution of human social behaviours? *Science* 324:1293–298.

Brandt, H., Hauert, C., and Sigmund, K. 2003. Punishment and reputation in spatial public goods games. *Proc. Roy. Soc. B* 270:1099–1104.

Darwin, C. 1871. The Descent of Man, and Selection in Relation to Sex. London: John Murray.

dos Santos, M., Rankin, D.J. and Wedekind, K. 2011. The evolution of punishment through reputation. *Proc. R. Soc. B* 278:371–377.

Flinn, M. and Alexander, R.D. 1982. Culture theory: the developing synthesis from biology. *Human Ecol.* 10(3): 383–400.

Flinn, M. and Alexander, R.D. 2007. Runaway Social Selection in Humans. Pp. 249–255 In: Gangestad, S. W. and Simpson, J.A. (eds). *The Evolution of Mind. Fundamental Questions and Controversies*. New York: Guilford Press.

Ginges, J. and Atran, S. 2011. War as a moral imperative (not just practical politics by other means). *Proc. Roy. Soc. B* 278:2930–2938.

Gürerk, Ö., Irlenbusch, B., & Rockenbach, B. 2006. The competitive advantage of sanctioning institutions. *Science* 312:108–111.

Hill, K.R., Walker, R., Bozicevic, M., Eder, J., Headland, T., Hewlett, B., Hurtado, A. M., Marlowe, F., Wiessner, P. & Wood, B. 2011. Coresidence patterns in hunter-gatherer societies show unique human social structure. *Science* 331:1286–89.

Irons, William. 2001. Religion as a Hard-to-Fake Sign of Commitment. Pp. 292–309 *In*: R.M. Nesse (ed.), *Evolution and the Capacity for Commitment*. New York: Russell Sage Foundation.

Jacquet, J., Hauert, C., Traulsen, A. and Milinski, M. 2011. Shame and honour drive cooperation. *Biol. Lett* 7:899–901.

Lahti, D.C. 2003. Parting with illusions in evolutionary ethics. *Biol. Phil.* 18:639–651.

Marlowe, F.W., Bebesque, J.C., Barrett, C., Bolyanatz, A., Gurven, M. and Tracer, D. 2011. The "spiteful" origins of human cooperation. *Proc. Roy. Soc. B* 278:2159–2164.

Matthew, S. and Boyd, R. 2011. Punishment sustains large-scale cooperation in pre-state warfare. *Proc. Natl. Acad. Sci. USA* 108:11375–11380.

Milinski, M., Semmann, D. & Krambeck, H.-J. 2002. Donors to charity gain in both indirect reciprocity and political reputation. *Proc. R. Soc. B* 269:881–883.

Nelissen, R.M.A. 2008. The price you pay: Cost-dependent reputation effects of altruistic punishment. *Evol. Human Behav.* 29:242–48.

Nowak, M.A. and Sigmund, K. 2005. Evolution of indirect reciprocity. *Nature* 437:1291–98.

Panchanathan, K. and Boyd, R. 2004. Indirect reciprocity can stabilize cooperation without the second-order free rider problem. *Nature* 432:499–502.

Price, M. 2005. Punitive sentiment among the Shuar and in industrialized societies: Cross-cultural similarities. *Evol. Hum. Behav.* 26:279–287.

Puurtinen, M. and Mappes, T. 2009. Between-group competition and human cooperation. *Proc. Roy. Soc. B* 276:355–360.

Rockenbach, B. and Milinski, M. 2006. The efficient interaction of indirect reciprocity and costly punishment. *Nature* 444:718–723

Semmann, D., Krambeck, H.-J. and Milinski, M. 2004. Strategic investment in reputation. *Behav. Ecol. Sociobiol.* 56:248–52.

Strassmann, B.I. 1981. Sexual selection, paternal care, and concealed ovulation in humans. *Ethol. Sociobiol.* 2:31–40.

Vehrencamp, S. 1983. A model for the evolution of despotic versus egalitarian societies. *Anim. Behav.* 31:667–682.

PART } I

General Foundations

Insect Behavior and Social Evolution

Phantasm
Soft staccato lisps of a bush katydid
Rising through cool mists of dusk
In a leftover lonely glacial spruce bog
High in the Appalachian Mountains

Quick insistent chirpings of brown field crickets
Across the moonlit landscape
Of a tiny hill prairie overlooking
Endless Mississippi River bottomlands

Eerily whining buzz of a great green
Grasshopper wafting on the wind
From far across the night-shrouded dunes
Alongside the lake named Michigan

Silvery tinkle of a miniature yellow cricket
Hidden in sun-speckled undergrowth
Of shadowy swamps across
The Everglades of Florida

Crashing synchrony of numberless cicadas
Chorusing deafeningly in the warm brightness
Of a June afternoon in open oak forests
Across the southeastern hills of Ohio

Lazy deep chir-ruping of a solitary beach cricket
Rising ghost-like in crashes of surf
Along a desolate rocky stretch
Of Atlantic coast at midnight

In such atmospheres I have imagined
I am the only human on earth, alone
A thousand centuries ago.

R. D. Alexander, *2011, p. 213*

INTRODUCTION

From Cricket Taxonomy to a Darwinian Philosophy of Man
Mary Jane West-Eberhard

Figure 1.1 Singing male field cricket, courtesy of R. D. Alexander

Dick Alexander once recounted the story of a conversation with his maternal grandfather, Noble Porter Heath II, an Illinois farmer, about his plan to go to Australia to pursue taxonomic studies of the singing Orthoptera. "That's a long ways to travel on a cricket," his grandfather quipped. This volume tells of an intellectual voyage that went from a doctoral thesis on the taxonomy, sound production, and classification of field crickets and cicadas, through studies of the social behavior of insects and mole-rats, to seminal influential work on human culture and morality—also a long ways to travel on a cricket.

Alexander's path-breaking contributions to the study of human evolution did not come out of the blue. They grew from an unusually acute understanding of Darwinian evolution, that is, adaptive evolution by natural selection. He was already a well respected evolutionary biologist—"distinguished" if we can judge by several prominent awards—when he jumped wholeheartedly into the perpetual fray of speculation about human evolution with a 1969 lecture on "The search for an evolutionary philosophy of man" (Alexander 1971). His ability to see the adaptive significance of major patterns in human behavior owes to his preparation for this as a pioneer in Darwinian studies of nonhuman behavior. So looking at his earlier work is important for understanding his innovative thinking about the social behavior of humans and other animals.

The pioneer aspect of Alexander's work, and the beginnings of a broad interest in evolution, were evident in his graduate research in entomology at Ohio State University. As a student of Donald Borror, Alexander was introduced to insect sounds as tools for the identification of species. But Borror used the characteristics of insect songs as if they were morphological characters with no attention to their functions or evolution, the aspects of most interest to Alexander even then: "Function – behavior, and what it accomplished for the animal— was my first interest" (RDA unpbl ms Autobiography IIIE, 2003).

Systematics has many virtues for understanding evolution. Among them is the need for close attention to phenotypic variation and to the importance of genetics, in particular reproductive (genetic) isolation, for producing the observed variation. Taxonomic research and fieldwork also promotes deep understanding of a particular group of organisms, including its morphology, behavior, and natural history. Alexander's earliest publications included articles on taxonomy that revealed large numbers of cryptic species, and of previously unappreciated morphological characters, based on studies of their acoustical communication (1957a), as well as papers on sound production itself (1957b) and arthropod communication in general (1960; 1964; 1967). The early publications grew increasingly broad in scope, as they treated an expanding list of behavioral and life history phenomena as factors in speciation (e.g., Alexander and Moore 1962; Alexander and Bigelow 1960) and evolution (e.g., Alexander and Brown 1963).

Many of Alexander's early papers appealed broadly to students of behavior and evolution, including especially graduate students. My favorite is the now-classic and still valuable monograph on cricket behavior, "Aggressiveness, territoriality, and sexual behavior in field crickets (Orthoptera: Gryllidae)" (Alexander, 1961). That paper documents the effects of male isolation, history of wins and losses, age, size, and copulatory success on social rank, and it describes comparative studies of mating behavior and sound communication, illustrating these phenomena with 63 figures, including photographs of behavior and audiospectographs of the calling, fighting, and courting sounds of eight different species. The monograph ends with a discussion of the evolution of social behavior in insects (see especially p. 212), something that is surprising given the title of the paper, but not given the author. It specifies how phenomena observed in crickets—evolutionary change in length of adult life, and the transfer to females of communication morphology that originated in males—could affect the evolution of sociality in general: "Comparison of the behavior of different species of field crickets and related Gryllinae suggests some of the probable intermediate stages in the evolution of social behavior in insects through an initial isolation of breeding pairs and their attachment to and modification of particular localities" (p. 217).

Dick Alexander was a leader in promoting behavior as a taxonomic tool, but went beyond that to emphasize the importance of behavior for understanding adaptive evolution in general: "To paraphrase an old adage, it is not what one *has* that counts in evolution, it is what one *does* with what one has—and what one *does* is not always entirely clear from what one apparently has," meaning morphology alone (Alexander 1962a). He once brought an antique apple peeler to class and asked what we guessed its function might be. The gadget had a number of puzzling complicated parts, but their functions were clear once an apple was attached and the handle turned.

A 1962 article in *Evolution* on "Evolutionary change in cricket acoustical communication" (Alexander 1962b) could serve as a model of how insightful, detailed comparative study of phenotypes—behavior, ontogeny, sex differences and ecology—can be used to make deductions regarding evolution without fossils and without explicit genetic data, using indirect evidence of genetic divergence. These early publications emphasized the use of behavior as the secret to understanding adaptive significance, and the search for adaptive significance as the secret to understanding behavior. It is obvious that such lines of thinking were important for developing ideas about human evolution. They show how Alexander's bold breach of the boundary between biology and the social sciences grew from a background in taxonomy, a field where orderly and essentially conservative practices prevail and expertise is often deep but circumscribed within what has been called the "comfortable separateness" of a specialized field (Anonymous, 2011). Notable for reaching beyond such boundaries, the 1962 paper on "The Role of Behavioral Study in Cricket Classification" (Alexander 1962a), presented at a 1961 meeting of the American Association for the Advancement of Science, won the AAAS Newcomb Cleveland Prize for an "outstanding contribution to science."

The scientific values that grew out of Dick Alexander's early work deeply influenced his later research and that of his students, who, like me, took those values as their own. One of them is a consuming passion for work on a particular group of organisms, with a determination to learn all that is known about it—a commitment to taxon-centered research whose merits I have discussed elsewhere (West-Eberhard 2001). Concentrating initially on evolutionary questions in a particular group of organisms always reveals something of general interest. And it quickly makes a beginner into an expert ready to tackle larger questions, with a concrete basis for critical thinking on almost any major topic in evolutionary biology. It also fosters confidence in one's ability to make an original contribution of some importance. In Dick's case the originality and breadth of his comparative studies of the singing insects made him quickly recognized as an expert on the evolution of animal communication and its role in systematics and speciation. From there it is not so difficult to embark on comparably original work on the evolution of humans.

Still, one might wonder: What would motivate a scientists trained as an entomologist to turn from successful work on six-legged creatures to confront the complex and controversial world of research on human evolution? This question has been answered by Alexander himself, in a biographical essay (Alexander 2009, p. 23):

> How and why did I make the transition, initiated around 1967, from studying the singing insects as a systematist and behaviorist to eventually writing

some 50 articles and two books about how evolution applies to humans and, in particular, human behavior? I decided in 1954, the day after I passed the written and oral preliminary examinations for the doctorate, that I wanted to be an *evolutionary* biologist. A few years later I realized I would like to think of myself (grandly!) as trying to falsify the hypothesis that everything about life is a result of evolution,... The approach I set for myself would require that I proceed eventually to the most difficult of all traits (behavior) and the most difficult of all species (humans).

The excerpts reprinted here are from "Comparative animal behavior and systematics," (1969), the published version of a 1967 lecture given at an international conference on systematic biology convened by the National Academy of Sciences in Ann Arbor. It represents a clear bridge between the early papers on behavior and systematics and the later ones on humans. This is the paper that Alexander himself recognizes as the turning point in his commitment to put human evolution at the center of his research (Alexander 2009). Even though the lecture was presented at a symposium on the classification of organisms, it is really an essay on adaptive evolution and the urgency of extending modern Darwinian thought to studies of human behavior. Following some brief introductory remarks on behavior and systematics, the paper moves abruptly to a section headed "Behavior and Man's Evolution," with a single transitional sentence: "The facts I have outlined above suggest some of the problems and possibilities in using behavior to understand the history of life. I think one of the best illustrations of these problems and possibilities comes from the evolution of man himself" (p. 495).

With that sentence Alexander began more than four decades of research devoted to understanding human behavior and the evolution of sociality. The excerpts reprinted here emphasize the general principles of adaptive evolution that inspired Dick Alexander, and through him others, to give studies of human evolution a solid Darwinian base.

References

Alexander, R.D. 1957a. The taxonomy of the field crickets of the eastern United States (Orthoptera: Gryllidae: Acheta). *Ann. Entomol. Soc. Amer.* 50(6):584–602.

Alexander, R.D. 1957b. Sound production and associated behavior in insects. *Ohio J. Sci.* 57(2):101–113.

Alexander, R.D. 1960. Sound communication in Orthoptera and Cicadidae. In: W. E. Lanyon and W.H. Tavolga (eds.), *Animal Sounds and Communication*, AIBS Publications 7:38–92.

Alexander, R.D. 1961. Aggressiveness, territoriality, and sexual behavior in field crickets (Orthoptera: Gryllidae). *Behaviour* 17:130–223.

Alexander, R.D. 1962a. The role of behavioral study in cricket classification. *System. Zool.* 11(2):53–72.

Alexander, R.D. 1962b. Evolutionary change in cricket acoustical communication. *Evolution* 16(4):443–467.

Alexander, R.D. 1964. The evolution of mating behavior in arthropods. *Symp. R. Entomol. Soc. Lond.* 2:78–94.

Alexander, R.D. 1967. Acoustical communication in arthropods. *Annu. Rev. Entomol.* 12:495–526.

Alexander, R.D. 1969. Comparative animal behavior and systematics. In: *Systematic Biology. Proceedings International Conference on Systematics* (Ann Arbor, Michigan, July 1967). National Academy of Sciences Publication 1962: 494–517.

Alexander, R.D. 1971. The search for an evolutionary philosophy of man. *Proc. R. Soc. Victoria* 84:99–120.

Alexander, R.D. 2009. Understanding ourselves. In L.C. Drickamer and D.A. Dewsbury (eds.), *Leaders in Animal Behaviour: The Second Generation*. Cambridge: Cambridge University Press, pp. 1–37.

Alexander, R.D. 2011. *The Mockingbird's River Song: Poems, Essays, Songs and Stories, 1946-2011*. Manchester, MI: Woodlane Farm Books.

Alexander, R.D. and Bigelow, R.S. 1960. Allochronic speciation in field crickets, and a new species *Acheta veletis. Evolution* 14(3):334–346.

Alexander, R.D. and Brown, W.L. 1963. Mating behavior and the origin of insect wings. *Univ. Mich. Occas. Pap.* 628:1–19.

Alexander, R.D. and Moore, T.E. 1962. The evolutionary relationships of 17-year and 13-year cicadas with three new species (Homoptera: Cicadidae: Magicicada). *Univ. Mich. Mus. Zool. Misc. Pub.* 121:1–59.

Anonymous, 2011. Editorial statement. *Daedalus* 140(1):1.

West-Eberhard, M.J. 2001. The importance of taxon-centered research in biology. In M. J. Ryan (ed.), *Anuran Communication*. Washington, DC: Smithsonian Institution Press, pp. 3–7.

COMPARATIVE ANIMAL BEHAVIOR AND SYSTEMATICS

Excerpt from Alexander, R. D. 1969. Comparative animal behavior and systematics. In: *Systematic Biology. Proceedings of the International Conference on Systematics* (Ann Arbor, Michigan, July 1967). National Academy of Sciences Publication 1962: 494–517.

Behavior is probably the most diverse aspect of the animal phenotype—at least, as William Morton Wheeler (1905) put it, "in the field of possible observation." On this basis alone, behavior should be fascinating to the systematists because they are always looking for characters. Furthermore, among biologists, systematists are, more than any other group, the real students of diversity. Comparison is their chief method of exploration, and the comparative method, of course, depends upon and thrives upon diversity.

On the other hand, to some extent the diversity of behavior results from its being, in general, more directly and probably more complexly related to the genotype than to any other aspect of the phenotype. This particular feature discourages the systematists. They are not interested in getting involved with phenotypic variations that might be due solely to variations in the developmental environment. After all, morphology is troublesome enough in that regard.

Behavior has some other special features. In general, it is more strongly selected—or perhaps I should say more directly selected—than morphology or physiology. By this I mean that in any representation of the chains of cause-effect relationships between gene action and selective action in animals, behavioral characteristics nearly always would be placed directly next to selective action.[1]

The systematist's concern with adaptation should prevent him from passing this off too lightly. On the other hand, behavior is often difficult to document or to communicate to others. As Dr. Wagner stressed in his paper, repeatability is the essence of science, and to many taxonomists this traditionally has meant that morphology alone is sacred. Very little behavior is evidenced by preserved specimens or fossils.

---

[1] A botanist asked me abruptly in a phone conversation recently, "What *is* animal behavior, anyhow?" I tried to answer him unhesitatingly, and my reply came out: "Behavior is what animals have interposed between natural selection and the other (morphological and physiological) aspects of their phenotypes." Even with an indefinite amount of reflection, I think it might be difficult to improve on the emphasis in that definition.)

Behavior and Man's Evolution

The facts I have outlined above suggest some of the problems and possibilities in using behavior to understand the history of life. I think one of the best illustrations of these problems and possibilities comes from the evolution of man himself. We surely would all agree that the most important thing we could possibly discover about man's transition from the nonhuman state to the human state would be how he behaved during that period—the details of what he did and how he lived while he was evolving into a man. We know positively that he did make the transition from ape to man. What we do not know is precisely *how* he did it. By that I mean we do not know what the selective forces were, and, for example, why such forces seem to have been relatively strong and unidirectional for a while—at least in regard to changes in size of the brain case—and then to have slacked off, perhaps rather abruptly, some tens of thousands of years ago. We speak (vaguely, I think) of tools and communication, and of growing food and fighting off predators, but the truth is we still have no really good notion how and why men with bigger brains once outreproduced those with smaller brains and then stopped doing so.

A wide range of possibilities still exists, and the answers could very well turn out to be more startling than most of us might suppose. As one example, we do not really know what kinds of predators, if any, might have been involved in the steady increase in man's brain size, and, as much as we may dislike the idea, I believe the possibility still exists that man himself is the only one that could have done the job.

Perhaps I can explain what I mean, and demonstrate some of our ignorance about man's evolution, by posing a question. Intraspecific competition, in connection with natural selection, may be said to occur in three possible forms. Sometimes different individuals simply compete indirectly, without direct interactions, for whatever food, mates, shelter, or other commodities may be in short supply. In other cases, some kinds of individuals may partially or completely exclude others from the best sources of food, mates, and shelter through territoriality of one sort or another. There is another possibility, less often recognized. Superior individuals might sometimes actually pursue and destroy competitors, or potential competitors, thus removing them and their descendants from the possibility of competing. Such a superior individual might, in addition to removing competition, actually derive direct benefit from the slaughter, through cannibalism. Which of these three kinds of intraspecific competition operated during the evolution of humans from nonhuman primates, and how significant was each? The question has certainly not been answered; I do not think it has even been clearly posed before. Yet the different possibilities could scarcely fail to produce widely different attitudes among men trying to understand themselves and their behavior through knowledge of history. [Since submission of this manuscript the ideas involved here have been discussed and extended in a book review coauthored by D. W. Tinkle (Bioscience 18:245–248).]

Sometimes I have thought that to understand the selective action that made a nonhuman primate into a man could be the most important question in all of biology. It could change man's attitude toward nearly everything he does or tries to do—in education, politics, religion, and all the rest—for it could tell him more precisely what he is, and therefore why, in one sense, he persists in doing some of the things he does, and why he still fails to accomplish some of the things he seems to want to do. Any adult who has tried to explain to a child the pre-eminence of things sexual in so much of human affairs (as well as in the lives of other organisms) without using natural selection in his explanation surely will understand what I mean. To use another example, it is possible that we should be taking the history of selective action upon man much more directly into account in our attempts to deal with overpopulation and its consequences.

In other words, we cannot learn how man became a man, and therefore, in a sense, what a man really is, without knowing some things about the history of his behavior. Yet, it seems that the only thing we can do about this problem is to dig and scrape around at a few fossils that reflect his morphology and represent a few indirect traces of his behavior.

The Comparative Method In Behavior

It *seems* as though this is all we can do; but my theme here is that such an idea about evolution is false. I suggest that we can find out how man's behavior evolved and the kinds of selective action that were involved. More fossils will help, of course, but we can do it without fossils if we have to; in any case, the most important advances in understanding man's history may not come from fossil evidence, and I consider it unlikely that satisfactory progress will come from the efforts of humanists who are not simultaneously first-rate evolutionary biologists. I believe that we will make the significant advances in this area in the same way that we eventually would have arrived confidently at the conclusion—even without the help of a single fossil—that man and the other living primates have diverged from common ancestors. We would have done this, of course, through extensive, intensive, and perceptive comparative study over a period of time long enough for us to have developed—on the side, from direct observation and experimentation—an understanding of the steps and the mechanics of the process of evolution.

It should be clear by now that I am not arguing simply about the role of behavior as a tool for taxonomists. I want to argue instead for the establishment of a reasonable relationship between those biologists interested primarily in behavior and those interested primarily in systematics in the broadest sense—a relationship that will result in the kind of reverberating feedback between these fields that both need, and have needed, for a long time. I think the key to this relationship—perhaps the only key—lies in applying the comparative method to behavior on a much wider scale than has been the case. I realize that I am one small voice in a long line of people carrying this particular argument to the zoologists. But I do think the point has not yet been properly made.

To some zoologists—though perhaps not to those here—to argue for a rejuvenation of comparative study must sound a little old-fashioned. Nowadays biologists are calling for precise, quantitative results and for more and more experimentation. Comparative study and the broad-scale, observational-descriptive work that undergirds it are often viewed as outdated, trivial pursuits. The need for more experimental work, however, and the possibility of more precise experimentation do not reduce the need for good, evolutionarily oriented, comparative investigations. Rather, though it may surprise some biologists, the need for comparative study is thereby increased, for it is a central role of comparison to tell us which experiments to do, and which ones to do first.

In his recent book on adaptation and natural selection, Williams (1966a) argued that systematists never will prosecute the study of adaptation the way it ought to be prosecuted. All of us will agree that there is a lot of shortsighted, narrow-minded systematic work going on but, contrary to Williams' argument, I believe that the methods of systematists represent a great potential contribution to the study of adaptation, beginning at the point where we find ourselves today. And I refer specifically to the comparative study of behavior, which Williams himself employed effectively in his book. Comparative study is the stock-in-trade of the systematist. It has never been the stock-in-trade of any other group of biologists in a very extensive, persistent, or pertinent fashion—least of all, perhaps, the behaviorists.

One of my favorite psychologists argued recently that molecularly oriented biologists are on the wrong track when they believe they can predict everything of significance about the biological world through a knowledge of structure and function at molecular and submolecular levels. He noted that, while theoretically this may be possible, it is unreasonable or impractical unless one knows beforehand what it is that he must be able to predict. I suggest that the same criticism can be leveled at many people studying behavior. They expect to be able to predict from precise, quantitative, laboratory experimentation without having any idea of the complexity, the variety, or even the nature of the things they will have to predict.

Zoologists left behavior largely to the psychologists, long past the time of knowing that psychologists in general do not answer the kinds of questions that zoologists must have answered. And systematists, in turn, have left zoological studies of behavior to the experimental zoologists, despite the fact that certain questions about behavior that are of importance to everyone are not going to be answered for a very long time using the methods generally conceded to experimental biology.

There is nothing mysterious about the comparative method. Yet I am convinced that many systematists and other biologists who use it all the time scarcely know what they are accomplishing with it, are not sufficiently prepared to explain and defend its problem-solving value, and, in any case, could not give a clear exposition of its usefulness to systematics or to biology in general.

The Problem of Instinct

So far, I have said nothing about what undoubtedly has been the knottiest problem in the study of animal behavior, and the one responsible more than any other,
I suppose, for slowing the advance of comparative study of behavior. If we call this
the "problem of instinct," most people have a good idea what is meant. It would
be more descriptive, however, to term it the problem of the extent and nature of
hereditary influences in behavioral variations, both between species and among
the individuals of each species.

Adaptation is a result of selective action on alternative genetic phenomena.
Therefore, it is critical for my topic to determine which behavioral variations correlate with genetic variations. Few people challenge the idea that certain species
differences, such as frog and insect calls, firefly flashes, or other behavioral characteristics identified as reproductive isolating mechanisms in any kinds of animals,
have genetic bases. And, we know from hybridization experiments that insect and
anuran call differences do indeed have genetic bases. In fact, results from crossing
experiments on crickets and frogs probably are cited more frequently in reviews
concerning transmissibility of behavioral variations than are any experiments with
other kinds of animals.

The systematist wants to know more about this. He wants to know whether he
can be sure that he is not examining some behavioral difference that has nothing to do with hereditary differences. Some systematists and other biologists have
gotten involved in long, bitter, and futile arguments about whether heredity or
environment has greater influence in determining the characteristics of particular
behavior patterns.

Concerning this topic, we are pursued now by a whole string of admonish-
ments: "To ask how much a given aspect of behavior depends upon genetic factors and how much upon environmental factors is like asking how much of the
area of a field depends upon its length and how much upon its width"; "Nothing
is inherited but the genotype and a little cytoplasm"; "Heredity is particulate; but
development is unitary"; "Instead of speaking of this or that trait as genetic or
environmental, the correct way is to ask yourself which, and the extent to which,
differences in characters are due to environment on the one hand and to heredity
on the other."

Konishi (1966) has recently written a paper that, I think, clarifies some issues
Involved in this problem. He points out that, as one of our shortcomings, we have
acted as though it is always true that, in behavior, stereotypy = species specificity =
inheritance = central coordination = spontaneity = self-differentiation. These factors are not strictly correlated and, as with learning, what has been called "instinctive" behavior really is not a single phenomenon, and it should not be treated as
if it were.

But not all these issues are of great or immediate concern to the systematist or
evolutionary biologist interested in behavior. What is of concern is predictability.

And it is very likely that significant increases in predictability, in many cases, can be attained sooner by insightful, properly directed, broad-scale (even superficial) comparisons than by detailed studies of development of specific patterns of behavior in individual animals or species. The comparative anatomists, as many ethologists have emphasized, have already shown this to be true. We systematists seem to have allowed ourselves to be overly concerned about precisely how individual patterns of behavior develop. We do not know a great deal about the development of morphology in a wide variety of animals; but we do know a very great deal about speciation, adaptation, and phylogenetic history—all of which knowledge was gained directly, almost solely, from comparisons of those very features of anatomy whose development we still do not understand.

When we will have carried out broad-scale comparative studies of behavior similar to those available in anatomy, and when we begin to acquire the glimmers of understanding that will come from predictiveness based on such studies, then those investigators concerned chiefly with the developmental bases of phenotypic differences will, indeed, have something to think about and work with. Because the question then would concern *how much* genetic variation is involved, a considerably sharper focus should be provided for the investigations of many biologists now skeptical that broad-scale comparisons can be made in the absence of extensive information on developmental pathways and stimuli for particular behavioral units.

This remark may raise some eyebrows, but I suggest that behavioral variations that at first glance appear useful to systematists—particularly to those working at and above the species level—rarely lack correlation with specific genetic variations. For example, is anyone here in a position to describe a species difference in behavior—any species difference in behavior—that he has cause to suspect does not have a genetic basis? [2]

Further, and of great importance, the extent and nature of correlations between behavioral variations and genetic variations—or their absence—is predictable to a large extent.

One aspect of such predictability can be exemplified by cricket calls. Examination of cricket biology soon reveals that in most temperate species only the eggs pass the winter, and that the auditory organs are not functional until maturation or near-maturation. With this information alone we can predict confidently that (at least usually) selection favors insulation from influences by environmental sounds in the establishment of the pattern of the call (R. D. Alexander, "Arthropods" *in* T. Sebeok, *Animal Communication,* to be published by Indiana University Press), for there can be no appropriate sounds available to copy.

---

[2] Alexander and Bigelow (1960) have given a possible example. Males of *Gryllus veletis* generally are much more aggressive than those of *G. pennsylvanicus.* They also occur more sparsely, and the difference can be erased, or even reversed, if males of *G. veletis* are crowded in the laboratory and males of *G. pennsylvanicus* are isolated.

On the other hand, even a limited knowledge of passerine-bird biology allows the reverse prediction: that most young birds probably have evolved specific ways of being influenced by their parents' song patterns. There are at least two reasons: the overlap of young and adults in each generation, and an apparent premium on individuality in song pattern; the latter is associated with the presence of specialized parental behavior and tendencies toward monogamy and is promoted by having part of the pattern learned.

In precocious birds both the critical periods of song learning and the imprinting of following behavior are predictable on the same general basis. Even the indiscriminateness of suitable stimuli for imprinting of following behavior is predictable, for the situation is such that unsuitable stimuli are not likely to be available, and selection, therefore, will have no chance to focus on a restricted group of stimuli. Likewise, we would predict that different populations of birds within a given species should sometimes have song differences that lack genetic bases, as is the case with human languages.

The psychologists who chastised ethologists for erecting dichotomies with regard to learned and unlearned behavior were right, for many ethologists had their dichotomies out of focus. But this argument became a source of confusion rather than clarification when it took the form of rejecting all implications of important dichotomies in the way behavior patterns develop. An important dichotomy can be identified in the examples I have just given: Does the selection favor the use of a given stimulus (sound, for example) in the establishment of a pattern in the same modality as the stimulus or does it specifically favor insulation from all stimuli in that particular modality? This is a dichotomy as to direction of selection, and it leads to extreme differences in certain relationships between genetic phenomena and behavioral characteristics. We identify it and discover its significance by studying adaptation and natural selection in relation to behavioral development.

I suggest that selection acting consistently on any kind of behavioral variation, regardless of its original basis, will usually result in the presence, ultimately, of genetic variation that relates directly to the behavioral variation. This would mean that the more ancient a behavioral difference, the more likely it is to have some genetic basis. By this I do not mean to imply any special kind of selection, I simply mean that selection will work on both genetic and non-genetic variations, but evolution will occur only when genetically based variations become available.

No broad-scale attempt has been made to study adaptation in behavioral terms by the kind of comparison and prediction that I have just described. Yet, if what I have argued is true, such attempts would be fruitful, even those dealing with the behavior of man, of all organisms the most "labile" in the functioning of his phenotype.

I will use a simple example cited by Williams (1966a), who notes that the females of many kinds of animals usually are described as being more "coy" or/

discriminating or reluctant in copulation than the males (or one could turn it around and say that the males are more "aggressive" in courtship), and he notes that this is predictable because in each copulation or fertilization the female invests a greater proportion of her total reproductive potential than the male invests of his. If this argument is correct, then, as Williams points out, the situation should be reversed in parental animals in which the male is solely responsible for the zygotes, or more involved in parental behavior. Such reversals have been reported in pipefishes in the genus *Syngnathus,* in which the males carry the fertilized eggs (Fiedler, 1954), though not in all such fish (Breder and Rosen, 1966; Straughan, 1960), and also in such birds, as some tinamous and phalaropes, in which the males incubate the eggs and protect the young (Bent, 1927; Tinbergen, 1935; Höhn, 1967).

Similar reversals as to which sex behaves territorially, fights off intruding individuals, and courts more aggressively have been reported in the ornate tinamou (Pearson and Pearson, 1955), red phalarope, northern phalarope, and Wilson's phalarope (Tinbergen, 1935; Höhn, 1967; Bent, 1927). Polyandry is more likely to be prominent in such animals, and strict polygyny ought to be rare, although both polygamy (or promiscuity on the part of both sexes) and monogamy have been reported (Lancaster, 1964; Höhn, 1967). Polygyny, on the other hand, is prominent among species in which the females carry most or all of the parental responsibility, and polyandry is almost nonexistent. Some of the disagreements in the literature (e.g., see Höhn, 1967) may result from differences in sex ratios among demes studied by different investigators. In some cases, what happens when sex ratios are locally or temporarily uneven may be important in understanding how selection has operated.

The relationship between reproductive effort and proportion of reproductive potential involved in any circumstance or event can be extended to include not only the proportion of eggs or sperm used per copulation but also the proportion of the breeding season used per clutch or pregnancy and the proportion of the total probable reproductive life involved in each season, as Dr. Tinkle demonstrated on this program. Such considerations of proportions must include also the likelihood of changes in reproductive possibility—such as improvements through learning about one's mate or about the food and predators in one's territory—and the likelihood of improvement in weather conditions. Williams (1966b) and Lack (1966) have pointed out that this means that longer juvenile lives will correlate roughly with longer reproductive lives, and that clutch sizes will increase with age. We should expect, especially in long-lived, monogamous animals with specialized parental behavior, that selection continually will maximize the slope of a line depicting the increasing reproductive ability of individuals and pairs.

Since man's plasticity in behavior seems for a long time to have been a major reason for our reluctance to discuss the general problem of behavior in relation to heredity and, therefore, a major reason for the reluctance of systematists and other biologists to use behavior in comparative work, it is appropriate that I conclude by referring to the possibility of heredity in an example of variation in man's behavior. I will use the previously mentioned theory of female "coyness,"; or difference between male and female, and suggest that what we frequently and sometimes jokingly refer to as the "double standard" in man's sexual behavior is, in part, a reflection of differences in selective action on male and female behavior during man's evolutionary history.

In general, in both polygynous and monogamous animals with specialized parental behavior, selection should favor females that promote monogamy and should favor males that promote polygyny. Even in evolutionary lines in which monogamy is never actually realized, tendencies toward it in females would be favored consistently if the male's cooperation in any way promoted the female's reproductive success. Likewise, even in a monogamous line polygynous tendencies in males often would be favored because in a species in which the female is responsible for the fertilized eggs a male is much more likely to benefit from, shall we say, "stealing" copulations with his neighbors' females than is the female who indulges in the same kind of behavior. Tendencies toward polyandry in man are evidently rare, but tendencies toward polygyny are not nearly so rare.

Is it reasonable to argue that there likely are no genetic correlates underlying even subtle intraspecific differences of this sort in an organism as plastic as man, when selection on man's breeding system has probably been consistent in the ways I have described all through man's evolutionary history? I think not.

Concluding Remarks

I have dealt in this paper with a few points that I believe will be useful in making the analysis of behavior more important in systematic work than it has been in the past, in searching for both similarities and differences among organisms. To bring these two fields into closer cooperation, I believe we need, chiefly, to be (1) more aware of the role of the comparative method in biology and in behavioral analysis, (2) more thoughtful in our searches for behavioral variations likely to be correlated with genetic differences, and (3) as systematists, more cognizant than we have been of the significance of studying adaptation directly, both by experimentation and by comparison.

Acknowledgments

I wish to thank Daniel Otte, Ann Pace, and Mary Jane West, graduate students at the University of Michigan, for assistance in developing the ideas presented here and for critical examination of the manuscript at various stages.

References

Alexander, R.D. and Bigelow, R. 1960. Allochronic speciation in field crickets, and a new species *Acheta veletis. Evolution* 14:334–336.

Bent, A.C. 1927. *Life Histories of North American Shore Birds. Part 1.* Washington, DC: US National Museum Bulletin 142.

Breder, C.M. and D.E. Rosen, 1966. *Modes of reproduction in fishes.* Neptune City, NJ: T.F.H. Publications.

Fiedler, K. 1954. Vergleichende verhaltenstudien an seenadeln, schlangennadeln und seepferdchen(Syngnathidae). *Z. Tierpsychol.* 11:358–416.

Höhn, E.O. 1967. Observations on breeding biology of wilson's phalarope (*Steganopus tricolor*) in central Alberta. *Auk.* 84:220.

Konishi, M. 1966. The attributes of instinct. *Behaviour* 27:316–327.

Lack, D. 1966. *Population Studies of Birds.* Oxford: Clarendon Press.

Lancaster, D.A. 1964. Life history of the Boucard Tinamou in British Honduras. Part II: Breeding Biology. *Condor* 66:253–276.

Pearson, A.K., and Pearson, O.P. 1955. Natural history and breeding behavior of the tinamou, *Nothoprocta ornata. Auk* 72:113–127.

Straughan, R. P. L. 1960. 100 seahorses spawn. *Aquarium J.* 31:302–308;325–326.

Tinbergen, N. 1935. Field Observations of East Greenland Birds. I. The Behaviour of the Red-Necked Phalarope (*Phalaropus lobatus* L.)in Spring. *Ardea* 24:1–42.

Wheeler, W. M. 1905. An interpretation of the slave-making instincts in ants. *Bull. Am. Mus. Nat. Hist.* 21:1–16.

Williams, G.C. 1966a. *Adaptation and Natural Selection.* Princeton, NJ: Princeton University Press.

Williams, G.C. 1966b. Natural selection, the costs of reproduction, and a refinement of Lack's principle. *Am. Natur.* 100:687–690.

2 }

Cooperation

About the Social Contract

Am I an outlaw, you ask? Am I?

Tell me which rules of this desperately
social beast whose genes I share must be
forgotten, ignored, or despised for me to qualify.

Tell me the shape of the particular horrors I
personally must plunge into a man's eye,

how I must shave my ideas or my
hair to stay out of that No-Baboon's Land,

and why.

Explain to me the times when my
personal conscience cannot fly
in the face of the social ones of this time
or that, concerning who must die,
and in what order.

Tell me whom I may or may not love, and why.
Say how much of me will be permitted to identify

with the unusual dreams of one whose eye
may seem to see it all differently. Apply
your (or is it "our"?) laws, and I
will give you my reply.

And why.

Or have I?

Alexander 2011, p. 20

A New Theory of Cooperation
Steven A. Frank

The function of laws is to regulate...the reproductive strivings of individuals and subgroups within societies, in the interest of preserving unity in the larger group....Presumably, unity in the larger group feeds back beneficial effects to those...that propose, maintain, adjust, and enforce the laws.

ALEXANDER 1979, P. 240

A corollary to reproductive opportunity leveling in humans may occur through mitosis and meiosis in sexual organisms....The leveling of reproductive opportunity for intragenomic components...is a prerequisite for the remarkable unity of genomes.

ALEXANDER 1987, P. 69

When I first read Alexander's (1987) *The Biology of Moral Systems*, I was stunned to find a truly new theory of biological cooperation. I had thought of the theory of social evolution as an already mature subject, fully developed with regard to fundamental concepts. Yet, in Alexander's book, I could see a completely new way to think about social evolution. In this essay, I explain that new theory and where it came from.

Dick Alexander, Bill Hamilton, and Bob Axelrod taught me the classical theory of social evolution at the University of Michigan in the late 1970s and early 1980s. That classical theory provided two explanations for cooperation: kin selection and reciprocity. At that time, the theory seemed mature and complete with regard to setting the foundation for the field. I hardly expected to come across a totally new way of thinking about the evolution of cooperation.

Yet the main weakness of the theory was also apparent. Extensive cooperation occurs between nonrelatives. Different genes in genomes are functionally integrated but not related. Larger human societies often have many highly cooperative but distantly related individuals. Some of this cooperation between nonkin can be explained by extensions of reciprocity to a general notion of mutual benefit for interacting partners (West-Eberhard 1975). In the early 1980s, kin selection plus these extended notions of reciprocity were the main conceptual tools.

Those limited conceptual tools led to blind spots about unsolved problems. Only rather forced theories of mutualism could work for the nearly complete integration of genes into cooperative genomes. Only a very enthusiastic belief in the scope of reciprocity could explain the broad social integration in larger groups of weakly related humans.

In the 1970s, Alexander had already seen the problem. Primitive human groups were perhaps something like chimpanzee troops, in which kin selection and reciprocity could be at least roughly matched to the level of cooperation and social integration. But human history has been characterized by a great expansion in group size. How could one account for the transitions from chimpanzee troop to village to nation-state? Group against group competition potentially explains the benefit of larger group size. But what prevented internal conflict in larger groups, which could no longer be bound by close relatedness or tight reciprocity?

Without some mechanism to control those internal conflicts, larger groups could not be sustained. Stating the puzzle in that way, the solution is clear. There must be some mechanism that suppresses internal conflict. If individuals cannot compete against members of their own group, then they can only increase their success by increasing the success of the group as whole. Suppression of internal competition unites all group members into a cohesive and cooperative unit.

That idea of internal suppression and group cohesion is the new theory of cooperation that I took away from *The Biology of Moral Systems*. I was truly surprised that so simple and so powerful an explanation could appear as a new idea in 1987. For Alexander, human moral systems were systems for regulating internal competition within groups to promote group cohesion and success in competition against other groups. The broader generality of regulating internal competition in biological evolution was clear to Alexander, and he had in fact reasoned about human evolution by extension of his understanding of the evolutionary history of life.

Of course, no truly deep and general idea appears out of nothing. The first clear description often coincides with others who soon express similar notions arising out of independent lines of study. In the early 1990s, I became increasingly interested in Alexander's theory of internal suppression as a way to achieve group integration. My own interest developed because I was trying to understand the principles by which early genomes arose near the origin of life. How did different replicating molecules come together to form larger, cohesive genetic systems? I felt that was an important question, because it seemed that we could not say we understood any aspect of sociality if we could not at least give a plausible theory for the origin of genomes.

The more I thought about it, the more I realized that, at that time in the early 1990s, we really did not have the conceptual tools to give a plausible theory of the origin of genomes. So I was brought back to Alexander's vision of how biological systems could achieve broader group integration in spite of limited relatedness and limited opportunities for reciprocity and synergism. Specifically, I had to understand how a theory of suppression of internal competition could work broadly in evolutionary theory.

In effect, I needed to work out the step-by-step aspects of Alexander's theory applied to the origin of genomes, in the absence of any advanced behavioral or cognitive abilities. My own work appeared in Frank 1995, but that is not the key point here. What matters is that I also had to understand the historical genesis of Alexander's thinking and the different parallel lines by which similar ideas developed through the 1980s and 1990s (Frank 2003).

Alexander cited two clear precedents to his own ideas. First, John Rawls's (1971) famous theory of justice from moral philosophy developed the notion of the "veil of ignorance." A just society establishes rules that individuals regard as fair from behind a veil of ignorance about their position within society. An individual may, in practice, end up on one end or the other of any particular social interaction. (Harsanyi [1953] developed a similar idea; see Skyrms [1996] for discussion of these ideas in an evolutionary context.)

It does not pay to argue the fine details of how precisely these humanistic thoughts presage current evolutionary understanding. These thoughts from moral philosophy do contain the following kernels: group cohesion returns benefits to individuals, and randomization of position levels individual opportunity and promotes group cohesion. In other words, given randomization of individual success within the group, an individual increases success only by increasing the success of the group as a whole.

The second precedent came from Leigh's (1971, 1977) work on Mendelian segregation in meiosis. In standard diploid genetics, each genetic locus has one allele from the mother and one from the father. Each gamete made by an individual has either the maternal or paternal allele. Mendelian segregation, or fair meiosis, gives an equal chance to maternal and paternal alleles of being in a successful gamete. Meiotic drive subverts fairness by giving one allele a greater chance of transmission. The pieces of chromosomes that can drive against their partners gain a reproductive advantage by increasing their chance for transmission to offspring. As driving chromosomes spread because of their transmission advantage, they often carry along deleterious effects that are partly protected from selection by being associated with transmission advantage (Zimmering et al. 1970).

Other parts of the genome lose when a driving chromosome carries with it deleterious effects into the majority of gametes. Suppression of drive has the immediate effect of reducing association with the deleterious effects of driving chromosomes; it has the long-term consequence of taking away the transmission advantage that protects the deleterious effects. Drive suppression thus helps to purge the genome of the deleterious effects carried by driving chromosomes. The many genes of the genome repress the drive "as if we had to do with a parliament of genes, which so regulated itself as to prevent 'cabals of a few' conspiring for their own 'selfish profit' at the expense of the 'commonwealth'" (Leigh 1977, p. 4543).

When meiosis is fair, randomization puts each allele behind a veil of ignorance with regard to its direct transmission (interests) in each progeny. Behind the veil, each part of the genome can increase its own success only by enhancing the total number of progeny and thus increasing the success of the group. However, discussing "interests" in arguments about how natural selection operates can be misleading. In this case, natural selection directly favors the immediate advantage of drive suppression, which reduces association with the deleterious effects that often hitchhike along with drive. The long-term advantage of purging the hitchhiked deleterious effects also contributes to favoring drive suppression when groups compete against groups, for example, species against species (Leigh 1977).

Leigh (1977) noted that alignment of individual and group interests shifts selection to the group level. However, meiosis was the only compelling case known at that time. Without further examples, there was no reason to emphasize repression of internal competition as an important force in social evolution and the formation of evolutionary units. From the conceptual point of view, it may have been clear that repression of internal competition could be important, but not clear how natural selection would favor such internal repression.

Alexander and Borgia (1978) joined Leigh in promoting the possible great potency of internal repression in shaping interests and conflicts in the hierarchy of life. From this, Alexander (1979, 1987) developed his theories of human social structure (see introductory quotes). In this theory, intense group-against-group competition shaped societies according to their group efficiencies in conflicts. Efficiency, best achieved by aligning the interests of the individual with the group, favored in the most successful group's laws that partially restricted the opportunities for reproductive dominance within groups. For example, Alexander (1987) argued that socially imposed monogamy levels reproductive opportunities, particularly among young men at the age of maximal sexual competition. These young men are the most competitive and divisive individuals within societies and are the pool of warriors on which the group depends for its protection and expansion. If these young men cannot compete against their neighbors within their groups, then they can increase their success only by cooperating with their neighbors in competition against other groups.

Rawls and Leigh directly influenced Alexander. With regard to biology, Leigh (1971, 1977) may have been the first to emphasize how repression of internal competition aligns individual and group interests. However, meiosis provided the only good example at that time, so the idea did not lead immediately to new insight. Alexander (1979, 1987) used Leigh's interpretation of meiosis as the foundation for his novel theories about human social evolution. I was aware of the discussion about meiosis in the 1970s, but I only realized the general implications for repression of competition as a powerful evolutionary force after reading Alexander (1987). With two examples—meiosis and the structuring of social groups—I could see how a simple idea could be applied to different contexts.

Around the same time, Buss (1987) was independently analyzing cellular competition in metazoans. Many multicellular animals are differentiated into tissues that predominantly contribute to gametes and tissues that are primarily nonreproductive. This germ-soma distinction creates the potential for reproductive conflict when cells are not genetically identical. Genetically distinct cellular lineages can raise their fitness by gaining preferential access to the germline. This biasing can increase in frequency even if it partly reduces the overall success of the group.

One way to control renegade cell lineages is to enforce a germ-soma split early in development (Buss 1987). This split prevents reproductive bias between lineages during subsequent development. Once the potential for bias has been restricted, a cell lineage can improve its own fitness only by increasing the fitness of the individual. This is another example of how reproductive fairness acts as an integrating force in the formation of units.

Buss stimulated Maynard Smith (1988) to consider how social groups became integrated over evolutionary history. Maynard Smith disagreed with Buss's particular argument about the importance of the germ-soma separation in metazoans. But in considering the general issues, Maynard Smith had in hand several possible examples of group integration, including meiosis and genomic integration and perhaps repression of cellular competition in metazoans. From these examples, Maynard Smith (1988, pp. 229-230) restated the essential concept in a concise and very general way:

> One can recognize in the evolution of life several revolutions in the way in which genetic information is organized. In each of these revolutions, there has been a conflict between selection at several levels. The achievement of individuality at the higher level has required that the disruptive effects of selection at the lower level be suppressed.

This view led to Maynard Smith and Szathmary's (1995) book *The Major Transitions in Evolution*, a popular account of the history of life based on the key events in which competition within groups became suppressed. Maynard Smith and Szathmary's synthesis arose independently from Alexander's work. But Alexander was the first to take Leigh's insight about meiosis and apply that view in a general and powerful way to propose a solution to the puzzle of progressive social integration of complex groups. Buss (1987) came soon after. In this light, *The Major Transitions in Evolution* can be seen as the great development and synthesis of Alexander's insight about suppression of competition in the evolution of cooperation.

Since 1995, suppression of competition has developed into a broad research topic applied to many puzzles of cooperative evolution. I list just three of the most recent applications taken from articles published during the past few years. These applications give a sense of the research that has grown out of Alexander's original insights. Frank (2003) reviews several additional examples.

Higginson and Pitnick (2011) summarize aspects of competition between sperm within the ejaculate of a single male. They conclude by noting the potential importance of mechanisms that repress such intra-ejaculate competition (p. 265):

> Competition between sibling sperm may reduce male reproductive fitness, even in monogamous systems, by reducing the number of

fertilization-competent sperm per ejaculate (e.g. killing of Y-chromosome-bearing sperm in the case of sex chromosome meiotic drive) or by displacing sibling sperm from the site of storage or fertilization. Male-level selection for adaptations that reduce intra-ejaculate competition in favour of improved whole-ejaculate success aligns the interests of males and the sperm they produce. When competition between sibling sperm is restricted or prevented, individual sperm fitness can only be maximized by enhancing inter-ejaculate competitive success (Frank 2003). We refer interested readers to two recent reviews for a more detailed discussion of sperm-level selection and male-sperm conflict (Immler 2008; Pizzari and Foster 2008).

Hoffmann and Korb (2011) studied replacement of reproductives in termite colonies. They emphasized the potential importance of mechanisms that repress competition during the replacement phase of the colony life cycle. They noted "Strong conflicts were predicted as all colony members (except soldiers) have the capability to become neotenic replacement reproductives. Our behavioural observations first implied that there was no overt conflict during replacement, but the killing of some neotenics suggested that conflict was not completely suppressed. Various mechanisms were found that may regulate conflict" (p. 270). Among possible mechanisms that may repress competition in the colony, they emphasize a possible form of randomization analogous to fair meiosis:

> One proximate mechanism that can reduce overt conflict is the sensitive period: a short period during the moulting interval when the developmental fate of an organism at the next moult is determined.... During this period, individuals are sensitive to environmental stimuli, such as the absence of reproductives.... Among the workers, only individuals in the sensitive period are able to respond to orphaning and become neotenic replacement reproductives. As all workers regularly pass through this period, they all have a fair chance of becoming neotenic, while at the same time the number of actually competing individuals is reduced. Thus, the sensitive period functions as a "fair lottery" mechanism similar to, for example, Mendelian segregation during gamete formation (Frank 2003).

The evidence remains ambiguous with respect to this argument for a fair lottery mechanism to repress competition. The point here is that Alexander's conceptual framing of cooperative evolution plays a key role in Hoffman and Korb's (2011) thinking about termite biology.

Many other aspects of social insect cooperation have been interpreted with respect to repression of competition within colonies (Ratnieks et al. 2006). Smith et al.'s (2009) recent study continues the development of this topic. They summarize their conclusions (p. 78):

> Cheaters are a threat to every society and therefore societies have established rules to punish these individuals in order to stabilize their social

system.... Recent models and observations suggest that enforcement of reproductive altruism (policing) in hymenopteran insect societies is a major force in maintaining high levels of cooperation.... In order to be able to enforce altruism, reproductive cheaters need to be reliably identified.... [Our data provide] the first direct evidence that cuticular hydrocarbons are the informational basis of policing behaviors, serving a major function in the regulation of reproduction in social insects. We suggest that even though cheaters would gain from suppressing these profiles, they are prevented from doing so through the mechanisms of hydrocarbon biosynthesis and its relation to reproductive physiology. Cheaters are identified through information that is inherently reliable.

The current literature contains many additional applications of Alexander's theory of cooperative evolution. Repression of competition has indeed joined kin selection as the second key process in the major evolutionarily transitions throughout the history of life.

Acknowledgments

Parts of this introduction were taken from Frank (2003). My research is supported by National Science Foundation grant EF-0822399, National Institute of General Medical Sciences MIDAS Program Grant U01-GM-76499, and a grant from the James S. McDonnell Foundation.

References

Alexander, R.D. 1979. *Darwinism and Human Affairs*. Seattle: University of Washington Press.

Alexander, R.D. 1987. *The Biology of Moral Systems*. New York: Aldine de Gruyter.

Alexander, R.D. 2011. *The Mockingbird's River Song: Poems, Essays, Songs and Stories, 1946-2011*. Manchester, MI: Woodlane Farm Books.

Alexander, R.D., and Borgia, G. 1978. Group selection, altruism, and the levels of organization of life. *Annu. Rev. Ecol. Syst.* 9:449–474.

Buss, L.W. 1987. *The Evolution of Individuality*. Princeton, NJ: Princeton University Press.

Frank, S.A. 1995. Mutual policing and repression of competition in the evolution of cooperative groups. *Nature* 377:520–522.

Frank, S.A. 2003. Repression of competition and the evolution of cooperation. *Evolution* 57:693–705.

Harsanyi, J. 1953. Cardinal utility in welfare economics and the theory of risk taking. *J. Pol. Econ.* 61:434–435.

Higginson, D.M. and Pitnick, S. 2011. Evolution of intra-ejaculate sperm interactions: do sperm cooperate? *Biol. Rev.* 86:249–270.

Hoffman, K. and Korb, J. 2011. Is there competition over direct reproduction in lower termite colonies? *Anim. Behav.* 81:265–274.

Immler, S. 2008. Sperm competition and sperm cooperation: the potential role of diploid and haploid expression. *Reproduction* 135:275–283.

Leigh, E.G., Jr. 1971. *Adaptation and Diversity*. San Francisco: Freeman, Cooper.

Leigh, E.G., Jr. 1977. How does selection reconcile individual advantage with the good of the group? *Proc. Natl. Acad. Sci. USA* 74:4542–4546.

Maynard Smith, J. 1988. Evolutionary progress and levels of selection. In: M.H. Nitecki (ed.), *Evolutionary Progress*. Chicago: University of Chicago Press.

Maynard Smith, J., and Szathmary, E. 1995. *The Major Transitions in Evolution*. San Francisco: Freeman.

Pizzari, T. and Foster, K. 2008. Sperm sociality: Cooperation, altruism, and spite. *PLoS Biol.* 6, e130.

Ratnieks, F.L.W., Foster, K.R., and Wenseleers, T. 2006. Conflict resolution in insect societies. *Annu. Rev. Entomol.* 51:581–608.

Rawls, J. 1971. *A Theory of Justice*. Cambridge: Harvard University Press.

Skyrms, B. 1996. *Evolution of the Social Contract*. Cambridge: Cambridge University Press.

Smith, A.A., Holldobler, B., and Liebig, J. 2009. Cuticular hydrocarbons reliably identify cheaters and allow enforcement of altruism in a social insect. *Current Biology* 19:78–81.

West-Eberhard, M.J. 1975. The evolution of social behavior by kin selection. *Q. Rev. Biol.* 50:1–32.

Zimmering, S., Sandler, L. and Nicoletti, B. 1970. Mechanisms of meiotic drive. *Annu. Rev. Genet.* 4:409–436.

HUMANS

Ultrasociality Based on Reciprocity

Excerpt from: Alexander, R. D. 1986. *The Biology of Moral Systems*. New York, Aldine Press.

Humans have taken a route to ultrasociality entirely different from that of the social insects. In the largest and evidently most unified or stable human groups (i.e., large, long-lasting nations) partial (rather than complete) restrictions on reproduction have the effect of leveling or equalizing opportunities to reproduce. Socially imposed monogamy and graduated income taxes are examples of such *reproductive opportunity leveling*. The tendency in the development of the largest human groups, although not always consistent, seems to be toward equality of opportunity for every individual to reproduce via its own offspring, rather than toward specializing baby production in one or a few individuals and baby care in the others. However humans specialize and divide labor, they nearly always insist individually on the right to carry out all of the reproductive activities themselves. One consequence is that the human individual has evolved to be extraordinarily complex (and evidently to revere individuality), and another is that the complexity and variety of social interactions among human individuals is without parallel. Because human social groups are not enormous nuclear families, like social insect colonies, a third consequence is that competition and conflicts of interest are also diverse and complex to an unparalleled degree. Hence, I believe, derives our topic of moral systems. We can ask legitimately whether or not the trend toward greater leveling of reproductive opportunities in the largest, most stable human groups indicates that such groups (nations) are the most difficult to hold together *without* the promise or reality of equality of opportunity (see also Alexander 1974, 1979a; Alexander and Noonan, 1979; Strate, 1982; Betzig 1986).

Other Special Cases of Cooperation

GENES IN GENOMES

A corollary to reproductive opportunity leveling in humans may occur through mitosis and meiosis in sexual organisms. It has generally been overlooked that these very widely studied processes are so designed as usually to give each gene or other genetic subunit of the genome (= the genotype or set of genetic materials of the individual) the same opportunity as any other of appearing in the daughter cells. Alexander and Borgia (1978) and Williams (1979) have speculated that this equality of opportunity came about because only alleles with equal (or better) likelihoods of being present in daughter cells have survived; possibly, more generalized mechanisms have come to be involved in modern forms. It is not inappropriate to speculate that the leveling of reproductive opportunity for intragenomic components—regardless of its mechanism—is a prerequisite for the remarkable unity of genomes, some of them composed of thousands or hundreds of thousands of recombining, potentially independent genes and other subunits (Leigh, 1983; Alexander and Borgia, 1978).

MONOGAMOUS PAIRS

To the extent that males and females (of any species) commit themselves to lifetime monogamy, the interests of two individuals in a pair approach being identical. This point is often confused by biologists and social scientists alike (e.g., Dawkins, 1976, and Sahlins, 1976, both thought that unrelated spouses necessarily disagree more than relatives). The reason is the same as that causing identity of interests in the different individual workers in a eusocial insect colony: the two different individuals realize their reproduction through identical third parties which each of them gain by helping a great deal. In the case of worker insects the third parties are the reproductive brothers and sisters produced by the queen, their mother. If the queen dies and is replaced by one of the workers' siblings, the situation may not be altered even though the workers are less closely related to a sister's offspring than to their sisters, and less closely related to a sister's offspring than is the sister herself. When a queen changeover occurs, unless workers retain some ability and likelihood of themselves becoming the queen (and in many modern species they have lost this ability), they can do no better than by cooperating fully with one another to produce reproductive nieces and nephews.

Given that the members of monogamous pairs are evolved to invest parentally, then, to the extent that (1) philandering is unlikely or too expensive to be profitable, and (2) the relatives of one or the other are not significantly more available for nepotistic diversions of resources, each member of the pair will profit from complete cooperation with the other to produce and rear their joint offspring. In humans this condition is most likely in (1) societies in which (a) families live and work separately and (b) husband and wife are in fairly close contact most of the time and (2) societies in which married couples are "neolocal," living in some new location apart from both sets of relatives but close enough to be affected by the interests of their kin networks in sustaining the marriage. In such societies (which historically have probably been most often agricultural), I predict that the devotion of husband and wife will be measurably most complete.

Aside from clones, social insects and humans have developed the largest known societies, measured by numbers of complexly interacting individuals. They are also the most complexly communicating organisms. They have both accomplished this by expanding confluences of interest and reducing conflicts of interests, and it is at least possible that monogamy was involved in both cases, early in the evolution of social insects and late in the evolution of the largest human societies. What we have to understand, for both social insects and humans, is how the situations develop in which workership and monogamy, respectively, come to be the rule or norm. I believe I am correct in saying that in neither case are the answers yet available.

Monogamy and Reproductive Opportunity Leveling

Our understanding of the manner in which monogamous pairs come to cooperate is not much better than our understanding of why social insect colonies sometimes become huge and sometimes do not. Because monogamy in large technological nations is imposed socially (meaning that the costs of its alternatives are imposed by the rest—or some part—of society), understanding its background becomes a part of the effort to understand moral systems. Alexander et al. (1979; see also Alexander, 1975) have argued that socially or legally imposed monogamy is a way of leveling the reproductive opportunities of men, thereby reducing their competitiveness and increasing their likelihood of cooperativeness. The imposition of monogamy by custom or law has the interesting effect of reducing both male-male and male-female conflicts to a minimum, especially when clans are discouraged (as in nation states: see Alexander, 1979a, pp. 256-259), and when married couples do not have differential access to their respective relatives (e.g., when they are "neolocal" or reside in a new locality rather than becoming a part of one or the other extended family of relatives). Moreover, the combination of socially or legally imposed monogamy, neolocality, and close association of the married couple in work not only leads to minimizing of philandering and conflict of interest between husband and wife, but also characterizes the largest (and perhaps the most unified—or durable—of all large) human societies. Young men at the age of maximal sexual competition are the most divisive and competitive class of individuals in human social groups; they are also the pool of warriors. It is not trivial that socially imposed monogamy (and the concomitant discouragement of clans as extended families that control members) correlates with (1) justice touted as equality of opportunity; (2) the concept of a single, impartial god for all people; and (3) large, cohesive, modern nations that wage wars and conduct defense with their pools of young men (Alexander, 1979a). To a large extent socially imposed monogamy has spread around the world by conquest. The social imposition of monogamy thus simultaneously (1) inhibits the generation of certain kinds of within-group power dynasties that might compete with government and lead to divisive within-group competition and (2) promotes those activities and attitudes that generate and maintain success in the wielding of reciprocity as the binding cement of social structure (honesty, sincerity, trust).

Humans almost certainly began to evolve their social tendencies and capabilities in small kin groups. If so, during that process they incidentally acquired the capability to maintain social organization in ever larger and more complex social groups through systems of reciprocity rather than nepotism per se. In such groups there is only one way to approach an equalization of reproductive opportunity, and that is by sets of rules or moral systems. In humans the laws and mores of larger and larger groups seem increasingly to (1) guarantee to every individual the right to produce and rear its own offspring and (2) restrict the amount and likelihood of variation in reproduction among families. China is currently an extreme in both size (over one billion) and regulation of reproduction (Keyfitz, 1984). Until 1981 or 1982, government assistance was given for a first child, but funds were withdrawn if a second was born. More recently it was reported (e.g., Ann Arbor, Michigan, *News*, 1982; "Nova" and "60 Minutes" television programs, 1984) that enormous pressure for sterilization followed the birth of a single child. Not long ago, India briefly attempted to require sterilization after three children were born to any person; the government of India now pays individuals who submit to sterilization. More subtle, but also more widespread, are laws that reduce variance in access to resources, such as graduated income taxes, the vote, representative government, elected (not hereditarily succeeding) officials, and universal education.

MacDonald (1983) discusses the "leveling" effect of monogamy, although he seems to find it puzzling, in evolutionary terms, perhaps because he does not consider the significance of equality of opportunity as a basis for social unity in the face of extrinsic threats. Once this factor is weighed in, one sees that the real puzzle is not, as MacDonald supposes, to account for leveling processes, but to account for the maintenance of despotic societies, within which the greatest disparities in opportunities for individuals occur (e.g., Betzig, 1986); or, rather, to explain why some sizes and kinds of societies involved huge disparities in individual opportunity (those intermediate in size; Alexander, 1979a), while others (large and small) have leveled them to extreme degrees. I think the answer will come from comparing the histories of interaction between neighboring societies, effects of physical or physiographic barriers on their sizes, and separation of warriors (soldiers) from their families.

Despite their obvious and dramatic differences from one another, then, the most extremely ultrasocial systems of humans and other species are apparently all based on reproductive opportunity leveling. The essential difference is that in (some) clones and eusocial forms all individuals realize their reproduction through the same sets of babies and have specialized baby production and baby care in different individuals, and humans have done neither of these things.

Literature Cited

Alexander, R.D. 1974. The evolution of social behavior. *Ann. Rev. Ecol. Syst.* 5:352–383.

Alexander, R.D. 1975. The search for a general theory of behavior. *Behav. Sci.* 20: 77–100.

Alexander, R.D. 1979a. *Darwinism and Human Affairs.* Seattle: University of Washington Press.

Alexander, R.D. and G. Borgia. 1978. On the origin and basis of the male-female phenomenon. In *Sexual selection and reproductive competition in insects*, M. F. Blum and N. A. Blum (eds.). Academic Press, pp. 417–440.

Alexander, R.D., J.L. Hoogland, R.D. Howard, K.L. Noonan, and P.W. Sherman. 1979. Sexual dimorphism and breeding systems in pinnipeds, ungulates, and humans. In *Evolutionary Biology and Human Social Behavior: An Anthropological Perspective*, N.A. Chagnon and W. G. Irons (eds.). North Scituate, Mass.: Duxbury Press, pp. 402–435.

Alexander, R.D. and K.L. Noonan. 1979. Concealment of ovulation, parental care and human social evolution. In *Evolutionary Biology and Human Social Behavior: An Anthropological Perspective*, N. A. Chagnon and W. G. Irons (eds.). North Scituate, Mass.: Duxbury Press, pp. 436–453.

Betzig, L. 1986. *Despotism and differential reproduction: A Darwinian view of history.* Hawthorne, N.Y.: Aldine.

Dawkins, R. 1976. *The Selfish Gene.* Oxford: Oxford University Press.

Keyfitz, N. 1984. The population of China. *Sci. Amer.* 250:38–47.

Leigh, E.G. 1983. When does the good of the group override the advantage of the individual? *Proc. Natl. Acad. Sci. U.S.* 74:2985–89.

MacDonald, K. 1983. Population, social controls and ideology: Toward a sociobiology of the phenotype. *J. Social Biol. Struct.* 6:297–317.

Sahlins, M.D. 1976. *The Use and Abuse of Biology: An Anthropological Critique of Sociobiology.* Ann Arbor, : University of Michigan Press.

Strate, J.M. 1982. An evolutionary view of political culture. Ph.D. dissertation, University of Michigan.

Williams, G.C. 1979. The question of adaptive sex ratio in outcrossed vertebrates. *Proc. Royal Soc. London B* 205:567–580.

Eusociality in Naked Mole-Rats

Heroes Are Works of Art

Inspirations decorating the walls of minds,
invitations to previously impossible reaches,
barely glimpsed peaks of uniqueness,
standouts in endless seas of possibilities.

Heroes stretch the imagination,
promising itineraries of unfolding expansion;
they proffer guides to the triumphs of synthesis,
declare the meaning of life.

Heroes model reality and security, sanction
scenarios, activate and reactivate those
billions of our cortical neurons said to be
of hundreds of kinds, enabling their million
billions of connections comprising the most
elaborately complex and precious machinery
conceivable anywhere
any time
yet.

* * *

Heroes are for all times, all ages
they help us to know what we
hadn't even known we wanted to know,
to facilitate all that we discover
can be discovered, all that
can be absorbed into ourselves.

Alexander *2011, p. 52*

INTRODUCTION

Richard Alexander, the Naked Mole-Rat, and the Evolution of Eusociality
Paul W. Sherman

Eusociality is an entomologically derived term that identifies animal social systems with three characteristics: (1) multigenerational groups, (2) alloparental care of young (helpers), and (3) restriction of reproduction to a fraction of the individuals in each group (i.e., reproductive "skew"). Eusociality has been an important evolutionary puzzle ever since Darwin (1859: 268) highlighted worker ants as presenting "one special difficulty which at first appeared to me insuperable, and actually fatal to the whole theory." Because natural selection favors increased reproduction, how can individuals evolve gradually to reproduce less and less or not at all? And how can specialized morphological, physiological, and behavioral characteristics, including spectacular acts of suicidal altruism, evolve among individuals that do not directly reproduce?

Until the late 1970s eusociality was thought to occur only in two orders of insects: the Hymenoptera (ants, bees, and wasps) and the Isoptera (termites). Then in 1981 Jennifer Jarvis announced the discovery of eusociality in a mammal, the naked mole-rat (*Heterocephalus glaber*, order Rodentia, family Bathyergidae; Jarvis 1981). This was a watershed event in studies of social evolution, and Richard Alexander played a pivotal role. Several years previously Alexander had started thinking seriously about the evolution of eusociality. Contrary to prevailing wisdom at the time, Alexander believed the ecological factors that favored extended maternal care of offspring and natal philopatry, especially predation and limitation of suitable nest sites, were more important than extraordinarily close genetic relatedness between helpers and beneficiaries in predisposing organisms to eusociality. Accordingly, he hypothesized an imaginary eusocial mammal that lived like a termite in a safe, expandable tunnel system deep underground in rock-hard soil, where it fed on subterranean plant parts. Alexander presented his ideas in professional seminars and after one of these a colleague told him that his hypothetical mammal sounded a lot like the naked mole-rat of eastern Africa.

Alexander was astonished—and energized. He corresponded with Jarvis, then the world's naked mole-rat expert, and in the fall of 1979 he and I and our spouses visited her at the University of Cape Town in South Africa. We will never forget the thrill of entering Jarvis's laboratory and seeing those remarkable little furless, buck-toothed rodents scurrying through yellow plastic tunnel systems backward and forward, head-to-tail, like tube trains. A round box served as a nest, and colony members congregated there, snuggling together in a jumbled pile atop which sprawled a particularly large, elongate individual: the queen (breeding female). Instantly we realized that before us was the connection between vertebrate and invertebrate social systems—a new clue to understanding the puzzle of eusociality!

Soon after our arrival, Jarvis presented Alexander and me with a draft of her now famous 1981 *Science* paper. We commented extensively on it, and a friendship born in shared excitement and scientific curiosity quickly developed. We invited Jarvis to join our expedition to Kenya the following week. She agreed and, working together, we collected six partial colonies of naked mole-rats near Mtito Andei, a little village 233 km southeast of Nairobi. These animals were sent to Cornell University and the University of Michigan for studies of the animals' social and reproductive behaviors.

Results of the first decade of mole-rat research were summarized in a book entitled *The Biology of the Naked Mole-Rat* (Sherman et al. 1991). The excerpt that follows is from the book's first chapter, which Alexander had begun writing some 15 years previously. The chapter is comprehensive, and too long to be included in its entirety in this volume. However, the portions that had to be omitted contain important insights and arguments and, to place the reprinted excerpt in context, I will highlight several of these.

First, Alexander et al. (1991) noted that because naked mole-rats and termites are diploid, their social systems confirm W. D. Hamilton's (1964) original argument that although close relatedness facilitates the evolution of altruism, haplo-diploidy is neither necessary nor sufficient to account for eusociality. However, failure of the specific "¾ relatedness hypothesis" does not cast doubt on the importance of kin selection in the evolution of eusociality (West-Eberhard 1975; Queller & Strassmann 1998; Foster et al. 2006; Abbot et al. 2011), contrary to some recent assertions (e.g., Wilson 2005, 2008; Nowak et al. 2010). Instead, it suggests that factors in addition to kinship were evolutionary antecedents of eusociality. Alexander et al. argued that extended maternal care of offspring requiring progressive food provisioning ("subsociality" in entomological jargon) and incomplete metamorphosis (no pupal stage between immature and adult) were the key prerequisites in termites and mole-rats. Andersson (1984) came to similar conclusions in another important paper on the evolution of eusociality that was being developed at roughly the same time as Alexander's manuscript.

Second, Alexander et al. (1991) noted that naked mole-rats prove eusociality is not unique to insects, refuting explanations that hinge on peculiarities of insect physiology, genetics, or evolutionary history. Indeed eusociality has now been discovered in a diversity of organisms including other African mole-rats (Jarvis et al. 1994; Wallace and Bennett 1998), another class of invertebrates (Crustacea:

sponge-dwelling shrimp [Duffy et al. 2000]), and other orders of insects, some of which are haplodiploid (e.g., Hemiptera: gall-nesting thrips [Crespi et al. 1997]) but others not (Coleoptera: ambrosia beetles [Kent and Simpson 1992; Smith et al. 2009]; Homoptera: gall-forming aphids [Stern and Foster 1996]). These convergences raise the question: what common selective pressures favored eusociality in such phylogenetically and physiologically divergent creatures?

Third, Alexander et al. (1991) used the naked mole-rat to draw parallels between social evolution in vertebrates and invertebrates. Cooperative breeding and eusociality occur in birds, mammals, arthropods, and a few fishes when ecological pressures preclude independent dispersal and reproduction, groups cooperate to forage and defend themselves, and individuals can raise their inclusive fitness sufficiently by rearing siblings to compensate for foregoing personal reproduction (see also Andersson 1984). Thus both "extrinsic" (ecological) and "intrinsic" (genetic and developmental) factors are important in favoring cooperative breeding and eusociality (Evans 1977). The significance of both types of factors is captured in Hamilton's Rule, which posits that altruistic acts are favored by natural selection when $rB - C > 0$, where "C" is the personal fitness cost to the altruist, "B" is the fitness benefit to the recipient, and "r" is the relatedness between them.

Among the extrinsic factors, Alexander et al. (1991) believed that protection from predation was paramount. They argued that the evolution of eusociality was facilitated by availability of safe, long-lasting, expansible nest sites. Such "fortresses" (Queller and Strassmann 1998) would have to be defensible by small numbers of armed individuals (e.g., possessing potent stings, chemical weaponry, or sharp teeth), expandable to accommodate the growing family, and surrounded by sufficient food resources to feed the entire group. All are true for naked mole-rats (Brett 1991) and single-site nesting termites (Shellman-Reeve 1997). Among the intrinsic factors, kinship is essential because all cooperatively breeding and eusocial species live in family groups. Ancestrally, most of these were monogamous (for social insects, see Hughes et al. 2008; for birds, see Cornwallis et al. 2010) so recipients of help usually were full siblings.

The behavior of a young helper mole-rat or termite conforms to Hamilton's Rule because the cost to the helper (C) for staying in the safety of the natal burrow system is low since risks of independent dispersal are substantial; the benefit (B) to younger siblings is high because food provisioning by helpers promotes their rapid growth and helpers' fortress defense behaviors effectively protect them from predators; and relatedness (r) is high because helpers and recipients typically are full siblings, with inbreeding increasing their relatedness more (for naked mole-rats, see Reeve et al. 1990; for subterranean termites, see Vargo and Husseneder 2009). Developmental factors also are important because a helper can begin enhancing its inclusive fitness long before reaching adult size, thereby gaining a head start in reproduction (and therefore senescence) over individuals that wait until they are large and old enough to rear their own offspring. Queller (1989) was stimulated by Alexander's eusociality manuscript to expand the latter idea to cases where the reproductive head start occurs because helpers can rapidly bring juvenile siblings to independence.

One question that Alexander et al. (1991) sidestepped was whether cooperatively breeding species with helpers that differ morphologically and behaviorally from breeders (i.e., a worker "caste") are qualitatively different from species in which helpers and breeders are similar morphologically and equally totipotent. The issue is important for deciding whether cooperative breeding and eusociality should be lumped together (Sherman et al. 1995, Hardisty and Cassill 2010) or split (Crespi and Yanega 1995) theoretically and terminologically. Both proposals have been endorsed but consensus has not been achieved (Lacey and Sherman 2005, Beekman et al. 2006), probably because each approach is useful for addressing questions on different levels of analysis (Sherman 1988). Lumping focuses on similarities in personal reproduction among taxa and facilitates ultimate analyses (i.e., evolutionary history and fitness levels) of variations in social structures, whereas splitting focuses on behavioral and morphological specializations and facilitates proximate analyses (mechanistic and ontogenetic levels) of reproductive differences among individuals.

The last section of Alexander et al.'s (1991) excerpt implies that cooperative breeding and eusociality form a continuum of fundamentally similar social systems whose main differences lie in group sizes and degrees of reproductive restriction. The evolution of specialized helper phenotypes, loss of behavioral and reproductive totipotency, and intragroup breeding conflict are related to group size, which itself is determined by body size and ecological factors (the benefits of philopatry and costs of attempting to disperse). When organisms are small and ecological constraints are severe—for example, predation is heavy and safe, expandable nests sites are limited, or food is patchily distributed and hard to find but locally abundant in quantity—offspring remain in the natal nest and large groups form. In such groups the likelihood that any youngster will ever have an opportunity to breed is vanishingly small, so individuals can transmit their genes most effectively by focusing on rearing collateral kin. Over evolutionary time this favors phenotypic specializations that enable helpers to function as super-efficient food gatherers, food preparers and provisioners, and colony defenders, even at the expense of loss of reproductive competence.

When body sizes are larger and ecological constraints are less severe the reverse occurs: Groups are small enough that everyone has a good chance of eventually breeding so all colony members retain reproductive and behavioral totipotency and phenotypes do not diverge because the tasks that parents and helpers perform are similar; the phenotypic plasticity of vertebrates also reduces the likelihood that alternative morphotypes will evolve. Breeding-age helpers forgo reproduction only temporarily, so conflicts over breeding are frequent among small-group cooperative breeders (e.g., wolves, wild dogs, meerkats, naked mole-rats, acorn woodpeckers, and paper wasps—see Hart and Monnin 2006; Clutton-Brock 2009), whereas among large-colony cooperative breeders (most ants and many bees and wasps) conflict over reproduction is sporadic and occurs primarily over sex allocation and production of males (Ratnieks et al. 2006).

Cooperative breeding and eusociality thus differ in degree but not in kind. Classical definitions of eusociality (Hölldobler and Wilson 1990, p. 638) do not require that some colony members have lost behavioral or reproductive totipotency—only that there is "reproductive division of labor." Arraying species that breed cooperatively along a common axis that represents the distribution of lifetime reproductive success (Keller and Perrin 1995) unites studies of vertebrate and invertebrate societies, and facilitates identification of common extrinsic and intrinsic selective factors that result in societal convergences. The social structure of each species results from life-history decisions that are made either over evolutionary time (for large-group species) or ecological time (for small-group species—e.g., Tibbetts 2007) about whether or not to disperse from home, attempt to breed, or help rear siblings (Cahan et al. 2002). At each stage, the evolutionarily stable decision rule (Darwinian algorithm) is the one that most consistently results in satisfaction of Hamilton's Rule.

The book chapter from which the following selection was excerpted is Richard Alexander's opus magnum on eusociality. It heightened interest in this social system and has steered our thinking about its evolution for the past two decades. In addition, by highlighting parallels between vertebrate and invertebrate social systems, Alexander provided a foundation for the recent surge of interest in human alloparenting (Crittenden and Marlowe 2008; Sear and Mace 2008), cooperative breeding (Hill and Hurtado 2009; Hrdy 2009; Kramer 2010, 2011), and eusociality (Foster and Ratnieks 2005; Betzig, 2012). Over the past 20 years, many of the specific hypotheses presented by Alexander et al. (1991) have been explored and confirmed. For example, Bourke (1999) and Anderson and McShea (2001) provided data that support Alexander et al.'s arguments about how large colony size affects the evolution of caste differentiation and reproductive conflict in social insects, Duffy and Macdonald (2010) showed how small body size, natal philopatry, kinship, and availability of safe, expandable nesting sites facilitated the evolution of eusociality in sponge-dwelling shrimp, and Purcell (2011) confirmed the importance of predation and food distribution in molding arthropod group sizes and social organizations.

Since 1991, no paper has been published that treats the evolutionary origins and adaptive significance of eusociality more comprehensively than Alexander et al. The most recent review of the evolution of eusociality (Nowak et al. 2010), although deeply flawed because it dismisses the importance of kin selection (see Abbot et al. 2011; Rousset and Lion 201; Bourke 2011), nonetheless contains some good hypotheses about the evolutionary antecedents of eusociality—and all of them can be found in Alexander et al. (1991). Clearly Alexander et al.'s chapter has been a guidepost to understanding the evolution of cooperative breeding and eusociality. It will remain so for many years to come.

Acknowledgments

Thanks to Kyle Summers and Bernie J. Crespi for inviting this contribution, to Bernie J. Crespi, David C. Queller, Janet Shellman Sherman, and Kyle Summers for helpful suggestions on the manuscript, and to Richard D. Alexander for inspiration.

References

Abbot, P. et al. (137 co-authors). 2011. Inclusive fitness theory and eusociality. *Nature* 471: E1–E4.

Alexander, R.D. 2011. *The Mockingbird's River Song: Poems, Essays, Songs and Stories, 1946-2011*. Manchester, MI: Woodlane Farm Books.

Alexander, R.D., Noonan, K.M., and Crespi, B. J. 1991. The evolution of eusociality. In: P. Sherman, J.U.M. Jarvis, and R.D. Alexander (eds.), *The Biology of the Naked Mole-Rat*. Princeton, NJ: Princeton University Press, pp. 3–44.

Anderson, C. and McShea, D.W. 2001. Individual versus social complexity, with particular reference to ant colonies. *Biol. Rev. Cambr. Phil. Soc.* 76: 211–237.

Andersson, M. 1984. The evolution of eusociality. *Ann. Rev. Ecol. System.* 15:165–189.

Beekman, M., Peeters, C., and O'Riain, M.J. 2006. Developmental divergence: neglected variable in understanding the evolution of reproductive skew in social animals. *Behav. Ecol.* 17: 622–627.

Betzig, L. 2012. Darwin's question: how can sterility evolve? In: K. Summers and B.J. Crespi (eds.), *Human Social Evolution: The Foundational Works of Richard D. Alexander*. Oxford: Oxford University Press.

Bourke, A.F.G. 1999. Colony size, social complexity, and reproductive conflict in social insects. *J. Evol. Biol.* 12: 245–257.

Bourke, A.F G. 2011. The validity and value of inclusive fitness theory. *Proc. R. Soc. Lond. B* 278: 3313–3320.

Brett, R. A. 1991. The ecology of naked mole-rat colonies: burrowing, food, and limiting factors. In: P.W. Sherman, J.U.M. Jarvis, and R.D. Alexander (eds.), *The Biology of the Naked Mole-Rat*. Princeton, NJ: Princeton University Press, pp. 137–184.

Cahan, S.H., Blumstein, D.T., Sundstrom, L., Liebig, J., and Griffin, A. 2002. Social trajectories and the evolution of social behavior. *Oikos* 96: 206–215.

Clutton-Brock, T. 2009. Structure and function in mammalian societies. *Phil. Trans. R. Soc. Lond.* 364: 3229–3242.

Cornwallis, C.K., West, S.A., Davis, K.E., and Griffin, A.S. 2010. Promiscuity and the evolutionary transition to complex societies. *Nature* 466: 969–972.

Crespi, B.J., Carmean, D.A., and Chapman, T.W. 1997. Ecology and evolution of galling thrips and their allies. *Ann. Rev. Entomol.* 42: 51–71.

Crespi, B.J. and Yanega, D. 1995. The definition of eusociality. *Behav. Ecol.* 6:109–115.

Crittenden, A. and Marlowe, F.W. 2008. Allomaternal care among the Hadza of Tanzania. *Hum. Nat.* 19: 249–262.

Darwin, C.R. [1859] 1962. *The Origin of Species*. New York: Collier.

Duffy, J.E. and Macdonald, K.S. 2010. Kin structure, ecology and the evolution of social organization in shrimp: a comparative analysis. *Proc. R. Soc. Lond. B* 277: 575–584.

Duffy, J.E., Morrison, C.L., and Rios, R. 2000. Multiple origins of eusociality among sponge-dwelling shrimps(Synalpheus). *Evolution* 54: 503–516.

Evans, H.E. 1977. Extrinsic versus intrinsic factors in the evolution of insect sociality. *BioScience* 27: 613–617.

Foster, K.R. and Ratnieks, F.L.W. 2005. A new eusocial vertebrate? *Trends Ecol. Evol.* 20: 363–364.

Foster, K.R., Wenseleers, T. and Ratnieks, F.L.W. 2006. Kin selection is the key to altruism. *Trends Ecol. Evol.* 21: 57–60.

Hamilton, W.D. 1964. The genetical evolution of social behaviour, I and II. *J. Theoret. Biol.* 7:1–52.

Hardisty, B.E. and Cassill, D.L. 2010. Extending eusociality to include vertebrate family units. *Biol. Phil.* 25: 437–440.

Hart, A.G. and Monnin, T. 2006. Conflict over the timing of breeder replacement in vertebrate and invertebrate societies. *Insectes Sociaux* 53: 375–389.

Hill, K. and Hurtado, A.M. 2009. Cooperative breeding in South American hunter-gatherers. *Proc. R. Soc. Lond. B* 276: 3863–3870.

Hölldobler, B. and Wilson, E.O. 1990. *The Ants*. Cambridge: Harvard University Press.

Hrdy, S.B. 2009. *Mothers and Others: The Evolutionary Origins of Mutual Understanding.* Cambridge: Harvard University Press.

Hughes, W.O.H., Oldroyd, B.P., Beekman, M., and Ratnieks, F.L.W. 2008. Ancestral monogamy shows kin selection is key to the evolution of eusociality. *Science* 320: 1213–1216.

Jarvis, J.U.M. 1981. Eusociality in a mammal: cooperative breeding in naked mole-rat colonies. *Science* 212: 571–573.

Jarvis, J.U.M., O'Riain, M.J., Bennett, N.C., and Sherman, P.W. 1994. Mammalian eusociality: a family affair. *Trends Ecol. Evol.* 9: 47–51.

Keller, L. and Perrin, N. 1995. Quantifying the level of eusociality. *Proc. R. Soc. Lond. B* 260: 311–315.

Kent, D.S. and Simpson, J.A. 1992. Eusocality in the beetle*Austroplatypus mampertus* (Coleoptera: Curculionidae). *Naturwissenschaften* 79:86–87.

Kramer, K L. 2010. Cooperative breeding and its significance to the demographic success of humans. *Annu. Rev. Anthropol.* 39: 417–436.

Kramer, K.L. 2011. The evolution of human parental care and recruitment of juvenile help. *Trends Ecol. Evol.* 26: 533–540.

Lacey, E.A. and Sherman, P.W. 2005. Redefining eusociality: concepts, goals and level of analysis. *Annal. Zool. Fennici* 42: 573–577.

Nowak, M.A., Tarnita, C.E., and Wilson, E.O. 2010. The evolution of eusociality. *Nature* 466: 1057–1062.

Purcell, J. 2011. Geographic patterns in the distribution of social systems in terrestrial arthropods. *Biol. Rev. Cambr. Philos. Soc.* 86: 475–491.

Queller, D.C. 1989. The evolution of eusociality: reproductive head starts of workers. *Proc. Nat. Acad. Sci.* 86: 3224–3226.

Queller, D.C. and Strassmann, J.E. 1998. Kin selection and social insects. *BioScience* 48: 165–175.

Ratnieks, F.L.W., Foster, K.R., and Wenseleers, T. 2006. Conflict resolution in insect societies. *Annu. Rev. Entomol.* 51: 581–608.

Reeve, H.K., Westneat, D.F., Noon, W.A., Sherman, P.W., and Aquadro, C.F. 1990. DNA "fingerprinting" reveals high levels of inbreeding in the eusocial naked mole-rat. *Proc. Nat. Acad. Sci.* 87:2496–2500.

Rousset, F. and Lion, S. 2011. Much ado about nothing: Nowak et al.'s charge against inclusive fitness theory. *J. Evol. Biol.* 24: 1386–1392.

Sear, R., and Mace, R. 2008. Who keeps children alive? A review of the effects of kin on child survival. *Evol. Hum. Behav.* 29: 1–18.

Shellman-Reeve, J.S. 1997. The spectrum of eusociality in termites. In: J.C. Choe and B.J. Crespi (eds.), *Social Behavior in Insects and Arachnids*. Cambridge: Cambridge University Press.

Sherman, P.W. 1988. The levels of analysis. *Anim. Behav.* 36: 616–619.

Sherman, P.W., Jarvis, J.U.M., and Alexander, R.D. (eds.) 1991. *The Biology of the Naked Mole-Rat*. Princeton, NJ: Princeton University Press.

Sherman, P.W., Lacey, E.A., Reeve, H.K., and Keller, L. 1995. The eusociality continuum. *Behav. Ecol.* 6: 102–108.

Smith, S.M., Beattie, A., Kent, D.S., and Snow, A.J. 2009. Ploidy of the eusocial beetle *Austroplatypus incompertus* \(Schedl)(Coleoptera, Curculionidae)and implications for the evolution of eusociality. *Insectes Sociaux* 56: 285–288.

Stern, D.L. and Foster, W.A. 1996. The evolution of soldiers in aphids. *Biol. Rev. Cambr. Phil. Soc.* 71: 27–79.

Tibbetts, E.A. 2007. Dispersal decisions and predispersal behavior in *Polistes* paper wasp "workers." *Behav. Ecol. Sociobiol.* 61: 1877–1883.

Vargo, E.L. and Husseneder, C. 2009. Biology of subterranean termites: insights from molecular studies of Reticulitermes and Coptotermes. *Annu. Rev. Entomol.* 54: 379–403.

Wallace, E.D. and Bennett, N.C. 1998. The colony structure and social organization of the giant Zambian mole-rat,*Cryptomys mechowi. J. Zool.* 244: 51–61.

West-Eberhard, M.J. 1975. The evolution of social behavior by kin selection. *Q. Rev. Biol.* 50: 1–33.

Wilson, E.O. 2005. Kin selection as the key to altruism: its rise and fall. *Soc. Res.* 72: 159–166.

Wilson, E.O. 2008. One giant leap: how insects achieved altruism and colonial life. *BioScience* 58: 17–25.

THE EVOLUTION OF EUSOCIALITY

Excerpt from R.D. Alexander, K.M. Noonan, B.J. Crespi. The Evolution of Eusociality. In P.W. Sherman, J.U.M. Jarvis, R.D. Alexander (eds.). *The Biology of the Naked Mole-Rat*: 27-32. Princeton, NJ: Princeton University Press.

Safe or Defensible, Long-Lasting, Initially Small, Expansible, Food-Rich Nest Sites

In addition to gradual metamorphosis, termites and naked mole-rats have the advantage of a safe niche (microhabitat, nest) from which there is no necessity to exit because food is abundant within the site and because the niche is both long-lasting and expansible to accommodate a growing social group. Thus, many termites live within log fortresses, which are also their food. The nest or niche expands as the termites excavate the log, and they may also locate additional logs by burrowing underground and enhance defensibility by thickening or reinforcing walls with mud. Many species have evolved the ability to construct mud tunnels to reach additional food sources; some also live underground and forage outside on grasses (evidently secondarily; Wilson 1971). Naked mole-rats live underground, feeding primarily on large tubers, which must be approached and located by digging but which provide continuing food sources that do not require exit from the relative safety of underground tunnels (see Brett, chap. 5). At least in termites, nests typically begin small and, in some cases, can be expanded to accommodate thousands or millions of individuals, with abundant food still available locally.

These conditions are unlike those of virtually all social and solitary (nest-building) Hymenoptera, which must locate and transport food back to the nest, often by flying. We suggest that the peculiar combination of nest-site attributes shared by termites and naked mole-rats represents an important contribution to the likelihood of their evolving eusociality, compared with the Hymenoptera and with cooperatively breeding birds and mammals. For the most part, subterranean mammals either do not have abundant food supplies that can be located and used without emerging from the safety of the underground tunnels, or their food is distributed such that, even if they forage underground, the formation and maintenance of groups larger than a parent and its offspring are inhibited (e.g., moles that feed on insects, earthworms, or small subterranean parts of dispersed plants). Similarly, most birds and nonsubterranean mammals live or nest in locations that either are not defensible across generations or cannot be expanded to accommodate large social groups and still be defensible. A few species, such as hunting dogs, beavers, dwarf mongooses, and hole-nesting birds, produce offspring in relative safety and have evolved ways of moving significant amounts of food back to the den (transport, regurgitation, helper lactation). These are the vertebrate forms that most closely approach eusocial (see also Lacey and Sherman, chap. 10). Presumably, if their niches were expansible and their food supplies sufficiently abundant and localized around the nest site, some of them would have continued to evolve toward large-colony eusociality.

Four conditions can therefore be postulated that might lead to incipient eusociality. All depend on a safe, maintainable, or improvable (and costly or unlikely) nest site. (The third condition assumes monogamy and haplodiploidy; the others assume monogamy but do not require closer relatedness between siblings than between parent and offspring.)

> 1. Young are produced faster in the incipient eusocial colony even though all or virtually all emigrating nonsocial parents find suitable nest sites and produce viable young. In other words, expanding and improving a particular kind of nest site after it has been located and started is better (for the mother, as manipulator, or for the mother and all participating individual offspring) than distributing descendants among an adequate number of nest sites suitable for the raising of a single brood.
>
> 2. Young are produced faster in the incipient eusocial colony, but only because most emigrating nonsocial founders fail to reproduce. In other words, nest sites (or suitable nest sites) are severely limiting (Emlen 1981, 1984; Koenig and Pitelka 1981).
>
> 3. Young are not produced faster or saved in higher proportions in the incipiently eusocial colonies, but they are more closely related to helpers than are offspring. Thus, staying home and helping is genetically more profitable than starting a new family if the two alternatives produce the same number of descendants.
>
> 4. Young are not produced faster in the incipiently eusocial colony, but they are saved and helped enough to cause their producers to outreproduce noncolonial competitors. In other words, one must imagine that per capita reproduction becomes increasingly effective with three, four, or even up to hundreds of thousands of caretakers (parents and alloparents) as compared with one or two parents.

Nest sites meeting one or more of the above requirements must continue to be safe for multigenerational periods. If new colonies are initiated by individuals or pairs, as in most eusocial forms, nest sites may initially be hidden or inconspicuous or simply not valuable enough as food sources to attract certain kinds of predators. If eusocial colonies continue to increase in size, however, the nest must become physically or behaviorally more defensible because larger colonies of organisms with many juveniles are more attractive and detectable to parasites and predators. Structural defensibility can be enhanced by extending tunnels and making them more complex (enabling flight or delaying predators), minimizing sizes and numbers of openings into the nest, and enhancing the strength of walls. Behavioral defensibility can be enhanced by evolving tendencies and abilities of helpers to ward off attackers and by increasing the numbers of such defenders. Structural and behavioral defensibility can evolve together as access to a nest is restricted to passages defensible by individuals or small numbers of individuals (e.g., the enclosed paper nests of bald-faced hornets) and as individuals evolve increasingly effective defenses (Wilson 1971) for the particular kinds of structures they defend (e.g., enlarged heads and jaws; expellers of toxic substances as in squirt-gun termites, *Nasutitermes*). There is a sense here in which eusociality is indeed a continuation of parental care of offspring hidden or otherwise made safe in a nest.

Most eusocial forms live in the soil. Underground nests can be relatively invulnerable and also difficult to locate. Aside from army ant colonies (up to 700,000 individuals), the largest eusocial colonies (ants, termites; up to 10 million) either live primarily in the soil or extend their nests into it (Wilson 1971). Moreover, most eusocial forms that maintain nests in the open (primarily wasps) live in the smallest and least permanent colonies. Their relatives with large colonies (e.g., tropical wasps, honey bees, and stingless bees) invariably enclose the nest, either in a cavity or an enveloping structure (West-Eberhard, pers. comm.). In addition, they have evolved the ability to eliminate the small-colony vulnerable stages from their nesting cycle by swarming to found new colonies, and they are particularly aggressive and feared by humans (and probably other vertebrates). Army ants, which are nomadic and fearsome even to large vertebrates, also fission to start new colonies. Fallen tree trunks appear to rank next to soil as nesting sites meeting the above requirements.

Nesting sites that promote eusociality must also be places where a single female can monopolize the production of offspring and the use of helpers during the early stages in the evolution of eusociality. If our scenario emphasizing such origins is appropriate, these requirements appear to rule out locations, such as caves, where multiple safe and proximal sites for single-female or pair nesting prevent such monopolization.

It seems to follow from the argument thus far that small animals are more likely than large ones to evolve eusociality. We speculate that large animals, such as birds and mammals, may not be able to increase the value of logs and tree trunks sufficiently to allow them to evolve eusociality in such places and that nest-site limitations were thus crucial in such forms. Several predictions about vertebrate sociality follow. First, the most nearly eusocial vertebrates should be expected to live in the soil, in large hollow trees or logs, or in constructed dens with similar characteristics (as do beavers). Second, if, for example, giant hollow trees and, say, hole-nesting social woodpeckers or king-fishers coexisted long enough, our argument would predict the evolution of eusociality. Third, if caves typically had structures in them such as hollow spheres with small openings (spheres that could be expanded) then either birds or bats might have become eusocial.

Many small organisms live in apparently suitable sites yet have not evolved eusociality. Some may have failed to do so because parental care is of little or no value to them. Others, such as subsocial Embioptera, Gryllidae, Dermaptera, Hemiptera, Coleoptera, Scorpionida, and Arachnida that live subsocially in seemingly appropriate sites (but which, for one reason or another, may be too short-lived), may lack the ability to initiate evolution of adequate defense of a nest site or may not have been subsocial long enough. Many of these small forms are semelparous, and it seems obvious that the ancestors of all eusocial forms were iteroparous. Semelparous adults are not likely to improve nesting sites significantly or to create conditions leading their offspring to tarry at the nest. Moreover, even if some offspring did tarry, there would be no younger siblings to help unless the parents were iteroparous.

It may seem that eusociality should evolve much more easily in the tropics, because it is easier to establish there the kind of more or less continuous breeding that accompanies increasing colony size and continued nest defense. The life cycle of temperate insects may usually be so set by the seasons as to make it quite difficult to initiate continuous breeding as an aspect of the initiation of eusociality. This speculation seems to predict that persistent subsociality in the soil and in wood may be more prevalent in temperate regions than in the tropics (when it occurs in the tropics it is more likely to change to eusociality) and that eusocial insects evolved in the tropics. However, the possibility of seasonality yielding the selective situation that would lead to obligate workership in first broods without altering life spans in workers or queens, as described above for *Polistes fuscatus*, represents a counterargument.

Further Comments on Vertebrate Eusociality

It may be an oversimplification to assume that there are no eusocial vertebrates except naked mole-rats (see also Lacey and Sherman, chap. 10). African hunting dogs and wolves live in packs that hunt cooperatively. In some cases, one female and one male have pups, and their offspring from the last season or two help them rear the young, carrying back meat that they regurgitate for the pups and probably protecting them and their parents from some kinds of danger (Lawick-Goodall and Lawick-Goodall 1970; Mech 1970, 1988). Surely, helping in some of these species regularly causes helpers to produce no offspring. But the social groups are smaller than those of the eusocial insects, and there is no evidence yet of morphological divergence of parental and helper phenotypes.

Some cooperatively breeding birds behave like the social canines (Emlen 1984; J. L. Brown 1987) and, possibly, beavers (Wilson 1975), dwarf mongooses (Rood 1978), and naked mole-rats (Jarvis 1981; Lacey and Sherman, chap. 10; Jarvis et al., chap. 12; Faulkes et al., chap. 14). Some of these mammals and birds are similar to some wasps and bees, in which groups are small; phenotypes have diverged little or not at all among castes; obvious competition occurs among potential breeders; and high proportions of helpers seem to be waiting and watching in case they get the chance to breed.

In contrast to mammals, birds would appear to be significantly hampered because they cannot simultaneously expand nest sites to accommodate large numbers of individuals and defend them in stationary locations on a multigenerational basis. They do not possess sting equivalents to deal with the kinds of predators that wasps and bees are able to deter, and, as a consequence, they are not able to construct and use expansible nests equivalent to the exposed paper and mud nests of Hymenoptera.

Helper and parental phenotypes may also have failed to diverge in vertebrates because the jobs that parents and helpers do are very similar. Vertebrate workers may not have the same opportunities as eusocial insects for magnificently reproductive (family-saving) suicidal acts (probably in defense against vertebrates) and the specializations improving the ability to do them (West-Eberhard 1975). Canines probably lack the kinds of predators that could guide such evolution. Birds may have the predators but nothing paralleling the venomous sting of female Hymenoptera. One hymenopteran worker can deter either a huge predator (like a human or a bear) that can destroy its whole family (of hundreds or thousands) in one swipe, or a bumbler that could do it only by accident. By plugging a break in the nest fortress, one termite can also deter a predator. It is more difficult for most vertebrates to be such heroes, though such opportunities may exist for naked mole-rats when predatory snakes enter their burrows (see Jarvis and Bennett chap. 3; Brett, chap. 4; Braude, chap. 6).

Mammalian and avian social groups (other than "selfish herds ") never get as big as those of the eusocial insects, and this also restricts the opportunities for superreproductive heroism. The ultimate heroes among eusocial forms are the polistine wasp and honey bee soldier-workers whose barbed stings cannot be extracted, making their attacks on predators irreversibly suicidal. One predicts that barbed stings will be used for defense only in species that form new colonies in swarms, such as honey bees and some tropical wasps. In very small colonies, workers are too valuable for suicidal attacks to be beneficial. The only other barbed stings are those of some ants, which evidently use them to kill prey (A. Mintzer, pers. comm.), and those of the wasp genus *Oxybelus,* which uses them to carry prey (Evans and West-Eberhard 1970); the prediction thus seems to be met.

Another reason why the vertebrate reproductive and worker failed to diverge sufficiently could be the relatively great behavioral plasticity of vertebrates, which reduces the likelihood of the evolution of alternative phenotypes (separate and discontinuous; behavioral, physiological, and/or morphological). (Environmentally determined alternative phenotypes have evolved thousands of times in insects, not merely in connection with social life, but much more frequently in regard to dispersal in species in short-lived habitats, e.g., the phases of migratory locusts, alary morphs in Orthoptera and Hemiptera, alternative phenotypes in successive generations or on different hosts in aphids.) Assuming that vertebrate helpers at the nest improve the reproduction of their parents or siblings, their failure to evolve sterile castes may result from the absence of long-term predictable fluctuations in the reproductive value of helping versus reproducing directly. Again, the reversible flexibility of the individual vertebrate phenotype may be partly responsible for damping the effective severity of such fluctuations, and the relatively long lives and the iteroparity of vertebrates may have reduced the number of such fluctuations.

Literature Cited

Braude, S.H. 1991. Which naked mole-rats volcano? In *The Biology of the Naked Mole-Rat.* P.W. Sherman, J.U.M. Jarvis, R.D. Alexander, eds., pp. 185–194. Princeton, N.J.: Princeton University Press.

Brett, R.A. 1991. The ecology of naked mole-rat colonies: Burrowing, food and limiting factors. In *The Biology of the Naked Mole-Rat.* P.W. Sherman, J.U.M. Jarvis, R.D. Alexander, eds., pp. 137–184. Princeton, N.J.: Princeton University Press.

Brown, J.L. 1987. *Helping and Communal Breeding in Birds.* Princeton, N.J.: Princeton University Press.

Emlen, S.T. 1981. Altruism, kinship, and reciprocity in the white-fronted bee-eater. In *Natural Selection and Social Behavior.* R.D. Alexander and D.W. Tinkle, eds., pp. 217–230. New York: Chiron Press.

Emlen, S.T. 1984. Cooperative breeding in birds and mammals. In *Behavioral Ecology: an Evolutionary Approach.* 2d ed. J.R. Krebs and N.B. Davies, eds., pp. 305–339. Oxford: Blackwell.

Evans, H.E. and M.J. West-Eberhard. 1970. *The Wasps.* Ann Arbor: University of Michigan Press.

Faulkes, C.G., D.H. Abbott, C.E. Liddell, L.M. George and J.U.M. Jarvis. 1991. Hormonal and behavioral aspects of reproductive suppression in female naked mole-rats. In *The Biology of the Naked Mole-Rat.* P.W. Sherman, J.U.M. Jarvis, R.D. Alexander, eds., pp. 426–445. Princeton, N.J.: Princeton University Press.

Jarvis, J.U.M. 1981. Eusociality in a mammal: Cooperative breeding in naked mole-rat colonies. *Science (Wash., D.C.)* 212:571–573.

Jarvis, J.U.M. and N.C. Bennett. 1991. Ecology and behavior of the family Bathyergidae. In *The Biology of the Naked Mole-Rat.* P.W. Sherman, J.U.M. Jarvis, R.D. Alexander, eds., pp. 66–96. Princeton, N.J.: Princeton University Press.

Jarvis, J.U.M, M.J. O'Riain and E. McDaid. 1991. Growth and factors affecting body size in naked mole-rats. In *The Biology of the Naked Mole-Rat.* P.W. Sherman, J.U.M. Jarvis, R.D. Alexander, eds., pp. 358–383. Princeton, N.J.: Princeton University Press.

Koenig, W.D. and F.A. Pitelka. 1981. Ecological factors and kin selection in the evolution of cooperative breeding in birds. In *Natural Selection and Social Behavior.* R.D. Alexander and D.W. Tinkle, eds., pp. 261–280. New York: Chiron Press.

Lacey, E.A. and Sherman, P.W. 1991. Social organization of naked mole-rat colonies: evidence for division of labor. In *The Biology of the Naked Mole-Rat*. P.W. Sherman, J.U.M. Jarvis, R.D. Alexander, eds., pp. 274–357. Princeton, N.J.: Princeton University Press.

Lawick, H. van, and J. van Lawick-Goodall. 1970. *Innocent Killers*. London: Collins.

Mech, L.D. 1970. *The Wolf: The Ecology and Behavior of an Endangered Species*. New York: Natural History Press.

Mech, L.D. 1988. *The Arctic Wolf: Living with the Pack*. Stillwater, Minn.: Voyageur Press.

West-Eberhard, M.J. 1975. The evolution of social behavior by kin selection. *Q. Rev. Biol.* 50:1–33.

Wilson, E.O. 1971. *The Insect Societies*. Cambridge, Mass.: Belknap Press of Harvard University Press.

Wilson, E.O. 1975. *Sociobiology: The New Synthesis*. Cambridge, Mass.: Belknap Press of Harvard University Press.

Parent-Offspring Conflict and Manipulation

Life Effort

dogs
crossing streets
as if they had some place to go

people
hurrying
as if time were short

people
crossing streets,
as if they had some place to go

dogs
hurrying
as if time were short

Alexander, *2011, p. 18*

INTRODUCTION: THE EVOLUTION OF SOCIAL BEHAVIOR

David C. Queller

Richard Alexander's 1974 paper, "The evolution of social behavior" (Alexander 1974) probably marks the watershed publication in his career. Most of his previous papers are about acoustical communication or systematics, particularly in orthopterans. Most of his later ones are on questions of social evolution. Perhaps Alexander's orthopterans were like Darwin's barnacles, the empirical foundation for bigger ideas. This is not to trivialize the earlier work; it earned him election to the National Academy of Sciences. It must be very unusual for an Academy member to do his or her best work after they are elected, because of age, complacency, or just regression on the mean. But Alexander did.

The 1974 paper is an early attempt to describe a general theory of social behavior. It follows on Williams (1966) critique of loose group selection thinking, and on the work of Hamilton (1964a, b, 1966, 1967, 1972) and Trivers (1971, 1972) in building an individual-centered theory. Alexander was well ahead of the curve here; a wider recognition of these issues would begin to come only later, with the publication of *Sociobiology* (Wilson 1975) and *The Selfish Gene* (Dawkins 1976). Alexander argued that the fabric of social evolution is woven from three classes of behavior: nepotism, reciprocity, and parental manipulation. He applies these concepts to why groups form (primarily for protection against predators) and to why social behavior evolves within groups (primarily from increased reproductive competition). He then moves to several particular topics including parental manipulation in humans and the evolution of social insects, the latter of which is the portion reprinted here.

So there is a special pleasure in introducing this paper that broke ground in our understanding of many topics in social evolution. Less pleasurable is the need to mention a rather large error in the paper, but I know Alexander won't mind. In the early 1980s I remember him cheerfully telling everyone that the reason this paper had been so heavily cited was that it included this mistake. That's not an accurate assessment of the paper, but it gives on an accurate view of Alexander's spirit of inquiry and intellectual honesty. If you don't make an occasional mistake, you are probably not thinking as creatively as you ought to be.

The error concerns the last of Alexander's trio of social mechanisms, parental manipulation. From inclusive fitness theory, and from a not-yet-published article of Trivers (Trivers 1974), he knew that parents and offspring could have conflicting interests about parental care. An offspring that gets more parental resources decreases resources available to other offspring. Parents and offspring view this trade-off differently because of different relatednesses. A parent is equally related to all offspring but each offspring is most related to itself.

Alexander thought that parents would win in conflicts with their offspring, for two reasons. The first was that parents were more powerful, not just in the sense of being larger, but also because parents are the dispensers of resources and can withhold those resources if that is in their interests. I will return to this important theme shortly.

The second reason is the erroneous one. Alexander reasoned that offspring could not evolve to act selfishly against the interests of their parents because, when they later become parents, the roles are reversed; they will pass this selfish gene on to their offspring and suffer as a result. Qualitatively this role-reversal argument is quite correct, but it takes mathematics to show whether it holds quantitatively. Does the selfish offspring lose just as much when it becomes a parent as it gained earlier in life? It turns out that relatedness captures the quantitative aspect. Yes, a selfish offspring passes on its selfish genes to its offspring, but they are diluted by the half of the offspring's genes that come from elsewhere, and it is this half that makes the difference. The first mathematical model directly addressing this point was by a graduate student at Michigan, James Blick (Blick 1977), stimulated and encouraged by Alexander, who promptly changed his mind about the question. Subsequent models quickly confirmed this finding (Macnair and Parker 1978; Parker and Macnair 1978; Stamps et al. 1978). In some sense that had to be true if inclusive fitness theory was correct. If offspring have the same interests as their parents, and their parents have the same interests as the grandparents, and so on, then individuals would be following the collective interests of innumerable overlapping remote ancestors.

The error was not too serious for the rest of the paper for a number of reasons. First, despite his role-reversal argument that implies that offspring must follow parental interests, Alexander sometimes argues in a way that implies that their offspring's inclusive fitness does matter. For example, he notes that, except in clonal groups, the interests of individuals in a group are never identical. Similarly he argues that the amount of genetic overlap (relatedness) determines the amount of parental molding necessary to get offspring to cooperate. This seems difficult to reconcile with his argument that parents must win. But is more like how we would put the argument today based on parental manipulation of offspring inclusive fitness options. Parents can do things to offspring that change the offspring's inclusive fitness payoffs and therefore offspring behavior.

Second, even though the role-reversal argument was not correct, parental manipulation is real and important for the other reasons Alexander gave, chiefly the relative power of the parent. It seems significant that morphological castes in social insects are normally determined by nutrition, over which parents would likely have strong control. A mother might benefit from feeding some of her offspring less food if that makes them less able to reproduce but still able to be effective helpers (West-Eberhard 1978; Craig 1979). However, this idea was not supported in two experimental tests in primitively eusocial wasps (Queller and Strassmann 1989; Field and Foster 1999) so it is still an open question how important this was for the origin of eusociality. Still, at some point in the elaboration of eusociality nutrition becomes important so that there are opportunities for parents, and for older workers, to manipulate the caste of younger workers. The importance of this control is shown by the consequences of its absence in the stingless bees in the genus *Melipona* (Ratnieks 2001; Wenseleers and Ratnieks 2004). Here all female offspring are fed equally, and a large fraction of them selfishly opt to develop as queens even though few queens can be supported in this swarm-founding group. Most are therefore killed by the workers.

Finally, Alexander was able to focus on important empirical points that were valid even if they were motivated in part by a mistaken theory. In arguing that kin selection on offspring was not the primary force in eusociality, Alexander questioned the evidence for Hamilton's (1964b, 1972) haplodiploid hypothesis. This idea, that eusociality with female workers is particularly easy to evolve in haplodiploids because females were more related to their sisters (3/4) than to their offspring (1/2), seemed to be the great success of kin selection theory. It appeared to explain why so many origins of eusociality occurred in the haplodiploid Hymenoptera, why their workers were always female (while diploid termites had workers of both sexes), and why workers never lay female eggs but sometimes laid male eggs.

Alexander questioned all three of these points, pointing out that there were strong alternative explanations. The Hymenoptera are the most parental of insects, so it is not surprising that they would easily evolve allo-parental care; all it requires is a shift in who received the care. In addition, parental care in the Hymenoptera is almost entirely done by mothers, not fathers, so it is not surprising that females would most easily evolve allo-parental care. Finally, the restriction of worker reproduction to sons might have the simple proximate explanation that sons are all they can produce unless they undergo the risks of mating, and that the ability is useful if the queen dies. These are solid arguments that stand today.

On some important points, there was a lack of sufficient data. Alexander pointed out that the haplodiploid hypothesis depends on single mating by the queen, while Alexander's parental manipulation hypothesis did not. While noting the paucity of information on this important point, he was misled by the example of the highly promiscuous honeybee into thinking that multiple mating might be common. In the years since, with the aid of molecular methods, mate number has been determined for many species of social insects. A recent phylogenetic analysis of mate number data for 267 eusocial species (Hughes et al. 2008) showed that, although multiple mating occurs, it is always derived. Every origin of eusociality appears to have occurred in a singly mated species and multiple mating may have evolved only after workers were locked into their roles.

On another point with little data, Alexander was correct. He noted the absence of evidence for nepotism within colonies. For example, why do workers not favor full sisters over half sisters? This generalization has become stronger over the years (Keller 1997), and remains something of a puzzle.

But other data would eventually show that relatedness certainly does matter, and that offspring can sometimes win in conflicts with parents. The key results came from sex ratios (Trivers and Hare 1976). In haplodiploids with singly mated queens, female workers are three times more related to sisters than to brothers and are therefore selected to favor three times as much investment in sisters. Queens, in contrast, favor the standard 1:1 investment ratio in sons and daughters. Trivers and Hare (1976) supported the prediction of worker control using comparative analyses. These were initially questionable (Alexander and Sherman 1977), but support for the predictions has become very strong over the years (Nonacs 1986; Queller and Strassmann 1998; Chapuisat and Keller 1999). Today, key questions for sex ratio conflicts and other queen-worker conflicts concern who controls what levers of power (Beekman and Ratnieks 2003), and how conflicts get resolved (Ratnieks et al. 2006).

Much of the importance of Alexander's analysis comes from a shift in emphasis from relatedness to ecological benefits. Although we know that relatedness is essential, differences in relatedness had been overemphasized compared to the equally essential differences in ecological costs and benefits that form the rest of Hamilton's rule. Arguing from the example of termites, where haplodiploidy plays no role, Alexander suggested that the key to eusociality was a nest that constituted a valuable resource that could be utilized over multiple generations. This insight could be true under either parental manipulation or kin selection and Alexander later developed this idea further under a kin selection framework (Alexander et al. 1991). He argued that eusociality was particularly favored when nest sites were safe or defensible, long-lasting, initially small, and expansible, and food rich.

In general, there is a great deal of continuity between the thinking in the 1974 and 1991 papers, despite the shift from a parental manipulation emphasis to a kin selection emphasis (though with parental manipulation still involved). To understand eusociality we need not just relatedness, but ecological benefits and costs. We need to understand why alloparental care evolves not just in the Hymenoptera but in termites and vertebrates. In addition, we need to understand not just the workers, but also the reproductives.

I have focused on eusociality in order to match the section of the paper reprinted, but the contributions of the paper are much broader, including not only the major themes mentioned earlier, but also some very significant points mentioned only briefly. His comments on disease being a nearly unavoidable consequence of grouping predated the now generally acknowledged importance of pathogens in ecology and evolution (Schmid-Hempel 2011). Alexander's argument on page 350 explains how a dominant individual might concede some reproduction to a subordinate in order to keep it in the group, a foreshadowing of skew theory (Keller and Reeve 1994). On page 353, Alexander discusses how helping a relative may not be as easy to evolve as Hamilton's equations implied if the relatives are also particularly close competitors, again anticipating a lot of later theory (Taylor 1992; Wilson et al. 1992; Queller 1994; Van Dyken 2010). Most important, significant parts of the paper are devoted to human behavior and presage much of Alexander's pioneering work on this topic.

References

Alexander, R.D. 1974. The evolution of social behavior. *Annu. Rev. Ecol. Syst.* 4:325–383.

Alexander, R.D. 2011. *The Mockingbird's River Song: Poems, Essays, Songs and Stories, 1946-2011*. Manchester, MI: Woodlane Farm Books.

Alexander, R.D. and Sherman, P.W. 1977. Local mate competition and parental investment: patterns in the social insects. *Science* 196:494–500.

Alexander, R.D., Noonan, K.M., and Crespi, B.J. 1991. The evolution of eusociality. In: P.W. Sherman, J.U.M. Jarvis, and R.D. Alexander (eds.), *The Biology of the Naked Mole-Rat.* Princeton, NJ: Princeton University Press, pp. 3–44.

Beekman, M., and Ratnieks, F.L.W. 2003. Power over reproduction in social Hymenoptera. *Phil. Trans. R. Soc. Lond. B.* 358:1741–1753.

Blick, J. 1977. Selection for traits which lower individual reproduction. *J. Theoret. Biol.* 67:597–601.

Chapuisat, M., and Keller, L. 1999. Testing kin selection theory with sex allocation data in eusocial Hymenoptera. *Heredity* 82:473–478.

Craig, R. 1979. Parental manipulation, kin selection, and the evolution of altruism. *Evolution* 33:319–334.

Dawkins, R. 1976. *The Selfish Gene.* Oxford: Oxford University Press.

Field, J., and Foster, W. 1999. Helping behavior in facultatively eusocial hover wasps: an experimental test of the subfertility hypothesis. *Anim. Behav.* 57:633–636.

Hamilton, W.D. 1964a. The genetical evolution of social behaviour. I. *J. Theoret. Biol.* 7:1–16.

Hamilton, W.D. 1964b. The genetical evolution of social behaviour. II. *J. Theoret. Biol.* 7:17–52.

Hamilton, W.D. 1966. The moulding of senescence by natural selection. *J. Theoret. Biol.* 12:12–45.

Hamilton, W.D. 1967. Extraordinary sex ratios. *Science* 156:477–488.

Hamilton, W.D. 1972. Altruism and related phenomena, mainly in the social insects. *Annu. Rev. Ecol. Syst.* 3:193–232.

Hughes, W., Oldroyd, B., Beekman, M., and Ratnieks, F. 2008. Ancestral monogamy shows kin selection is key to the evolution of eusociality. *Science* 320:1213–1216.

Keller, L. 1997. Indiscriminate altruism: unduly nice parents and siblings. *Trends Ecol. Evol.* 12:99–103.

Keller, L., and Reeve, H.K. 1994. Partitioning of reproduction in animal societies. *Trends Ecol. Evol.* 9:98–102.

Macnair, M.R., and Parker, G.A. 1978. Models of parent-offspring conflict, II. Promiscuity. *Anim. Behav.* 26:111–122.

Nonacs, P. 1986. Ant reproductive strategies and sex allocation theory. *Q. Rev. Biol.* 61:1–21.

Parker, G.A., and Macnair, M.R. 1978. Models of parent-offspring conflict. I. *Anim. Behav.* 26:97–110.

Queller, D. 1994. Genetic relatedness in viscous populations. *Evol. Ecol.* 8:70–73.

Queller,D.C., and Strassmann,J.E. 1989. Measuring inclusive fitness in social wasps. In: Breed, and R. E. Page (eds.), *The Genetics of Social Evolution.* Boulder, CO: Westview Press, pp. 103–122.

Queller, D.C., and Strassmann, J.E. 1998. Kin selection and social insects. *Bioscience* 48:165–175.

Ratnieks, F., Foster, K.R., and Wenseleers, T. 2006. Conflict resolution in insect societies. *Annu. Rev. Entomol.* 51:581–608.

Ratnieks, F.L.W. 2001. Heirs and spares: caste conflict and excess queen production in melipona bees. *Behav. Ecol. Sociobiol.* 50:467–473.

Schmid-Hempel, P. 2011. *Evolutionary Parasitology: The Integrated Study of Infections, Immunology, Ecology, and Genetics.* Oxford: Oxford University Press.

Stamps, J., Metcalf, R., and Krishnan, V. 1978. A genetic analysis of parent-offsping conflict. *Behav. Ecol. Sociobiol.* 4:369–392.

Taylor, P.D. 1992. Altruism is viscous populations—an inclusive fitness model. *Evol. Ecol.* 6:352–356.

Trivers, R.L. 1971. The evolution of reciprocal altruism. *Q. Rev. Biol.* 46:35–47.

Trivers, R.L. 1972. Parental investment and sexual selection. In: B. Campbell (ed.), *Sexual Selection and the Descent of Man.* Chicago: Aldine.

Trivers, R.L. 1974. Parent-offspring conflict. *Am. Zool.* 14:249–264.

Van Dyken, J.D. 2010. The components of kin competition. *Evolution* 64:2840–2854.

Wenseleers, T., and F.L.W. Ratnieks. 2004. Tragedy of the commons in Melipona bees. *Biol. Lett.* 271:S310–S312.

West-Eberhard, M.J. 1978. Polygyny and the evolution of social behavior in wasps. *J. Kansas Ent. Soc.* 51:832–856.

Williams, G.C. 1966. *Adaptation and Natural Selection: A Critique of Some Current Evolutionary Thought.* Princeton, NJ: Princeton University Press.

Wilson, D.S., Pollock, G.B., and Dugatkin, L.A. 1992. Can altruism evolve in purely viscous populations? *Evol. Ecol.* 6:331–341.

Wilson, E.O. 1975. *Sociobiology: The New Synthesis.* Cambridge: Harvard University Press.

THE EVOLUTION OF SOCIAL BEHAVIOR

Excerpt from Alexander, R.D. 1974. The Evolution of Social Behavior, Annual Review of Ecology and Systematics 5:357-367.

Evolution of Sociality In Insects

The social insects have been a central theme in every major publication on natural selection for one important reason: they are apparently alone among all organisms in having evolved obligately sterile individuals. Darwin (41. p. 236) referred to the sterile castes of insects as the "one special difficulty, which at first appeared to me insuperable, and actually fatal to my whole theory." Darwin effectively solved the problem of how different kinds of sterile castes can evolve within a single species by realizing that selection can operate through the "family." As he put it (41, p. 238) "a breed of cattle, always yielding oxen with extraordinarily long horns, could be slowly formed by carefully watching which individual bulls and cows, when matched, produced oxen with the longest horns; and yet no one ox could ever have propagated its kind."

Nevertheless, as Hamilton (67) indicates, the problem of precisely how obligate sterility has evolved in the various social insects is still with us. As the most extreme form of altruism known, its relationship to everything said about social behavior up to this moment is obvious. Indeed, the social insects are probably the best example available for distinguishing the predictions and correlates of the three general systems of selection in social groups.

Several lengthy and detailed discussions of the probable selective backgrounds of insect sociality have been published recently (11, 50, 60, 67, 101, 109, 110, 166, 168, 169, 171, 173- 175). The following account is by comparison a brief and sketchy effort in which I shall attempt to distinguish in certain specific regards the predictions and correlates of theories principally invoking 1. reciprocity, 2. kin selection, and 3. parental manipulation of progeny.

Since Hamilton's (60) paper, with the principal exception of Michener (110) and Lin & Michener (101), only kin selection, in which each individual worker or soldier caste is expected to secure an overcompensating genetic return for its altruism, has been invoked to explain altruism in eusocial insects. Across the past several years: however, I have become convinced that kin selection is not a sufficient explanation for such behavior in insects, and that it may be only feebly and infrequently involved. Kin selection, I suggest, will prove ultimately to be most relevant to the kinship and breeding systems of primate and human societies, for only there does clear evidence exist of keen ability to discriminate among many different relatives within social groups. [Curiously. Hamilton (67) makes the same suggestion for reciprocity.] The broad applicability of Hamilton' s (60, 67) papers, and the changes in approach that they have caused, place them among the most important theoretical contributions to evolutionary biology since Fisher. But I believe that in some respects Darwin was more nearly correct than Hamilton, and that a form of parental manipulation of progeny in the interests of the parent best explains the sterile castes of insects. The difference between these two arguments can be clarified by referring to Hamilton's (60) summary statement (p. 29) "If a [hymenopteran] female is fertilized by only one male all the sperm she receives is genetically identical. Thus, although the relationship of a mother to her daughters has the normal value of 1/2, the relationship between daughters is 3/4. ..other things being equal, [a newly adult daughter would prefer] returning to her mother's [nest] and provisioning a cell for the rearing of an extra sister to provisioning a cell for a daughter of her own. From this point of view therefore it seems not surprising that social life appears to have had several independent origins in this group of insects."

If, however, other things are indeed equal, then queen offspring of the above monogamous female cannot maximize their inclusive fitnesses by their devotion to producing offspring only half like themselves. Only if we assume that the parent has evolved to mold or manipulate her offspring phenotypically so as to maximize her own reproduction can both worker and queen offspring maximize their respective inclusive fitnesses. This they can do because of the particular phenotypes with which the mother endows each of them as a result of the distribution of parental benefits and influences. Such an idea does not detract from the underlying significance of kin selection in sexual organisms. The amount of genetic overlap of different individuals must still determine the amount of parental molding necessary to effect cooperation. Nevertheless, it is clear that individual offspring consistently appearing in the same situations are unlikely on the basis of kin selection alone to evolve dramatically different roles in which one is a sterile helper at the nest and one reproduces in the normal fashion. Furthermore, so long as it is parental manipulation that brings about sibling cooperation, genetic relationships among siblings indicate only the amount of parental molding necessary, not whether or not it will be able to yield a given result. Alternative explanations for the prevalence of eusociality among Hymenoptera, and for its presence in male-diploid termites, are thus given more credibility.

The eusocial insects actually have two distinctive attributes: sterile castes and overlap of the mother's reproductive life with that of her offspring. Social groups with these attributes appear to derive from two different precursors: l. groupings of subsocial (parental) females (eventually including their offspring) and 2. extended families of single mothers. In either case extended parental care precedes eusociality and sterile castes. This apparent dichotomy has long puzzled students of insect social behavior, and is in fact responsible for much of the disagreement in recent theoretical arguments (101, 109, 110). The similarity of the two groups is greatest if the groups of subsocial females, in cases that lead to eusociality, are always sibling groups. Then, as Lin & Michener (101) note, the only difference would be that in one case the mother is present and in the other she is not. As a result, the problems of selection during evolution of sterile castes become essentially the same in the two cases. Because *(a)* single-queen colonies are vastly preponderant in eusocial insects, *(b)* facultative sterility has not been unequivocally demonstrated among nonsiblings, and *(c)* for reasons already indicated it is much easier to evolve sterile castes among siblings, I here suggest that the burden of proof may be upon the investigator who argues that sterile castes have evolved other than within broods of single mothers.

In this light we can begin our comparison by considering Michener's (109) proposal that groups of cooperating unrelated female bees evolved through stages in which differences in reproduction among them came to be actual division of labor in reproduction, and then led directly to the evolution of sterility in some of the females. Lin & Michener (101) defended this idea, but with the critical modification that the cooperating females may (sometimes!) be siblings.

As was shown above, in systems of reciprocity, each individual is continually gambling that his investment will improve both his phenotypic and his genotypic fitness; indeed, what is going on is a form of mutual exploitation under the benefits of group living. There is in fact no altruism except in a temporary sense that benefits may be given at one time and received only at a later time. Should systems of pure reciprocity exist, evolution will tend to reduce fitness shifts to zero—that is, to equalize investments and benefits to individuals. There is no alternative, and this is the precise opposite of what has to happen in the evolution of obligate sterility. As a result we can dismiss reciprocity as being the central factor in the evolution of sterile castes.

Bees or other parental insects may have interacted reciprocally in groups prior to the evolution of sterile castes, and they may have done so subsequent to the evolution of sterile castes. Different families of social insects in a single large group of the sort that are sometimes called "multiple-queen colonies" may even use sterile individuals now as their social donations, or their contributions to reciprocity in the maintenance of the entire group of families, as can also be proposed for facultatively sterile individuals in human religious sects. Group living among competing reproductives may have evolved among subsocial bees for any of the reasons for group living given earlier, and such group living may have [as Lin & Michener (101) suggest] somehow "primed" siblings in the direction of forming groups within which sterile castes could evolve. When such groups are composed of closely related nonsiblings, eusociality could feasibly evolve through kin selection; but this route seems less likely than the route of parental manipulation, for reasons out lined below. In no other sense can the interactions of unrelated competing reproductive females lead to evolved sterility.

In distinguishing the predictions and the correlates of kin selection and parental manipulation in accounting for sterile insect castes let us first consider the genetic relationship of altruist and beneficiary. A principal difference between kin selection and parental manipulation is that kin selection, as formulated by Hamilton (60), requires that each individual secure genetic returns for its altruism greater than the cost of the altruism to its own personal reproduction, this return deriving from the likelihood that given relatives will carry a gene for altruism carried by the altruist. To the extent that the evolution of parental care has placed parents in the position of being able to use their investments in some offspring to increase their total reproduction via other offspring, this requirement is nullified.

Genes for altruism among siblings that benefit the parent can spread regardless of their distribution in the brood with respect to dispensation of altruism. I believe that this fact may largely solve the problem of initially saving and spreading genes causing their bearers to be altruistic, advantages to parents thus perhaps providing a major source of genes leading to altruism in all contexts (including the temporary altruism of reciprocity). The significance of this explanation in accounting for phenomena such as aposematic coloration is obvious (see also 53). Thus the parent with a few brightly colored offspring in a poisonous brood may be both more likely to lose the brightly colored offspring and more likely to produce a bigger brood after predation. The allele for brighter color, assumed for this example to be recessive and present in other individuals in broods having a few homozygous bright individuals, may as a result be selected downward within broods while simultaneously being selected either downward or upward in the species or population as a whole. (See also Figure 1, Brood 1-2c-3c.)

While this situation continues (meaning until the alleles for aposematic coloration have spread widely), the selection that is going on will favor the parent who produces at least a few homozygous bright offspring, disfavor homozygous bright offspring, and either favor or disfavor alleles for brightness, depending upon the intensity and kind of selection. It may also favor parents whose offspring tend to cluster around the few bright offspring, probably to the added detriment of those individuals since they will likely be more obvious to predators in the middle of a group of moving caterpillars than when alone. If a mutant for brightness is not entirely recessive then its initial spread may be inhibited more than in the above example, except that a heterozygous parent will produce an entire partly bright brood. This example purposely omits the possibility of predators with a generalized ability (of whatever origin) to avoid brightly colored potential prey. In such cases alleles for brightness will be favored in all circumstances.

Let us now consider the genetic relationships of altruists and beneficiaries among social insects. Hamilton and others have emphasized the 3/4 average genetic relationship of sisters in a hymenopteran social colony, given the haplo-diploid sex-determining mechanisms of all Hymenoptera and a monogamous mother. This emphasis is misleading for three reasons. First, the termites, which have also evolved eusociality, have normal diploid males. Second, as pointed out by Trivers

(158), only the females are considered but brothers are also reared by workers, and in haplo-diploid species they are only 1/4 like [and 1/4 unlike!] their sisters. Third, eusocial hymenopteran queens at least frequently mate with more than one male. Considering its importance, relatively little attention has been paid to the mating of social Hymenoptera. Astonishingly, it was only recently discovered (Parker, ref. 124, lists references) that multiple inseminations (as many as 7-12 per queen) are evidently the rule in honeybees (each male can mate only once). Single mating has been established for few social hymenopteran females, but multiple insemination is apparently common (101. p. 141; 124; 175, p. 330).

There is no evident correlation between monogamy and eusociality or tendencies toward eusociality. It is possible, but not convincing in view of the generally polygynous or promiscuous hymenopteran background, to postulate 1. brief periods of monogamy in each line that became eusocial and 2. that once sterile castes had evolved, multiple matings could become the rule even if monogamy were critical in the appearance of sterility. Wilson (175, p. 33), after noting that multiple insemination "is not favorable to Hamilton's thesis" suggests 1. the above explanations and 2. the possibility that males are often closely related. But the necessarily dangerous mating flights of queen honeybees (compared to mating on the comb), even though drones have access to hive interiors, suggests selection favoring outbreeding; the appearance under inbreeding of useless diploid males that are killed in the pupal stage by the workers (83, 135, 178) suggests a long history of outbreeding. Hymenopteran siblings may average a closer relationship than siblings in species with diploid sexually produced males, but a 3/4 average relationship is yet to be demonstrated. A point which detracts from the hypothesis suggested here is the tendency of the sperm of different honeybee males to clump inside the queen (152). This phenomenon reduces the variation in genetic relationships among the hive members at most times. But this phenomenon may be widespread (124), and there seems to be no evidence that it has been elaborated in honeybees because of a value in regard to kin selection (also, see below).

Monogamy in termites probably long preceded eusociality coinciding with extended parental care and ensconcement in burrows or crevices. Such behavior is widespread among orthopteroid insects; Alexander (2) has provided a hypothetical scheme indicating some of the steps by which this behavior could lead to eusociality. The nesting cavity of termites (as well as the nests of wasps and bees) is a resource possibly of value to breeding offspring. Parents could gain if adult offspring sometimes remained in the cavity because of the opportunity of taking it over from the parents when they died. Parents could gain further by 1. keeping such offspring from engaging in deleterious competition over the nest resource and 2. causing them to use their parental behavior in the parent's interest when healthy parents and adult offspring overlap. Long-lasting nests and overlap of parents and offspring serving as facultative workers would in turn select for longer parental life, and ultimately perhaps, for obligately sterile offspring. Abilities of parents to make their offspring helpers would often tend to increase the duration of the nest as a reproductive resource and reinforce the entire process. I believe that this hypothetical scheme may be generally applicable in accounting for insect eusociality, and for at least some cases of extended families in vertebrates.

The central role of the duration of the nest resource in the above hypothesis focuses interest on the manner in which nests are founded or pass from one generation to another. In this connection, West-Eberhard (168, pp. 66-67) has described a series of intense conflicts across several weeks among potential queens of the tropical paper wasp, *Polistes canadensis,* for possession of a 22-cell nest that had five foundresses when first observed. These queens may or may not have been sisters. Likewise, West-Eberhard describes as "offspring" three queens that fought for the nest for three weeks after the dominant queen was removed; she does not term them sisters, although they likely were. The evolutionary background of this kind of conflict can only be understood through knowledge of the frequency with which *P. canadensis* nests are usurped by nonsibling queens. West-Eberhard describes several usurpations, but with little knowledge of the relationships of the contending queens.

A parallel to the parent-offspring interactions in the above evolutionary situation can be drawn with long-lived trees. So long as the insect nest, as a reproductive resource, persists longer than the incipiently social queen, there will be selection for longer adult life; with trees, a similar effect accrues from persistence of the resource of a place in the sun and soil. Both the tree and the insect are then in competition with offspring for the resource, but the evolution of offspring that compete with a healthy parent will be thwarted in either case. One predicts, as a result, that seedlings will be less able to grow up under their own parents (e.g., 165) than will seedlings of other species, which can evolve to compete; and within-species allelopathy should be viewed as a parent-offspring interaction, rather than simply intraspecific competition leading (for example) to some kind of population regulation.

A healthy tree with a long life ahead of it gains only from offspring that germinate somewhere other than beneath it, and it loses from those that germinate beneath it. The extent to which parental poisoning of young will evolve, if it is not genotype specific within species, will be determined by the frequency with which seedlings germinating beneath conspecific adult trees do so beneath their own parents; if the adult is often enough a nonparent, competitive ability will evolve in the seedlings too. Trees should also evolve so as to maximize their likelihood of replacement by an offspring, however, and with certain combinations of lengths and predictabilities of juvenile and adult life, the result will be greater likelihood, at least at certain times, of seedlings succeeding under their own parents. The production of suckers or sprouts from roots of dying or afflicted trees must reflect a history of success in trees replacing themselves, in this case by genetically identical offspring.

With trees there is no obvious capability of evolving to use some juvenile seedlings to produce and rear others, so the competition can be clarified (partly) in terms of Hamiltonian kin selection: The tree is more interested in producing further offspring of its own (1/2 like it) than in giving up the resource to its offspring so that they can produce grandchildren (1/4 like it), particularly if the replacement is likely to be a single offspring. The same is true of the insect (and it is particularly important that in each case the resource is suitable for a single reproductive individual). But this description does not specify why the parent wins in the competition, nor does it explain the evident "altruism" of the offspring. The social insect differs from the tree in that, being already parental and with parentally inclined offspring ready to assume ownership of the next resource, it is evidently only small steps away from the capability of using those offspring as effective parental investment contributing to the reproduction of other offspring. Once assistance of parents is seen in this light, the step to obligate sterility in some offspring is easy to envision. It is possible that, in explaining insect eusociality, more attention should be given to the effects of evolving the potential for producing a reliable and persistent (homeostatic) environment useful to a single adult, which in turn selects for longer adult life (as in trees), causing particular kinds of parent-offspring competition.

The hypothesis that sterile insect castes evolved in the context of assisting the reproduction of their parents thus leads to predictions somewhat different from those of Hamiltonian kin selection. Sterile offspring may in this hypothesis be totally altruistic, for no genetic return is required except to the parent (or, to say it another way, to the brood as a whole). The sterile offspring are only a part of the mother's parental investment, and genetic relationships among the brood, sex determining mechanisms, and numbers of matings by the mother may all be more or less irrelevant. The reason is that the correlation is not with altruism being directed at close relatives, but with altruism being directed at siblings, whose relationships to the mother (for each sex) are always the same.

Supporting the hypothesis that eusociality in the Hymenoptera derives from the prevalence or extensive parental care, which has no great relevance in itself to male haploidy, is the fact that parasitic Hymenoptera and the plant feeders of the suborder Symphyta possess the male-haploid system of sex determination, no extensive parental care, and no social behavior. Likewise the termites evolved eusociality without male haploidy.

It appears that the critical factor in the evolution of eusociality is overlap of breeding parents with adult offspring or extensive parental care of siblings in an environment favoring cooperative nest-founding (therefore genes in the parent causing sibling offspring to cooperate in the parent's interests whether or not the parent is present). One of the difficulties experienced by entomologists in applying their usual precise definitions has involved the question of whether or not "true" social life (eusociality) should require only that parents tend their offspring to adulthood or that there be in addition sterile castes. The reason the problem has existed is that no parental insects are known to tend their offspring to adulthood, and overlap with them, without having sterile castes. The closest, perhaps, is *Halictus quadricinctus*, in which the mother remains in the nest "and is still present when the first of her offspring emerge" (175; see 101, pp. 146-47, for other doubtful cases). This virtual absence of parents in the same nest overlapping adult offspring without sterile workers further argues that it is parent-offspring interactions and not selection on sibling interactions as such that is involved in eusociality. Eusocial insects are unusual in having parents that overlap the total adult life of some offspring. Even of successions of offspring.

Regarding the relationships of sister workers in the Hymenoptera it is also relevant that with, say, two matings by the mothers the sisters may average 50% genetic overlap (or more: see 67), but with two haploid fathers their relationships actually vary more than they would with a single diploid father because the haploid sperm contributions of two different fathers cannot recombine. Some pairs of workers will overlap genetically much more than others; this point has never been made clear, and one result is that efforts to apply kin selection, (e.g., 27. 28, 50) have considered only average relationships and thus between-group selection. The lack of evidence of within-colony discrimination in single-queen species, even given two or more fathers, calls forth the spectacle of nurse bees sometimes tending young queens with whom they share relatively few genes. Again, it is to the mother's advantage (although not to the fathers', in the case of multiple mating) that sisters treat each other alike. Although Hamilton (60, 67) believes that worker laying indicates worker reluctance to raise the queen's male offspring, discrimination by workers against the queen's male offspring has apparently not been reported, and other explanations for worker laying are likely (see below). If bees and other social insects can discriminate offspring of different mothers within multiple-queen colonies (evidence of aggression among workers in multiple-queen colonies would represent the critical datum), and if kin selection is the main force in the evolution and maintenance of worker altruism, it is legitimate to wonder why workers with different fathers have not evolved the ability to discriminate full and half siblings. If altruism is a matter of queens manipulating their parental investments, this problem ceases to exist.

A second problem involves the question of why there are no male workers in the Hymenoptera. The kin selection argument is that, because of their haploidy, they are less closely related to one another and to their sisters, hence have less stake in the colony (60). But there is another much more compelling explanation. First, males are rarely parental in the Hymenoptera, social or not (see 175 for the possibility of specialized exceptions among ants), although the females are more parental than perhaps any other insects. More importantly, the hymenopteran female controls the sex ratio of her brood by fertilizing or not fertilizing her eggs. As a parent she can therefore produce whatever proportion of females (thus, whatever proportion of workers) is most advantageous to her in the immediate situation. Under these conditions it is scarcely necessary to invoke kin selection to explain the absence of the genetic revolution necessary to make hymenopteran males parents. As Trivers & Willard (159) point out, the altruism from female progeny toward male progeny, in this case without compensating genetic return, will favor parents able to produce appropriately greater proportions of the more altruistic sex; Michener (110) has compiled sex ratios for social bees that seem to support this argument. Such altruism could not affect primary sex ratios if it occurred beyond the period of parental care (53), and were thus solely a matter of kin selection.

The history of the situation in regard to sex of workers (and soldiers) is again quite different in termites. Young termites are not helpless maggot-like offspring tended from hatching to adulthood in cells as hymenopteran offspring were before social behavior; termite sterility was not preceded by such extreme parental behavior. And termite females evidently do not have the kind of immediate and precise control over the sex ratio of their broods possessed by hymenopteran females. Thus, to the extent that they are now extremely parental, male and female termite workers probably became so more or less together.

A third point involves the production of males parthenogenetically by the worker females. Some consider this tendency support for kin selection (60, p. 31), some have considered it evidence against kin selection (101, pp. 153-55), and some consider it evidence that offspring may evolve so as to compete directly against their parents—in other words, they may "break out" of the clutches of manipulative parents. But there are other ways to view this phenomenon, at least in some cases. When a queen dies or is lost for whatever reason she has only one way to reproduce further if there are no larvae that can still be made into queens. Her final blaze of reproductive glory will be to have her workers make as many males as they can before the colony is dead. I suggest that queens have been favored whose workers begin frantically to make males with the slightest waning of her influence. (Hamilton, 67, refers to such behavior by workers as "selfish," apparently referring to the "race" by the workers to see which will reproduce most. But such behavior matches the mother's wishes, and in some eusocial insects leads to the production of a new queen, whereupon the workers "altruistically" kill the incipient queens that didn't make it).

The above explanation is insufficient to account for all of the varying reports on the phenomenon of male production by workers (67, 101, 175); but, perhaps owing to the fragmentary nature of current information, so is every other single explanation. Perhaps more relevant than parent-offspring competition in the problem of male production by workers is father-mother competition. Since all males are produced parthenogenetically, fathers will gain from producing both worker daughters that make males, in competition with their mothers, and, paradoxically, queen daughters that do so while suppressing their worker daughters' male production; the effect through a male's worker daughters is perhaps more immediate (see also 101). Queens, on the other hand, will gain from worker daughters that do not make males and queen daughters that do and that suppress male production by their worker daughters. When the queen is alive and healthy it is solely in the male's interest that worker females make sons, and this may be the only clear competition between male and female parents in colonies of social Hymenoptera. Coupling this conflict with the value to the queen of her daughters making males when she is dead or waning may provide explanations for many of the confusing variations reported in this phenomenon. Moreover, a mechanism can be postulated whereby the male may to some extent compete successfully against his mate in this regard; this by constantly evolving sperm that are somehow able to thwart tendencies by the queen to lay unfertilized eggs, while producing daughters that tend to lay if they are phenotypically channeled into becoming workers.

The three points outlined above all seem to support the idea that the parents of sterile insects have made them so in their own interests, and have made them altruistic beyond the possibilities of kin selection as so far formulated. This theory, which has not previously been proposed, is also supported by the fact that the pheromonal influence of the queen is in every case either directly, or indirectly through the existing castes, the determiner of sterility. And, as noted earlier, it provides a solution to the question of why the queen honeybee has evolved a special sting apparently used only against her sexual sister. Hamilton (67) was so puzzled over this phenomenon as to suggest that *Apis* queens return from their nuptial flight into strange colonies often enough for the queen's sting to evolve as a result. If the colony is largely a manipulation of the queen's parental investment, then both extremely altruistic and extremely selfish adaptations among offspring are easily explained so long as they contribute to the queen's reproduction. Not only could a sting evolve solely because it efficiently dispatched the closest relative of the stinging individual at appropriate times, but the "quacking" of young queens still in pupal cells, in answer to the "piping" of an emerged sister queen whose response may be to sting them to death (68), can also be understood in this light.

The queen's sting, then, may be analogous to the necrotic tip of the proximal embryo of the pronghorn, the graded sizes of owl nestlings, and the various other determiners of clutch or litter size in different animals: it is a device that prevents partitioning of the parental investment (measured in honeybees largely in terms of available workers) beyond the point at which it is maximally reproductive to the parent. That this circumstance is not clear from the arguments so far provided on this topic (as Hamilton's puzzlement would imply) indicates that it is insufficient to argue that k in Hamilton's (60) formula somehow includes variations in intensity or directness of reproductive competition.

Obviously parental manipulation of progeny is not restricted to physical coercion or pheromonal control. It means chiefly that parents with one kind of offspring outreproduce those with another. Whether the offspring are selfish or altruistic, and the exact manner in which they are caused to be selfish or altruistic, is another problem. A good example with which to illustrate these points is the paper wasp, *Polistes fuscatus* (68), often considered to represent an intermediate stage in the evolution of sociality in insects because founding females are facultatively sterile. One reason for its illustrative value is that founding females, which sometimes cooperate and may (frequently or always) be siblings, begin reproduction long after their mother's death.

Queens of *P. fuscatus* begin nests in spring, build up a population of workers during the summer, and in fall produce new queens and males. The new queens both mate and overwinter apart from the old nest site. In spring they found nests singly or in groups. When they found nests in groups only one queen lays, the others serving as workers for her even though they too are fertilized. West-Eberhard considered that the individual subordinates may gain genetically by cooperating to help their most fit sister. But at least two potentially alternative explanations exist and have not previously been discussed.

First, subordinate females may get to take over the nest (because the original queen is somehow lost) often enough before the reproductive brood is produced. (Production of queen daughters only near the end of summer after producing solely diploid worker females also raises interesting questions about the fate of sperm provided by different males). Second, queens may gain by producing daughters that sometimes cooperate at the individual expense of all but the dominant, actual queen and thus build fewer nests more swiftly. Obviously the old queen need not be present at nest-founding for this altruistic tendency to evolve. Whether such altruism evolves depends solely on whether the parents carrying the genes responsible for it outreproduce. This outcome in turn will depend chiefly on two things: 1. Is it more reproductive to build fewer nests more swiftly? or 2. Is there a high likelihood that subordinates will accidentally direct their altruism at nonsiblings? If new queens tend to return to the old nest site to start nests, altruism may rarely be misdirected. If they generally nest in new sites the possibility for error may be increased. A testable difference in predictions is thus provided between the behavior of individuals either of the same species or of different species with different dispersing tendencies; unfortunately, without consistent differences in inbreeding coefficients or number of matings per queen between populations it will not help us in this case to distinguish between kin selection and parental manipulation.

We may ask, finally, why the sterility of subordinate *Polistes* queens remains facultative. If the old queen is really producing, in effect, a brood of queens and workers, why not obligate sterility? Three categories of environmental uncertainty may combine to help explain this situation: 1. varying queen mortality or incapacity before production of the sexual brood in autumn, 2. varying availability of nest sites, and 3. varying likelihood of siblings reliably nesting together without interlopers. Sometimes, apparently, the queen gains if all her surviving daughters found nests alone.

Literature Cited

2. Alexander. R.D. 1961. Aggressiveness, territoriality, and sexual behavior in field crickets (Orthoptera: Gryllidae). *Behaviour* 17:130–223.

11. Altmann, S. A., Altmann, J. 1972. *Baboon Ecology. African Field Research.* N.Y.: Karger. vii+220pp.

27. Brown, J. 1970. Cooperative breeding and altruistic behavior in the Mexican Jay (*Aphelocoma ultramarina s*). *Anim. Behav.* 18:366–78

28. Brown, J. 1972. Communal feeding of nestlings III the Mexican Jay (*Aphelocoma uttramarina*): interflock comparisons. *Anim. Behav.* 20:305–403

41. Darwin, C. 1967. *On the Origin of Species. A Facsimile of the first Edition with an Introduction by Ernst Mayr.* Boston: Harvard Univ. Press. xviii + 502 pr. Orig. Publ. 1859

50. Eberhard, W. G. 1972. Altruistic behavior in a sphecid wasp: Support for kin-selection theory. *Science* 172: 1390–91

53. Fisher, R. A. 1958. *The Genetical Theory of Natural Selection.* N.Y.: Dover. xiv +291 pp.

60. Hamilton, W. D. 1964. The genetical evolution of social behaviour. I. II. *J. Theor. Biol.* 7:1–52

67. Hamilton, W. D. 1972. Altruism and related phenomena, mainly in the social insects. *Ann. Rev. Ecol. Syst.* 3:193–232

68. Hansson, A. 1945. Lauterzeugung und Lautauffassungersvermögen der Bienen. *Opusc. Entomol. Suppl.* 6:1–124.

75. Horn, H. S. 1971. Social behavior of nesting Brewer's Blackbirds. *Condor* 72:15–23.

83. Kerr, W. F., Nielsen, R. A. 1967. Sex determination in bees (Apinae). *J. Apicult. Res.* 6:3–9

101. Lin, N., Michener, C. D. 1972. Evolution of sociality in insects. *Quart. Rev. Biol.* 47:131–59

109. Michener, C. D. 1958.The evolution of social behavior in bees. *Proc. Tenth Int. Congr. Entomol. Montreal* 2:441–447

110. Michener, C. D. 1969. Comparative social behavior of bees. *Ann. Rev. Entomol.* 14:299–342

124. Parker, G. A. 1970. Sperm competition and its evolutionary consequences in insects. *Biol. Rev. Cambridge Phil. Soc.* 45:525–67

135. Rothenbuhler, W. 1957. Diploid male tissue as new evidence on sex determination in honeybees. *J. Hered.* 48:160–68

152. Taber. S. 1955. Sperm distribution in the spermathecae of multiple-mated queen honeybees. *J. Econ. Entomol.* 48:522–25

158. Trivers, R. L. Haplodiploidy and the evolution of the social insects. Manuscript

159. Trivers. R. L., Willard. D. E. 1973. Natural selection of parental ability to vary the sex ratio of offspring. *Science* 179:90–92

165. Webb, L.J., Tracey, J.G., Haydock, K. P. 1967. A factor toxic to seedlings of the same species associated with living roots of the non-gregarious subtropical rain forest tree *Grevillea robusta. J. Appl. Ecol.* 4, 13–25

166. West, M. J. 1967. Foundress associations in polistine wasps: Dominance hierarchies and the evolution of social behavior. *Science* 157:1584–85

168. West-Eberhard, M. J. 1969. The social biology of polistine wasps. *Univ. Mich. Mus. Zool. Misc. Publ.* 140:1–101

169. West-Eberhard. M.J. Toward an evolutionary theory of social behavior. *Quart. Rev. Biol.* In press.

171. Williams, G. C. 1966. *Adaptation and Natural Selection.* Princeton, N.J.: Princeton Univ. Press. x + 307 pp.

173. Williams, G. C., Williams. D. C. 1957. Natural selection of individually harmful social adaptations among sibs with special reference to social insects. *Evolution* 11:32–39

174. Wilson, E. O. 1963. The social biology of ants. *Ann. Rev. Entomol.* 8:345–68

175. Wilson, E. O. 1971. *The Insect Societies.* Cambridge, Mass.: Belknap. ix +548 pp.

178. Woyke. J. 1963. Drone larvae from fertilized eggs of the honeybee. *J. Apicult. Res.* 2:19–24

PART II

Human Social Evolution

Biology and Culture

In the beginning it was, "Why?"
curious, incredulous, or snickering.
And the answer came, quick and confident,
amid a swelling flood of pride.

And later there were no whys,
but respect, envy, association coveted.
Or perhaps the why was lost in a sea of shame
and smug, knowing nods –
no whys because they know now.

They know now!
And so it is I who wonders,
and the question bores, insistent,
and will not likely be answered.

Are the understanding and culture of all of ourselves
no more than feeble, stumbling fingers that
rip and tear in fruitless efforts to unravel
the complex fabric of cause and effect,
the too-intricately woven threads of truth?

Alexander, *2011, p. 22*

INTRODUCTION

Mark Flinn, University of Missouri

I do not know what Richard Alexander dreams about. But when he is awake, he seems endlessly driven to understand the most perplexing, difficult, important questions about the evolution of life. Like Kipling's Dingo running after Old Man Kangaroo, the chase is always on. Of all the scientific enigmas that captured Dick's persistent, dogged attention, the evolutionary foundation for human culture is perhaps best beloved[1] with major discussions in many of Alexander's articles and in both *Darwinism and Human Affairs* (1979a) and *The Biology of Moral Systems* (1987). His article "Evolution and Culture" in the classic volume *Evolutionary Biology and Human Social Behavior* (edited by Napoleon Chagnon and William Irons) is perhaps his singular most direct treatment of this difficult topic. It begins with a stark line-in-the-sand:

...all of life is subjected continually and relentlessly to a process of differential reproduction of variants...all of the aspects of life are owing, directly or indirectly, to the cumulative effects of this process. No significant doubt has ever been cast on the first part of this argument, and the only alternatives to the second, advocated since 1858, have been divine creation and culture. (R.D. Alexander 1979b: 59)

This is a bold challenge. Is culture explicable as an outcome of the organic evolutionary process, or is Culture something else, a separate and in some respects independent process more akin to divine creation? The question remains unresolved; most of Alexander's arguments are as salient today as they were over 30 years ago. In this introductory essay I summarize his key points about culture and try to push the chase on by examining the issue of why human culture can involve such astonishing levels of informational creativity.

Alexander starts by explaining several misconceptions and obstacles to understanding modern evolutionary theory. First he addresses the basic challenges

---

[1] Among my childhood memories are the wonderful times spent snuggled together with my siblings in front of a fireplace in my father's study listening to bedtime readings. One of our favorites was Kipling's "Just-so" stories. That the label was later used by S.J. Gould and others as a derogation of evolutionary analyses of human behavior seems small-minded. My own children prefer Alexander's books "*Thumping on Trees*" (2010) and "*Red Fox*" (2004) to Kipling.

Table 1 } Murdock's Organic/Cultural Evolutionary Scheme

Evolutionary Processes	Biological	Cultural
Inheritance; information	Gene (DNA)	Culture trait (idea; meme)
Source of novel variation	Mutation	Innovation; mistakes
Selection	Natural selection	Choice
Mode of transmission	Reproduction (genetics)	Social learning
Random effects	Drift	Sampling error
Separation/contact	Isolation/gene flow	Cultural diffusion

posed by those who deny that "microevolutionary" (small, cumulative) changes can produce major structures of organisms (such as the human brain) and differences among species. Creationists posit that the "macroevolutionary" gaps between humans and other life require divine intervention. Alexander notes that exceptions to Darwin's challenges (ways to falsify the general theory of evolution by natural selection) have yet to be found (see also Alexander 2012).

On the other end are those that accept the general proposition that humans are products of organic evolution, but see Culture as a distinct phenomenon, with its own evolutionary processes. George Peter Murdock (1949) described an analogous evolutionary scheme (table 1).

Alexander points out several key shortcomings of this analogy. First, it is not clear that the units of cultural inheritance have the necessary property of high-fidelity replication. "Genes" (DNA strands) can make exact physical copies of themselves; ideas and mental representations do not have a similar consistent basis for replication. Second, cultural innovations or inventions are usually directed at solving problems and are not random in the sense that mutations are. The remarkable creativity of human behavior is a special case that I examine more extensively below. Third, and most important, the cognitive mechanisms (brain; mind) that underpin cultural information are themselves products of human evolutionary history. Hence cultural evolution is not independent of organic evolution (see Alexander 1979:73).

Vygotsky (1978) observed that children are especially tuned to their social worlds and the information that it provides. Alexander posits a complementary adaptive logic: The social world is a rich, vital source of useful information for cognitive development. The human brain has been designed by natural selection to take advantage of this bonanza of data. Some, perhaps in a gesture of appeasement to the cultural *tabula rasa* old guard, would have culture running off in its own (second) evolutionary system with its own distinct (but linked!) inheritance mechanisms similar to Murdock's scheme above (for reviews, see Dawkins, 1982; Durham, 1991; Henrich & McElreath, 2003; Richerson & Boyd, 2005). Others advocate a more restrictive grounding in the biology of learning (Tomasello, 2011), viewing socially learned information or culture as a rather special type of phenotypic plasticity (Alcock, 2005; Flinn, 1997; Flinn & Alexander, 1982, 2007).

Alexander exemplifies this latter paradigm, modeling culture as a compilation of flexible responses by individuals to specific environmental contingencies, analogous to the biological concept of reaction norms and consistent with the basic premises of evolutionary psychology (e.g., Daly & Wilson, 1983).

Here I examine how the Alexander model of culture goes beyond the concept of "evoked culture" as constrained response to variable environments guided by specialized psychological modules. Advances in the understanding of the evolutionary basis of the phenotype, captured in part by the emergent field of "evo-devo" (evolutionary developmental biology) and its reemphasis of the complexity of ontogeny (West-Eberhard, 2003), have apparent relevance to this question of culture and its variants (e.g., Frankenhuis & Panchathan 2010; Heyes, 2003).

Alexander suggests that culture may be viewed as a highly dynamic information pool that is generated and filtered by the extensive information-processing abilities associated with our flexible communicative and sociocognitive competencies (Alexander, 1979b). With the increasing importance and power of information in hominin social interaction, culture and tradition may have become an arena of social cooperation and competition (Flinn, 2004; see also Baumeister, 2005; Sternberg & Grigorenko, 2004). The key issue is novelty. One of the most difficult challenges to understanding human cognitive evolution, and its handmaiden culture, is the unique informational arms race that underlies human behavior. The reaction norms posited by evolutionary psychology to guide evoked culture within specific domains may be necessary but insufficient. The mind does not appear limited to a predetermined Pleistocene set of options—such as choosing mate A if in environment X but choosing mate B if in environment Y—analogous to examples of simple phenotypic plasticity, such as the development of winged and wingless morphs in response to crowding and food availability in migratory locusts. The human jukebox does not just keep the same old selection of tunes; the Beatles displaced Elvis, and so forth. Humans do not have a limited, predetermined set of phenotypic trajectories. We build airplanes.

Keeping up in the hominin social chess game required imitation. Getting ahead favored creativity to produce new solutions to beat the current winning strategies. Random changes, however, are risky and ineffective. Hence the importance of cognitive abilities to hone choices among imagined innovations in ever more complex social scenarios. The theater of the mind that allows humans to "understand other persons as intentional agents" (Tomasello, 1999, p. 526) provides the basis for the evaluation and refinement of creative solutions to the never-ending novelty of the social arms race. This process of filtering the riot of novel information generated by the creative mind favored the cognitive mechanisms for recursive pattern recognition in the open domains of both language (Deacon, 1997; Pinker, 1994) and social dynamics. The evolutionary basis for these psychological mechanisms underlying culture appears rooted in a process of "runaway social selection" (Alexander, 2006; Flinn & Alexander, 2007; Flinn 2011).

Runaway Social Selection

Darwin (1871) recognized that there could be important differences between (a) selection occurring as a consequence of interaction with ecological factors such as predators, climate, and food, and (b) selection occurring as a consequence of interactions among conspecifics (i.e., members of the same species competing with each other over resources such as nest sites, food, and mates). The former is termed *natural selection* and the latter *social selection,* of which sexual selection may be considered a special subtype (West-Eberhard, 1983). The pace and directions of evolutionary changes in behavior and morphology produced by these two types of selection—natural and social—can be significantly different (Alexander, 1974, 2005; West-Eberhard, 2003).

Selection that occurs as a consequence of interactions between species can be intense and unending, for example with parasite–host red queen evolution (Hamilton, Axelrod, & Tanese, 1990). Intraspecific social competition may generate selective pressures that cause even more rapid and dramatic evolutionary changes.

Decreasing constraints from natural selection, combined with increasing social competition, can generate a potent runaway process. Human evolution appears characterized by such circumstances (Alexander, 2006; Flinn, Geary, & Ward, 2005). Humans, more so than any other species, appear to have become their own most potent selective pressure via social competition involving coalitions (Alexander, 1989; Geary & Flinn, 2001, 2002; Wrangham, 1999; e.g., Chagnon, 1988) and dominance of their ecologies involving niche construction (Deacon, 1997; Laland, Odling-Smee, & Feldman, 2000).

The primary functions of the most extraordinary human mental abilities— language, imagination, self-awareness, empathy, Theory of Mind (ToM), foresight, and consciousness—involve the negotiation of social relationships (Adolphs, 2003; Flinn et al., 2011; Geary, 2005; Siegal & Varley, 2003; Tulving, 2002). The multiple-party reciprocity and shifting nested subcoalitions characteristic of human sociality generate especially difficult information-processing demands for these cognitive facilities that underlie social competency (Flinn et al., 2012). Hominin social competition involved increasing amounts of novel information and creative strategies. Culture emerged as an increasingly potent selective pressure on the evolving brain.

Evolution of the Cultural Brain

The human brain is a big evolutionary paradox. It has high metabolic costs, it takes a long time to develop, it evolved rapidly, it enables behavior to change quickly, and it generates unusual levels of informational novelty. As noted earlier, its primary functions include dealing with other human brains (Adolphs, 2003; Gallagher & Frith, 2003). The currency is not foot-speed or antibody production but the generation and processing of data in the social worlds of the human brains' own collective and historical information pools. Some of the standout features of the human brain that distinguish us from our primate relatives are asymmetrically localized in the prefrontal cortex, including especially the dorsolateral prefrontal cortex and frontal pole (Rilling & Sanfey, 2011; for review see Geary, 2005). These areas appear to be involved with "social scenario building" or the ability to "see ourselves as others see us so that we may cause competitive others to see us as we wish them to" (Alexander, 1990, p. 7) and are linked to specific social abilities such as understanding sarcasm (Shamay-Tsoory, Tomer, & Aharon-Peretz, 2005) and morality (Moll, Zahn, de Oliveira-Souza, Krueger, & Grafman, 2005). An extended childhood seems to enable the development of these necessary social skills (Flinn, Ward, & Noone, 2005; Joffe, 1997). Learning, practice, and experience are imperative for social success. The information-processing capacity used in human social competition is considerable and perhaps significantly greater than that involved with foraging skills (Roth & Dicke, 2005).

Evolution of the Cultural Child

The altricial (helpless) infant is indicative of a protective environment provided by intense parental and alloparental care in the context of kin groups (Alexander, 1990; Chisholm, 1999; Flinn & Leone, 2006; Hrdy, 2005; Muehlenbein & Flinn 2011). The human baby does not need to be physically precocial. Rather than investing in the development of locomotion, defense, and food acquisition systems that function early in ontogeny, the infant can work instead toward building a more effective adult phenotype. The brain continues rapid growth, and the corresponding cognitive competencies largely direct attention toward the social environment. Plastic neural systems adapt to the nuances of the local community, such as its language (Bjorklund & Pellegrini, 2002; Bloom, 2000). In contrast to the slow development of ecological skills of movement, fighting, and feeding, the human child rapidly acquires skill with the complex communication system of human language (Pinker, 1999). The extraordinary information-transfer abilities enabled by linguistic competency provide a conduit to the knowledge available in other human minds. This emergent capability for intensive and extensive communication potentiates the social dynamics characteristic of human groups (Deacon, 1997; Dunbar, 1997) and provides a new mechanism for social learning and culture. The recursive pattern recognition and abstract symbolic representation central to linguistic competencies enable the open-ended, creative, and flexible information-processing characteristic of humans, especially of children.

Reconciling Domain-Specific Modularity with Informational Novelty

Humans are unique in the extraordinary levels of novelty that are generated by the cognitive processing of abstract mental representations. Human culture is cumulative; human cognition produces new ideas built on the old. To a degree that far surpasses that of any other species, human mental processes must contend with a constantly changing information environment of their own creation.

Cultural information may be especially dynamic because it is a fundamental aspect of human social coalitions. Apparently arbitrary changes in cultural traits, such as clothing styles, music, art, perceptions of beauty, food, dialects, and mate choice decisions, may reflect information "arms races" among and within coalitions.

The remarkable developmental plasticity and cross-domain integration of some cognitive mechanisms may be products of selection for special sensitivity to variable social context (e.g., Boyer, 1998; Carruthers, 2002; Sperber & Hirschfeld, 2004). Human culture is not just a pool or source of information; it is an arena and theater of social manipulation and competition via cooperation. Culture is contested because it is a contest.

The effects of coalition conformity and imitation of success may drive culture in directions difficult to predict solely on the basis of simple functional concerns or evolved psychological mechanisms. This social dynamic would explain the apparent lack of a simple biological utilitarianism of so much of culture and the great importance of historical context and social power (e.g., Wolf, 2001). Deconstruction is a complicated but necessary enterprise, for we are all players in the social arena. The twist is that we are evolved participants.

Alexander's analysis of culture may reconcile important gaps between the evolutionary psychological paradigm and the more history-oriented anthropological approaches (e.g., Richerson & Boyd, 2005) because it suggests an evolved human psychology that is creative, dynamic, and responsive to cultural context, rather than being more rigidly constrained by domain-specific modules, or by an independent system of cultural rules.

Understanding the evolutionary basis for human culture is far more than a difficult academic issue. Such an understanding is critical to solving humanity's great problems of environmental degradation, social injustice, overpopulation, and war. Alexander's ideas about culture have become increasingly influential and broadly accepted, and continue to provoke deeper analyses of how the human mind evolved in concert with an increasingly complex informational environment.

References

Adolphs, R. 2003. Cognitive neuroscience of human social behavior. *Nat. Rev. Neurosci.* 4:165–178.

Alexander, R.D. 1974. The evolution of social behavior. *Annu. Rev. Ecol. Syst.* 5:325–383.

Alexander, R.D. 1979. *Darwinism and Human Affairs*. Seattle: University of Washington Press.

Alexander, R.D. 1987. *The Biology of Moral Systems*. Hawthorne: Aldine.

Alexander, R.D. 1989. Evolution of the human psyche. In P. Mellars & C. Stringer (eds.), *The Human Revolution* (pp. 455–513). Chicago: University of Chicago Press.

Alexander, R.D. 1990. How did humans evolve? Reflections on the uniquely unique species. Ann Arbor: Univ. Mich. Mus. Zool. Spec. Pub. 1.

Alexander, R.D. 2006. The challenge of human social behavior. *Evol. Psych.* 4:1–28.

Alexander, R. D. 2011. *The Mockingbird's River Song: Poems, Essays, Songs and Stories, 1946-2011.* Manchester, MI: Woodlane Farm Books.

Alexander, R.D. 2012. Darwin's challenges and the future of human society. In F. Wayman, P. Williamson, & B. Bueno de Mesquita (eds.), *Prediction: Breakthroughs in Science, Markets, and Politics.* Ann Arbor: University of Michigan Press. (in press).

Baumeister, R.F. 2005. *The Cultural Animal: Human Nature, Meaning, and Social Life.* New York: Oxford University Press.

Bjorklund, D.F., & Pellegrini, A.D. 2002. *The Origins of Human Nature: Evolutionary Developmental Psychology.* Washington, DC: American Psychological Association.

Bloom, P. 2000. *How Children Learn the Meanings of Words.* Cambridge: MIT Press.

Boyer, P. 1998. Cognitive tracks of cultural inheritance: How evolved intuitive ontology governs cultural transmission. *Amer. Anthropol.* 100:876–889.

Carruthers, P. 2002. The evolution of consciousness. In P. Carruthers & A. Chamberlain (eds.), *Evolution and the Human Mind: Modularity, Language and Meta-cognition* (pp. 254–276). Cambridge: Cambridge University Press.

Chagnon, N.A. 1988. Life histories, blood revenge, and warfare in a tribal population. *Science* 239:985–992.

Chisholm, J. 1999. *Death, Hope, and Sex.* Cambridge: Cambridge University Press.

Darwin, C.R. 1871. *The Descent of Man and Selection in Relation to Sex.* London: Murray.

Deacon, T. W. 1997. *The Symbolic Species: The Co-evolution of Language and the Brain.* New York: Norton.

Dunbar, R. I. M. 1997. *Grooming, Gossip and the Evolution of Language.* Cambridge: Harvard University Press.

Durham, W. 1991. *Coevolution: Genes, Culture, and Human Diversity.* Palo Alto, CA: Stanford University Press.

Flinn, M.V. 1997. Culture and the evolution of social learning. *Evol. Human Behav.* 18:23–67.

Flinn, M.V. 2004. Culture and developmental plasticity: Evolution of the social brain. In K. MacDonald & R. L. Burgess (eds.), *Evolutionary Perspectives on Child Development* (pp. 73–98). Thousand Oaks, CA: Sage.

Flinn, M.V. 2011. Evolutionary anthropology of the human family. In: C. Salmon & T. Shackleford (eds.), *Oxford Handbook of Evolutionary Family Psychology.* Oxford: Oxford University Press, pp. 12–32.

Flinn, M.V., & Alexander, R.D. 1982. Culture theory: The developing synthesis from biology. *Human Ecol.* 10:383–400.

Flinn, M.V., & Alexander, R.D. 2007. Runaway social selection. In S. W. Gangestad & J. A. Simpson (eds.), *The Evolution of Mind* (pp. 249–255). New York: Guilford Press.

Flinn, M.V., Geary, D.C. & Ward, C.V. 2005. Ecological dominance, social competition, and coalitionary arms races: Why humans evolved extraordinary intelligence. *Evol. Human Behav.* 26:10–46.

Flinn, M.V. & Leone, D.V. 2006. Early trauma and the ontogeny of glucocorticoid stress response in the human child: Grandmother as a secure base. *J. Develop. Proc.* 1:31–68.

Flinn, M.V., Nepomnaschy, P., Muehlenbein, M.P., & Ponzi, D. 2011. Evolutionary functions of early social modulation of hypothalamic-pituitary-adrenal axis development in humans. *Neurosci. Biobehav. Rev.* 35:1611–1629.

Flinn, M.V., Ponzi, D., & Muehlenbein, M.P. 2012. Hormonal mechanisms for regulation of aggression in human coalitions. *Hum. Nature*, 22:68–88. DOI 10.1007/s12110-012-9135-y

Flinn, M. V., Ward, C. V., & Noone, R. 2005b. Hormones and the human family. In: D. Buss (ed.), *Handbook of Evolutionary Psychology*. New York: Wiley, pp. 552–580.

Frankenhuis, W.E. & Panchanathan, K. 2011. Balancing sampling and specialization: an adaptationist model of incremental development. *Proc. R. Soc. Lond. B* 2011 278: 3558–3565. doi: 10.1098/rspb.2011.0055

Gallagher, H.L., & Frith, C.D. 2003. Functional imaging of "theory of mind." *Trends Cog. Sci.* 7: 77–83.

Geary, D.C. 2005. *The Origin of Mind*. Washington, DC: American Psychological Association.

Geary, D.C., & Flinn, M.V. 2001. Evolution of human parental behavior and the human family. *Parenting: Science Pract.* 1:5–61.

Geary, D.C., & Flinn, M.V. 2002. Sex differences in behavioral and hormonal response to social threat. *Psychol. Rev.* 109:745–750.

Heyes, C. 2003. Four routes of cognitive evolution. *Psychol. Rev.* 110:713–727.

Hrdy, S. B. 2005. Evolutionary context of human development: The cooperative breeding model. In: C. S. Carter & L. Ahnert (eds.), *Attachment and Bonding: A New Synthesis*. Dahlem Workshop 92. Cambridge: MIT Press.

Joffe, T. H. 1997. Social pressures have selected for an extended juvenile period in primates. *J. Hum. Evol.* 32:593–605.

Laland, K.N., Odling-Smee, J., & Feldman, M.W. 2000. Niche construction, biological evolution, and cultural change. *Behav. Brain Sci.* 23:131–175.

Moll, J., Zahn, R., de Oliveira-Souza, R., Krueger, F., & Grafman, J. 2005. The neural basis of human moral cognition. *Nat. Rev. Neurosci.* 6:799–809.

Muehlenbein, M.P., & Flinn, M.V. 2011. Patterns and processes of human life history evolution. In T. Flatt & A. Heyland (eds.), *Oxford Handbook of Life History*. Oxford: Oxford University Press, pp. 153–168.

Pinker, S. 1994. *The Language Instinct*. New York: William Morrow.

Pinker, S. 1999. *Words and Rules: The Ingredients of Language*. New York: HarperCollins.

Rilling, J.K., & Sanfey, A.G. 2011. The neuroscience of social decision-making. *Ann. Rev. Psych.* 62:23–48.

Roth, G., & Dicke, U. 2005. Evolution of the brain and intelligence. *Trends Cog. Sci.* 9:250–257.

Shamay-Tsoory, S.G., Tomer, R., & Aharon-Peretz, J. 2005. The neuroanatomical basis of understanding sarcasm and its relationship to social cognition. *Neuropsych.* 19:288–300.

Siegal, M., & Varley, R. 2002. Neural systems involved with "Theory of Mind." *Nat. Rev. Neurosci.* 3:463–471.

Sperber, D., & Hirschfeld, L. 2004. The cognitive foundations of cultural stability and diversity. *Trends Cog. Sci.* 8:40–46.

Sternberg, R.J., & Grigorenko, E.L. 2004. Intelligence and culture: How culture shapes what intelligence means, and the implications for a science of well-being. *Phil. Trans. Roy. Soc. B* 359:1427–1434.

Tulving, E. 2002. Episodic memory: From mind to brain. *Annual Review of Psychology*, 53:1–25.

Vygotsky, L.S. 1978. *Mind in Society: The Development of Higher Mental Processes*. M. Cole, V. John-Steiner, S. Scribner, & E. Souberman (eds.). Cambridge: Harvard University Press. (Original work published in 1930, 1933, and 1935).

Walker, R.S., Flinn, M.V., & Hill, K. 2010. The evolutionary history of partible paternity in lowland South America. *Proc. Natl. Acad. Sci. USA* 107:19195–19200.

Walker, R.S., Hill, K., Flinn, M.V., & Ellsworth, R. 2011. Evolutionary history of hunter-gatherer marriage practices. *PLoS ONE* 6: e19066. doi:10.1371/ journal.pone.0019066

West-Eberhard, M.J. 2003. *Developmental Plasticity and Evolution*. New York: Oxford University Press.

Wolf, E.R. 2001. *Pathways of Power: Building an Anthropology of the Modern World*. Berkeley: University of California Press.

Wrangham, R.W. 1999. Evolution of coalitionary killing. *Yearbook Phys. Anthro.* 42:1–30.

EVOLUTION AND CULTURE[1]

Richard D. Alexander

Excerpt from *Evolutionary Biology and Human Social Behavior* (N.A. Chagnon and W. Irons, Eds.), 1979, Duxbury Press, North Scituate, MA: Duxbury Press, pp. 59-85.

The basic argument developed by Darwin, and destined to become the central principle upon which all of biology rests, was two-part in nature. The first part was that all of life is continually and relentlessly subjected to a process of differential reproduction of variants, which Darwin termed natural selection or "survival of the fittest." The second was that all of the attributes of life are owing, directly or indirectly, to the cumulative effects of this process. No significant doubt has ever been cast on the first part of this argument, and the only alternatives to the second, advocated since 1859, have been divine creation and culture.

Debates over the Scope of Selection

CONCERNING DIVINITY

The effects of accepting both parts of Darwin's argument are that (1) the traits of modern organisms are, in terms of the environments of history at least, assumed to be means of maximizing genetic reproduction and (2) the patterns of long-term change observable from paleontological data are assumed also to be owing to natural selection. Creationists believe that unfilled gaps in the paleontological record imply creation, hence did not involve change by natural selection; and most thoughtful people would agree that some extensive changes during human history, evidenced in the archaeological record, are likely to have been unaccompanied by genetic change, hence also did not involve change by natural selection.

A degree of importance for natural selection has been granted by both creationists and the most radical adherents to the idea that culture and biology have

---

[1] I thank Laura Betzig for allowing me to read an essay of hers which prompted me to begin immediately to develop and write down some of the ideas in this paper.

been independent throughout human history. Thus, the organized supporters of creation as an alternative to evolution, such as members of the Creation Research Society (see the *Creation Research Society Quarterly*), have found it necessary to accept the process of natural selection, which they refer to as "microevolution" (e.g., Moore and Slusher 1970). They have established their line of defense chiefly against "macroevolution," which is their name for natural processes, supposed by others to account for the formation of "major organs" and for the origin of large changes or differences among organisms, such as exist between species or "major groups." The creationists argue that because the formation of major groups or major organs cannot be observed directly or studied by experiment, its analysis is outside science; and they argue that this process is best explained as creation. Understandably, they have remained indefinite about the precise nature of major groups and major organs, or the levels at which divine creation is unavoidable.

Many lines of evidence indicate that creationists are wrong in their efforts to distinguish long-term and short-term changes in evolution. Darwin offered a devastating critique of this view, and anticipated the small, cumulative effects of gene mutations as well, when he noted that major organs are the products of large numbers of small changes of the observable kind attributed by creationists to "microevolution." This fact is easily demonstrated by crossing organisms with variant forms of a given major organ or attribute. Darwin went so far as to offer the challenge that "If it could be demonstrated that any complex organ existed, which could not possibly have been formed by numerous, successive, slight modifications, my theory would absolutely break down" (1967[1859]:189). A related class of evidence derives from laboratory or forced hybridization of different species or genera, a procedure which clearly shows that differences between such forms are also accumulations of small mutational changes of the directly observable kind (see also Alexander, 1978b).

CONCERNING HUMANITY

Creationists thus deny that the second part of Darwin's thesis applies to humans by denying that the *earliest* of human attributes—i.e., those actually responsible for the designation "human"—originated through natural selection. Students of culture, on the other hand, tend to deny that the most *recent* of human attributes can be understood by reference to natural selection—i.e., the details of cultural patterns and differences—because they feel that the advent of traditionally transmitted learning signaled the end of any necessary relationship between behavior and the differential reproduction of alternative genetic elements.

Some recent authors have developed arguments, often explicitly about human behavior and culture, as alternative to natural selection in ways that may seem to cast doubt upon even the first part of Darwin's argument. Three such arguments seem most prominent.

Is Selection Tautological?

The first argument is that the basic thesis of natural selection is tautological. Supporters of this view (e.g., Peters 1976) contend that we are unable to identify the "fittest" organisms or traits except retrospectively, and that, accordingly, we can only identify them as those which *have survived*. This argument ignores an enormous body of evidence confirming its falsity. Biologists, as well as plant and animal breeders, are continually able to identify as unfit individual organisms whose phenotypic attributes reveal ahead of time that their chances of reproducing are either nonexistent or relatively small (see also Ferguson 1976; Stebbins 1977). Success in such predictions is possible, as with the maintenance of adaptation, only to the extent that environments are predictable. But all environments of life have some predictable aspects. We can prove this directly, and the countless fashions in which organisms are marvelously and intricately tuned to their environments show that we are correct in assuming that the empirical evidence of environmental consistency is relevant to the process of evolution. Modern evolutionary biology depends upon an ability to generalize about adaptiveness, both across genetic lines and across generations, and remarkable success is being realized from such generalizations, especially with attributes common to most or all organisms, like sex ratios, senescence, and parental investment, and others for which the social environment is crucial, like group-living, nepotism, and sexual competition (see references in Alexander 1977a, 1977b, and Alexander et al., this volume, chapter 15). In the first case more effective comparisons are possible; in the second, the winning strategies are more stable and more easily identifiable.

Criticisms that statements about natural selection are tautological only concern their predictive value, but some detractors have supposed that they also cast into doubt the existence or universality of the entire process. Even retrospective judgments, however, are entirely sufficient to demonstrate the inevitability of differential reproduction, whether or not humans are aware of its workings or capable of assessing its consequences.

The argument that natural selection is tautological is often linked with statements by prominent evolutionists, such as Mayr (1963) or Simpson (1964), that evolution is not a particularly predictive or predictable phenomenon, to suggest that evolution does not even qualify as a scientific theory. The misapprehension involved is failure to see that Simpson and Mayr were talking about our inability to predict or give the adaptive reasons for ancient or long-term phylogenetic changes because we are necessarily ignorant of the environments of selection during geological time. We are not so ignorant of the current and recent environments of selection, and our understanding of them grows constantly.

Organic evolution leads to patterns of change in morphology, physiology, and styles of life. Some of these patterns are reflected by fossil remains and some by the array of organisms present at any given time. Evolution also involves speciation, which results in irreversible divergences of different patterns of life. Pattern changes and speciation together lead to phylogenies or family trees that presumably, if the record were complete, could be reconstructed to illustrate the whole history of life. But phylogenetic patterns, although they are outcomes of evolution, are not the essence of the process. The essence of the process is differential reproduction, or, as Williams (1966) put it, the maintenance of adaptation. Phylogenies are reconstructed with little understanding of the selective forces that produced their patterns because the environments of long-term history cannot be reconstructed with the precision necessary to reconstruct the generation-by-generation effects of natural selection. Efforts to "predict" phylogenies, or the nature of species (i.e., to presage what will be discovered about either the past or the future when more complete information is available) fail to the extent that we are ignorant about environments and the array of living organisms present at each time and place in history. They do not fail, as Peters (1976) suggested, because of the demonstrable independence between the causes of mutations and the causes of selection. Our inability to make long-term evolutionary predictions thus does not mean that the nature of the process yielding evolutionary patterns is itself to be doubted, or that evolutionary propositions are so tautological as to fail as scientific theory. Predictions about adaptiveness are most accurate when they concern short-term changes in the present, and there is every reason to believe that they fail increasingly with longer time spans, or when other eras are considered, simply because our information about environments is more incomplete in such cases.

Arguments about the relationship of long-term pattern changes during evolution to the process of natural selection, which is widely accepted as responsible for short-term changes, have relevance here because of the common failure to realize that long-term pattern-tracing (paleontology, archaeology) can be carried on without direct attention to or concern for the process responsible. In biologists' terms, the process of change is guided largely by natural selection. For anthropologists, it is probably fair to say that there is no universally accepted *guiding force* to account for the changes commonly called cultural evolution or for modern variations in culture. The question we must ultimately address is: what is the nature of this guiding force and to what extent has it been the differential reproduction of genes, realized through reproductive striving of individuals? In this question there is no implication that reproductive striving is consciously so directed.

Is Selection Often Impotent Because of Lack of Genetic Variation?

The second argument seeming to cast doubt on the universality of natural selection is that genetic variations relevant to selective forces are not always present, rendering selection ineffective. But this argument only specifies rare and temporary situations. As any plant or animal breeder knows, even in genetic lines on which directional selection has been practiced for a very long time, genetic variants and combinations now and then appear which are relevant to a desired direction of change. Because of their unpredictability, the only way to take advantage of such variants is to maintain selection. This realization causes humans to practice artificial selection in a way that parallels the differential reproduction of organisms induced by natural environments, which is also inexorable whether or not the variations involved are heritable (and also ineffective when they are not).

For this reason we may assume that in natural populations most novel variants which increase reproduction spread and become characteristic of the population, and that this is the usual process of evolutionary change.

The reasons why natural selection is regarded as the guiding force of evolution are not commonly discussed, but they are obviously crucial (see Alexander 1977a, for a fuller discussion). The most apparent ones are the following: (1) altering directions of selection alters directions of change in organisms (probably always, even if there is sometimes delay owing to specialization as a result of previous selection), (2) the causes of mutation (chiefly radiation) and the causes of selection (Darwin's "hostile forces" of food shortages, climate, weather, predators, parasites, and diseases) are independent, (3) only the causes of selection remain consistently directional for relatively long periods (thus, could explain directional changes), and (4) predictions based on the assumption that adaptiveness depends solely on selection are met (e.g., consider the history of sex-ratio selection: Fisher 1958[1930]; Hamilton 1967; Trivers and Willard 1973; Trivers and Hare 1976; Alexander and Sherman 1977; Alexander et al., Chagnon et al., this volume).

The greatest constraints on selection occur then, paradoxically, in two opposite situations: when the change of selective direction is very great and when unidirectionality persists for a very long time. In the first case, specializations as a result of previous selection reduce the likelihood of adaptive changes in certain directions; thus, moles are almost certainly less likely than squirrels to evolve wings. In the second case, alleles causing change in the favored direction are apt to be fixed by selection faster than mutants arise; thus, after generations of selection for increased milk production, dairy farmers know that "management" (i.e., environment) is crucial but, obviously, they will continue to favor breeding stock from their best producers in the expectation that once in a while the differences will be heritable. Unlike many aspects of phenotypes, cultural variations of humans, which lack correlation with genetic variations, may nevertheless be heritable because of traditional transmission of learned behaviors. Thus, as is well understood, culture has the unusual property of being able to evolve cumulatively in the absence of genetic change. Rates of cultural change within historical times are clear evidence that massive cultural change does indeed occur without genetic change, or in its virtual absence.

However, for cultural change to be independent of natural selection, the following hypotheses would also have to be true: (1) because directions and rates of cultural change are potentially independent of many or most human genetic changes, they are independent of the history of natural selection upon humans; and (2) genetic change through natural selection is not induced by cultural changes. Until very recently, both of these hypotheses have remained largely untested; should they eventually be rejected, as I believe likely, then even for human culture the second part of Darwin's argument will stand as stated above.

Does Evolutionary Theory Suggest Genetic Determinism?

The third argument about the relationship of biology and culture is exemplified by the newspaper announcement of the British Broadcasting Corporation film "The Human Animal," which identified sociobiology as "The field of study built on the theory that behavioral patterns in humans are inherited through genes." This definition is, perhaps innocently, a version of the argument that efforts to invoke biological explanations of human behavior are efforts to defend the notion that behavior is "genetically determined." It cannot be denied that some statements by biologists also suggest this kind of naivete, But there is equal naivete in supposing that merely to consider genes as influences upon behavior (or any other aspects of phenotypes) means that one is automatically excluding the environment or under-playing its role. To argue that behavior is a product of a history of natural selection, however, is in no way an argument that behavior is determined by the genes. It is almost the opposite—a declaration instead that the behavior of each organism is determined, not by the genes, but by the genes and the developmental and experiential environment together. That the mere introduction of genes as influences on behavior is construed as an unsupportable kind of genetic determinism is indicated by the tendency to contrast explanations which include genes with explanations invoking learning. For so many years we asked: "Is this behavior learned or genetic?" Finally, we are coming to realize that the answer is always "Both." The consequence of this realization is not the exclusion of biology from considerations about human behavior but its appropriate reintroduction into them. (For fuller discussions see Alexander 1978a, 1977b, 1977c.)

Traits, Learning, and Genetic Variation

To explain this paradox it is fruitful to consider still another question recently made prominent by self-professed critics of evolutionary approaches to the analysis of human behavior. What is a "trait"? This question seemingly has two aspects. First is the problem of how much or what part of the phenotype can be viewed as a unit in terms of function. In other words, upon what parts and amounts of the phenotype is selection acting in a given circumstance or environment? How much of the phenotype does a given kind or aspect of selection affect? The second part of the question involves how the genes work together during ontogeny to create the phenotype—in other words, how do the genes in the genotype relate to the identifiable components of the phenotype? A history of natural selection suggests that in some sense these two problems resolve into a single one: How do the units within genotypes (genes, supergenes, chromosomes, etc.) interact (through epistasis, pleiotrophy, linkage, etc.) to produce the functional units (appendages, sensory devices, reproductive organs, etc.) of the phenotype?

This question relates to the genes/learning dichotomy because of a common confusion between (1) whether or not a particular behavioral variant characterizes a particular genotypic variant and (2) whether or not a particular *set* of behaviors characterizes a particular genotypic variant. If a particular behavioral act was learned, then its individual presence as a variant is clearly not a result of genetic variation. But genes are necessarily causal (together with their environment) in the production of all behavior; the only problem is to understand how. To reach this understanding we must consider entire sets of learned behaviors in the set of different individuals possessing the same set of genes influencing that behavior in the collection of environments in which the set of individuals developed. Ideally, such entire sets of learned behaviors would be compared with other sets of learned behaviors in other sets of organisms which do differ genetically. In a species with a great deal of immediate-contingency learning in the behavioral repertoires of individuals, then, the effects of genes on learned behavior can only be understood by analyzing the behaviors expressed by *numerous* individuals who *collectively* have experienced the array of environments in which the behavior in question has evolved, or the environments in which it has usually been expressed. Traits can only be identified by examining the variations in learned activities in the normal range of environments of learning. Genetic change could shift the ease of learning in such a group of organisms one way or another along one or more axes, or reduce or abolish certain possibilities. The adaptive or evolved aspect of learning traits so identified will be the nature or range of expression correlated with the usual environments of history, with non-adaptive, maladaptive, or evolutionarily incidental aspects represented by those appearing in novel or rare environments. Obviously, such "incidental" or non-evolved aspects of learning are crucial in understanding human behavior because of the extraordinarily rapid changes induced by culture and technology. What would be left is a set of learning abilities, the range and relative ease of which have been tuned by natural selection acting on the genetic makeup of the population. It is difficult for me to conceive of *any other relationship* between learned behaviors and the process of natural selection.

Indeed, what I have just described is the general relationship between natural selection and all kinds of expressions of the phenotype, whether behavioral, physiological, or morphological. This is the reason why—even though learning, or variation resulting from environmental variation, is an explanation for observed behavioral variations which is alternative to an explanation based on genetic variation—cultural evolution is not an *alternative* to natural selection as a general explanation for the nature of human activities. Cultural patterns are, like all expressions of the phenotype, outcomes of different developmental environments acting on sets of genetic materials accumulated and maintained by natural selection. Culture differs from other aspects of phenotypes in the degree to which it can change without genetic change; but behavior in general differs from morphology and physiology in the same regard. This is the raison d'être of behavior. It is a way of responding to a greater proportion of the information available to the organism from immediate contingencies in its environments. Culture is a particular and elaborate system of behavior for doing the same. The questions we are led to ask about culture, after considering it in a biological context, are the same that would be asked from any other analytical approach: What forces influence its patterns? What do its expressions mean? The only distinctive aspect of a biological approach is that we are apt to ask these questions in relation to biological functions, or reproduction. This attitude may seem alien to social scientists, but the correct answers to questions about the significance of culture will be the same regardless of the manner in which they are approached.

CONCERNING CULTURE: NATURAL SELECTION AND CULTURE THEORY

Probably because anthropologists and others tended to identify the possession of culture and the capacity for culture as peculiarly human traits, the concept of culture has acquired and retained a certain singularity: hence, perhaps, efforts to seek general or singular theories of culture; and perhaps also the assertion by investigators reluctant to see culture in this way that "Culture is dead"- that is, that it does not possess the singularity attributed to it and that truly general theories of culture are therefore unlikely. Others, still convinced about the generality of the concept of culture, have come to the notion that, in the absence of acceptable functional explanations, culture can only be explained in terms of itself, or as a set of arbitrarily assigned meanings or symbolizations (White 1949; Sahlins 1976b), and specifically cannot be explained in terms of utilitarianism of any sort or at any level. White, for example, writes skeptically of man's "vaunted control of civilization" and of the fond belief that it "lies within man's power... to chart his course as he pleases, to mold civilization to his desires and needs."

Culture, such authors seem to be arguing, is something greater than humans collectively and almost independent of humans individually: It continues on courses perhaps unpredictable, and certainly swayed but slightly by the wishes of individuals, who are merely its "passive" transmitters. Sahlin; comes very close to describing culture as an aspect of the environment of humans about which they can do little but accept it in just those terms, thereby almost paralleling the biologists' concept of the genotype and the phenotype as parts of the environment of selection of the individual genes.

But, if human evolution, like that of other organisms, has significantly involved selection effective at genic levels, realized through the reproductive strivings of individuals, neither humans as individuals nor the human species as a whole have had "a" course to chart in the development of culture but rather a very large number of slightly different and potentially conflicting courses. In such event it would indeed be difficult to locate "a function for," or even "the functions of," culture; instead, culture would chiefly be, as Sahlins' view may be slightly modified to mean, the central aspect of the environment into which every person is born and must succeed or fail, developed gradually by the collections of humans that have preceded us in history, and with an inertia refractory to the wishes of individuals, and even of small and large groups. Culture would represent the cumulative effects of inclusive-fitness-maximizing behavior (i.e., reproductive maximization via all socially available descendant and non-descendant relatives) of the entire collective of all humans who have lived. I here advance this as a theory to explain the existence and nature of culture, and the rates and directions of its change.

If this theory is appropriate, then aspects of culture would be expected to be adversary to some of the wishes of each of us; few aspects of it would be viewed with equal good humor by all of us; and in just this circumstance we would not expect grand utilitarian views of culture, general theories of culture, or efforts at purposeful guidance of culture to succeed easily. These are exactly the kinds of failures that have always plagued culture theorists. Yet, by this theory, the inertia of culture would exist *because* individuals and groups did influence its directions and shape, molding it—even if imperceptibly across short time periods—to suit their needs, thereby incidentally increasing the likelihood that subsequent individuals and groups (a) could find ways to use it to their own advantages as well and (b) could not alter it so greatly or rapidly.

It would also be a source of confusion, in attempts to relate directions and rates of cultural change to utilitarian theories, that the reproductive efforts of individuals would not actually be directed at *changing* culture, as such; nor would such efforts lead to any particular directions of change in culture as a whole. The striving of individuals would be to *use* culture, not necessarily by changing it, to further their own reproduction. No necessary correlation would exist between success in the reproductive striving of an individual and the magnitude of the individual's effect on cultural change, or between the collective success of the individuals making up a group or society and the rate of cultural change. It would not matter if one were a legislator *making* laws, a judge *interpreting* them, a policeman *enforcing* them, a lawyer *using* them, a citizen *obeying* them, or a criminal *circumventing* them: Each of these behaviors can be seen as a particular strategy within societies governed by law, and each has some possibility of success.

Again, it would tend to be contrary to the interests of the members of society that cultural changes of any magnitude could easily be effected by any individuals except for inventions seen as having a high likelihood of benefiting nearly everyone. The reasons are that (1) changes, effected by individuals or subgroups in their own interests, would likely be contrary to the interests of others; and (2) once individuals have adopted and initiated a particular set of responses to the existing culture around their own interests, changes of almost any sort have some likelihood of being deleterious to them. These arguments not only suggest how anthropological interpretations of culture may be entirely compatible with the notion of reproductive striving principally effective at the individual (or genic) level, but also may explain the genesis of views that culture is somehow independent of individuals and groups and their wishes, and not easily explainable in utilitarian terms.

The Evolution of Culture

> It is a fundamental characteristic of culture that, despite its essentially conservative nature, it does change over time and from place to place. Herein it differs strikingly from the social behavior of animals other than man. Among ants, for example, colonies of the same species differ little in behavior from one another and even, so far as we can judge from specimens imbedded in amber, from their ancestors of fifty million years ago. In less than one million years man, by contrast, has advanced from the rawest savagery to civilization and has proliferated at least three thousand distinctive cultures.
>
> [GEORGE PETER MURDOCK, 1960B:247]

If long-term changes in human phenomena, as evidenced for example in the archaeological record, are cultural, and were not induced by natural selection or accompanied by genetic changes relating to cultural behavior, then we should be interested in answering two questions: First, what has guided cultural evolution? What forces can account for its rates and directions of change? Second, what degrees and kinds of correspondence exist today between the patterns of culture and the maximization of genetic reproduction of the individuals using, transmitting, and modifying culture? Are the degrees and kinds of correspondence, and of failure to correspond, consistent with the forces presumed to underlie rates and directions of cultural change?

At one end of a spectrum lies the possibility that all of the cultural changes during human history have been utterly independent of genetic change, neither causing such nor caused by it. At the other end is the possibility that changes in human behavior have correlated with genetic change to approximately the same degree as changes in the behavior of other species, such as non-human primates. Observations within recorded history are sufficient to show that neither of these extreme possibilities is likely. As examples, cultural changes, such as eyeglasses and treatments for diabetes, obviously influence genetic change; and cultural changes clearly have accelerated tremendously in recent decades without any evidence of parallel acceleration in genetic change. At least, then, cultural changes do influence genetic change although there is apparently no clear evidence that genetic changes are causing cultural changes, or that there is any close correlation between cultural changes and genetic changes that specifically influence behavior in relation to culture. Now, it is easy to understand, on theoretical grounds, how culture can change cumulatively without accompanying genetic changes that relate to the behaviors involved-and easy to argue that numerous such changes have occurred within recorded history when strikingly different cultures merged. Therefore the significance of the above two questions about the forces which change culture and the relationships of culture to maximization of reproduction by individuals is brought into an even sharper focus. We expect that the answers to these two questions will be complementary, and that the efforts to answer them should be conducted simultaneously and jointly.

Some changes in culture, such as those influenced by climatic shifts, natural disasters, and diseases, predators, and parasites (of humans and the plants and animals on which they depend), are beyond human control; others are explicitly under such control, although such control may be very direct (invention and conscious planning) or not so direct (resource depletion and pollution). The difficult question, in understanding the relationship between culture and our inevitable history of natural selection, is not in discovering the reasons behind cultural changes, as such, which are actually fairly obvious. Instead, it is in understanding exactly *how* such changes influence culture: What is done with them? What *direction* of change do they induce, and why? Those changes in culture which are consequences of human action appear to represent products of the striving of individuals and groups of individuals. Such changes, as with extrinsically caused changes, are also *responded to by changes* in the striving of individuals and groups of individuals. Inventions are seized upon. Pollution and resource depletion are lamented, and cause geographic shifts in population or efforts at inventions or practices which will either offset their effects on the lives of those showing the effort or allow them to take advantage of such effects. Attempts are made to predict and offset natural disasters and climatic shifts. All of these responses are easily interpretable as part of efforts by individuals, acting alone or in groups, to use culture to their own advantage in the fashion already suggested. But culture is not easily explainable as the outcome of striving to better the future for *everyone equally:* If that were the case, then surely conscious planning would quickly become the principal basis for cultural change, and it would be carried out with a minimum of disagreement and bickering (perhaps we shall actually be able to make our interests coincide to a greater degree by realizing that we have a background of competition in genetic reproduction, which may be less interesting to us once exposed to our conscious reflection).

It is possible to examine the problem of cultural change in a fashion parallel to that used for evolutionary change (e.g., Alexander 1977a). We can ask about the same five phenomena which characterize the process of genetic change (the most closely parallel argument is probably that of Murdock [1960b]).

1. *Inheritance:* Just as the morphological, physiological, and behavioral traits of organisms are heritable, given consistency in the developmental environment, the traits of culture are heritable through learning. They may be imitated, plagiarized, or taught.

2. *Mutation:* Like the genetic materials, culture is mutable, through mistakes, discoveries, inventions, or deliberate planning (Murdock's "variations," "inventions," and "tentations").

3. *Selection:* As with the phenotypic traits of organisms, some traits of culture reinforce their own persistence and spread; others do not, and eventually disappear for that reason (Murdock's "social acceptance," "selective elimination," and "integration"). (See also Campbell 1965, 1975.)

4. *Drift:* As with genetic units, traits of culture can also be lost by accident or "sampling error."

5. *Isolation:* As with populations of other kinds of organisms, different human societies become separated by extrinsic and intrinsic barriers; they diverge, and they may come into contact and remerge or continue to drift apart; items and aspects of culture may spread by diffusion (Murdock's "cultural diffusion" and "cultural borrowing").

Immediately, differences are apparent between the processes of change during genetic and cultural evolution. Unlike genetic evolution, the causes of mutation and selection in cultural evolution are not independent: Instead, there is a feedback between need and novelty. Most of the sources of cultural "mutation" are at least potentially related to the reasons for their survival or failure. Some culture theorists have tended to deny utilitarian connections between the sources or causes of cultural change and the reasons for their survival or failure. I suggest that the reason for these denials is that such theorists have never sought function both in terms of reproduction and at the individual level, as biologists now realize must be the case in organic evolution. Some, such as Franz Boas and Ruth Benedict, have emphasized the individual; others, such as Bronislaw Malinowski and A. R. Radcliffe-Brown, have emphasized function as survival value—even, sometimes, to the individual. None, however, has seen function as reproductive value. Instead, most functionalists have either sought group-level utilitarian effects or have regarded survival, not reproduction, of the individual as crucial (for reviews of such views, see Hatch 1973; Harris 1968).

Some recent investigators, such as Cloak (1976), Dawkins (1976), Durham (1976a), and Richerson and Boyd (1978), have concentrated on heritability of cultural traits (or cultural novelties or "instructions") and argued that their separate mode of inheritance thwarts the operation of natural selection of genetic alternatives. I regard this approach to the history of culture as similar to a view of the natural history of organisms that sees phenotypes in general (as opposed to no phenotypes) as essentially thwarters of natural selection. In one sense they are, since they necessarily render the action of selection on the genes less direct: Selection must now act through the phenotype. But this change had to have occurred because those genes that reproduced via phenotypes outsurvived their alternatives in the environments of history. So must it be with the capacity for culture, as the above authors for the most part acknowledge. Even if culture out-races organic evolution, creating blinding confusion through environmental novelties, to view the significance of its changes and its traits as independent of, or as mere thwarters of, natural selection of genetic alternatives, would be parallel to supposing that the function of an appetite is obesity.

The important question in cultural evolution is: Who or what decides which novelties will be perpetuated, and how is this decided? On what basis are cultural changes spread or lost? In other words, we are led to analyze exactly the same part of the process of cultural change as for genetic change. In cultural change the answer to this question of who decides, and how, actually determines the heritability of culture, since heritability of cultural items at least theoretically can vary from zero to 100 percent from one generation to the next, or even within generations. Any cultural trait, unlike a gene, theoretically can be suddenly cancelled and just as suddenly reinstated, in the population as a whole. Again, in theory at least, this can be done as a result of conscious decision based on what the involved parties see as their own best interests at the time. This reinforcing relationship among selection, heritability, and mutation in culture means that, unlike organic evolution, heritability of culture traits will not be steadily increased; nor will mutability be depressed because the majority of mutations are deleterious in the individuals in which they arise owing to the lack of feedback between mutational directions and adaptive value. Some cultural mutations appear (that is, are implemented, or translated from thought to action) because they are perceived to have value. Unlike evolutionary change, then, cultural change will acquire inertia to the extent that the interests of individuals and subgroups *conflict* (and have a history of conflicting), and whenever the distribution of power is such as to result in stalemates. In part this means that cultural change may be expected to continue accelerating, and this acceleration, I believe, will not only make it increasingly difficult to interpret human behavior in terms of history, but will also increasingly become apparent as the source of novel ethical problems, bound to increase in numbers and severity as cultural change accelerates, because ethical problems derive from conflicts of interest and these are bound to become more complex (Alexander, 1979).

Most recent and current efforts to relate genetic change and cultural change, then, seem really to be efforts to divorce them - to explain why and how culture and genes came "uncoupled" during human history. These arguments generally assume that the uncoupling is essentially synonymous with the appearance of culture—that culture is, by definition, an uncoupling of human behavior from gene effects.

I think these are the reasons why virtually all efforts to understand culture in biological terms have failed. We can easily assume that the capacity for culture *allowed* (as an incidental effect) various degrees of uncoupling of human behavior from reproductive maximization. In modern urban society, for example, such uncoupling is rampant. But to assume that uncoupling is the (historical, biological, evolutionary) function of culture, or its basic significance or attribute, is, as already suggested, like assuming that the function of an appetite is obesity.

There is enough evidence, even in everyday life, to indicate that in general human social behavior is remarkably closely correlated to survival, well-being, and reproductive success. If one accepts this assumption then it is easy to agree that the real question is: What forces could cause the continued coupling between culture and genes? In effect, we must discover, for cultural as well as genetic evolution, the nature of the "hostile forces" (paralleling Darwin's "Hostile Forces" or predators, parasites, diseases, food shortages, climate, and weather [see Alexander 1977a] responsible for natural selection's effects on gene frequencies) by which variations in human social behavior and capacity are selected, by the adjustment of strategies or styles of life, consciously and otherwise, by individuals and groups.

Few people would doubt that positive and negative reinforcement (learning) schedules relate, respectively, to environmental phenomena reinforcing (1) survival and well-being and (2) avoidance of situations deleterious to survival and well-being. With ordinary physical and biotic stimuli this relationship is easy to understand: We withdraw from hot stoves, avoid poisonous snakes, seek out tasty foods, appreciate warmth in winter, dislike getting wet in cold rains, etc. What about social stimuli? Should it not be the same? Should we not seek social situations that reward us and avoid those that punish us? Should not the actual definitions of reward and punishment in social behavior, as with responses to physical stimuli, identify for any organism those situations that, respectively, improve or insult its likelihood of social survival and well-being, with appropriate connotations for reproductive success? Is it possible that Sheldon (1961) was right in suggesting that "…the reason why many pleasures are wicked is that they frustrate other pleasures"? That evil consists "in frustrations, as the Thomist says, in privation of one good by another"? Is what is pleasurable, hence, "good" and "right," that which, at least in environments past, tended to maximize genetic reproduction?

ARBITRARINESS IN CULTURE

The symbolic or seemingly arbitrary nature of many aspects and variants of culture is commonly regarded as contrary to any functional theory, and especially to the notion that culture can somehow be explained by a history of differential reproduction by individuals. Of course, *seeming* arbitrariness may represent observer error based on failure to understand the significance of environmental variations. Arbitrariness may also be a consequence of the inertia to cultural change in the face of environmental shifts; of mistakes about what kind of behavior will best serve one's interests—especially in the face of the constant and accelerating introduction of novelty, primarily through technology. But, even if the assessment of arbitrariness is actually correct, it need not be contrary to a theory based on inclusive-fitness-maximizing, particularly if culture is explained as a product of the different, as well as the common, goals of the individuals and subgroups of individuals who have comprised human society during its history. Thus, however symbolism and language arose- say, because they were superior methods of communication—their existence, as the major sources of arbitrariness, also allowed the adjustment of messages away from reality in the interests of the transmitting individual or group. In other words, as abilities and tendencies to employ arbitrary or symbolic meanings increased the complexity and detail of messages, and the possibility of accurate transmission under difficult circumstances (e.g., more information per unit of time or information about objects or events removed in time or space), they also increased opportunities for deception and misinformation. It would be a consequence that arbitrariness could typify some of the different directions taken by cultural changes which were nevertheless crucial to their initiators and perpetuators.

Consider the relationship between status and the appreciation of fashion, art, literature, or music. What is important to the would-be critic or status-seeker is not alliance with a particular form but with whatever form will ultimately be regarded as most prestigious. If one is in a position to influence the decision he can, to one degree or another, cause it to become arbitrary. Fashion designers, the great artists, and the wealthy are continually using their status to cause such adjustments. In no way, however, does such arbitrariness mean that the outcomes are trivial or unrelated to reproductive striving. Precisely the opposite is suggested that arbitrariness may often be forced, in regard to important circumstances, because the different circumstances involved represent important alternatives and because forcing arbitrariness is the only or best way for certain parties to prevail.

These various suggestions may simultaneously explain the genesis of "great man" theories of culture and their failure as general explanations. Great men do appear, and their striving, almost by definition, is likely now and then to have special influences; but, for reasons given above, not necessarily great influences and not influences leading to particular, predictable, overall changes in culture.

The old saw that "one hen-pecked husband in a village does not create a matriarchy" also emphasizes not only that individuality of striving occurs within culture but that it does not necessarily lead to trends. Similarly the argument about status and arbitrariness is a variant of the adage that "when the king lisps everyone lisps," and it bears on the notion of a "trickle-down" effect in stratified or hierarchical social systems. But it indicates that the "trickle-down" effect, rather than being a societal "mechanism for maintaining the motivation to strive for success, and hence for maintaining efficiency of performance in occupational roles in a system in which differential success is possible for only a few..." (Fallers 1973) is a *manifestation* of such striving, and a *manifestation of degrees of success.*

As already noted elsewhere in this volume, several recent studies have suggested that many aspects of culture, involving such items as patterns of marriage, inheritance, and kinship behavior, and varying in expression among societies, are neither arbitrary nor independent of predictions from a theory dependent upon inclusive-fitness-maximizing by individuals (Alexander 1977a).

Like learning theory and other theories that stop with proximate mechanisms, Malinowski's "functional" theory of culture, which was couched in terms of satisfying immediate physiological needs, did not account for the existence of those needs (Alexander 1977b). Thus, Sahlins (1976a) was led to say that for Malinowski culture represented a "gigantic metaphorical extension of the digestive system." But Malinowski's theory would have made sense in the terms suggested here if it could only have been interpreted as seeing culture as a gigantic metaphorical extension of the *reproductive* system.

The ideas I have just suggested are alternative to recent efforts to explain the relationship of culture and genetic evolution—or, more particularly, their apparent lack of relationship—by suggesting that "cultural instructions" (Cloak 1975) or "memes" (Dawkins 1976) are selected in the same fashion as, and often in opposition to, genes or genetic instructions; or that two kinds of selection, often in opposition, are necessarily involved (Richerson and Boyd 1978). Arbitrariness, then, in fashion or any other aspect of culture, may not be contrary to the genetic reproductive success of those initiating and maintaining it, only to that of some of those upon whom it is forced, in particular those who are least able to turn it to their own advantage. To understand the reproductive significance of arbitrariness as a part of status-seeking, one need only understand the reproductive significance of status. One might suggest that there are *genetic instructions* which somehow result in our engaging in arbitrariness in symbolic behavior in whatever *environments* it is genetically *reproductive* to do so.

I'd suggest, then, that the rates and directions of mutability and heritability in culture are determined by the collectives and compromises of interest of the individuals striving at any particular time or place, together with the form and degree of inertia in the cultural environment as a result of its history; that the "hostile forces" that result in cultural change have tended increasingly to be the conflicts of interest among human individuals and subgroups in securing relief from Darwin's "Hostile Forces of Nature" (see above); and that, among these "Hostile Forces of Nature," increasingly prominent and eventually paramount have been what amounted to predators, in the form of other humans acting in groups or in isolation, with at least temporary commonality of interests (Alexander 1971, 1974, 1975a, 1977a).

By these arguments four outcomes are predicted: (1) a reasonably close correspondence between the structure of culture and its usefulness to individuals in inclusive-fitness-maximizing, (2) an even closer correlation between the overall structure of culture and those traits which benefit everyone about equally, or benefit the great majority, (3) extremely effective capabilities of individuals to mold themselves to fit their cultural milieu, and (4) tendencies for culture to be so constructed as to resist significant alteration by individuals and subgroups in their own interests and contrary to those of others. If these predictions are regarded as important we shall be led to analyze the variations in culture potentially as the outcomes of different strategies of inclusive-fitness-maximizing under different circumstances, and the proximate or immediate physiological and social mechanisms whereby inclusive fitness is maximized as potential explanations of degrees and directions by which cultural patterns diverge from actual inclusive-fitness-maximizing behaviors when technological change and other events create novel environments outside the limits of those in which earlier behaviors functioned.

Concluding Remarks

I think we may regard as settled the universality and inevitability of natural selection and the rarity of effective selection above the individual level, and as relatively trivial for social scientists the problem of the relative effectiveness of selection at the individual level as against some lower level. I also suppose that culture can evolve without genetic change, and that it does so frequently without diminution of inclusive-fitness-maximizing effects. It would appear that the immediate future in other areas of investigation will see concentration on two questions: (1) to what extent are cultural patterns actually independent of predictions from natural selection, and why, and (2) how could patterns of cultural behavior be consistent with natural selection in ways that do not do violence to our knowledge of the extent and nature of learning? The papers in this volume suggest this trend and indicate that in most cases the data, if they are to lead to convincing answers, will have to be gathered with these questions actually in mind.

The complexity of the picture developed by these arguments and conclusions indicates both the difficulty involved in extensive and thorough testing of an inclusive-fitness-maximizing theory of human sociality and the potential generality of such a theory. Such testing is the major challenge that lies ahead on the border between the social and biological sciences, together with the problem of dealing with the moral and ethical questions that arise along with any increase in understanding of human behavior and how to modify it. The tasks so identified are not likely to be easy or simple. But, then, no one who ever thought about human behavior in analytical terms is likely to have supposed that they would be.

References

Alexander, R. D. 1971. The search for an evolutionary philosophy of man. *Proceedings of the Royal Society of Victoria* 84:99–120.

Alexander, R. D. 1974. The Evolution of Social Behavior. *Annual Review of Ecology and Systematics* 5:325–383.

Alexander, R. D. 1975a. The search for a general theory of behavior. *Behavioral Science* 10:77–100.

Alexander, R. D. 1977a. Natural selection and the analysis of human sociality. In *Changing Scenes in the Natural Sciences*, 1776–1976. C. E. Goulden, ed. Pp. 283–337. Academy of Natural Sciences, Special Publication 12, Philadelphia: the Academy.

Alexander, R. D. 1977b. Evolution, human behavior and determinism. *Proceedings of the Biennial Meeting of the Philadelphia Science Association* (1976). Vol. 2:3–21.

Alexander, R. D. 1977c. Review of The Use and Abuse of Biology: An Anthropological Critique of Sociobiology. *American Anthropologist* 79:917–920.

Alexander, R. D. 1978a. Natural selection and societal laws. *In Science and the Foundation of Ethics. III. Morals, Science, and Society*. T. Engelhardt and D. Callahan, eds. Hastings-on-Hudson, New York: Hastings Institute of Society, Ethics and the Life Sciences. Pp. 249–290.

Alexander, R. D. 1978b. Evolution, creation and biology teaching. *American Biology Teacher* 4(2):91–104.

Alexander, R. D. and Sherman, P. W. 1977. Local mate competition and parental investment patterns in social insects. *Science* 196:495–500.

Alexander, R. D. 1979. Evolution, social behavior and ethics. *In Science and the Foundations of Ethics. IV*. Hastings-on-Hudson, New York: Hastings Institute of Society, Ethics and the Life Sciences.

Campbell, D. T. 1965. Variation and selection retention in sociocultural evolution. In *Social Change in Developing Areas: A Re-interpretation of Evolutionary Theory*. H. R. Barringer, G. L. Blankstern, and R. W. Mack, eds. Pp. 19–49. Cambridge, Mass.: Schenkman.

Campbell, D. T. 1975. On the conflicts between biological and social evolution and between psychology and moral tradition. *American Psychologist* 30:1103–1126.

Cloak, F. T. Jr. 1975. Is a cultural ethology possible? *Human Ecology* 3:161–182.

Darwin, C. 1967. *On the Origin of Species*. Boston, Harvard University Press (Facsimile of the 1859 edition).

Dawkins, R. 1976. *The Selfish Gene*. Oxford: Oxford University Press.

Durham, W. H. 1976a. The adaptive significance of cultural behavior. *Human Ecology* 4(2):89–121.

Fallers, L. A. 1973. Inequality: Social Stratification Reconsidered. Chicago: University of Chicago Press.

Ferguson, A. 1976. Can evolutionary theory predict? *American Naturalist* 110:1101–1104.

Hamilton, W. D. 1967. Extraordinary sex ratios. *Science* 156:477–488.

Harris, M. 1968. *The Rise of Anthropological Theory*. New York: T.Y. Crowell.

Hatch, E. J. 1973. *Theories of Man and Culture*. New York: Columbia University Press.

Mayr, E. 1963. *Animal Species and Evolution*. Cambridge, Harvard University Press.

Moore, J. N. and Slusher, H. S., eds. 1970. *Biology: A Search for Order in Complexity*. Grand Rapids, Mich.: Zondervan Publishing House.

Murdock, G. P. 1960b. How culture changes. In *Man, Culture and Society*. H. L. Shapiro, ed. Pp. 247–260. New York: Oxford University Press.

Peters, R. H. 1976. Tautology in evolution and ecology. *American Naturalist* 110:1–12.

Richerson, P. J. and Boyd, R. 1978. A dual inheritance model of the human evolutionary process I: Basic postulates and a simple model. *Journal of the Social Biological Structure* 1:127–154.

Sahlins, M. D. 1976a. *The Use and Abuse of Biology: an Anthropological Critique of Sociobiology*. Ann Arbor: University of Michigan Press.

Sahlins, M. D. 1976b. *Culture and Practical Reason*. Chicago: University of Chicago Press.

Sheldon, W.H. 1961. The criterion of the good and the right. In *Experience, Existence and the Good*. I.C. Lieb, ed. Pp. 275–284. Carbondale: Southern Illinois University Press.

Simpson, E. E. 1964. *This View of Life: the World of an Evolutionist*. New York: Harcourt, Brace and World.

Stebbins, G.L. 1977. In defense of evolution: tautology or theory? *American Naturalist* 111:386–390.

Trivers, R. L. and Willard, D. E. 1973. Natural selection of parental ability to vary the sex ratio of offspring. *Science* 179:90–92.

Trivers, R. L. and Hare, H. 1976. Haplodiploidy and the evolution of the social insects. *Science* 191:249–263.

White, L. A. 1949. *The Science of Culture.* New York: Farrar, Straus & Giroux.

Williams, G. C. 1966. *Adaptation and Natural Selection.* Princeton, Princeton University Press.

Intergroup Competition and Within-Group Cooperation

THE OTHER SIDE OF THE SOCIAL COOPERATION COIN

A hydrogen bomb is an example of mankind's enormous capacity for friendly cooperation. Its construction requires an intricate network of human teams, all working with single-minded devotion toward a common goal.
Let us pause and savor the glow of self-congratulation we deserve for belonging to such an intelligent and sociable species.

— ROBERT S. BIGELOW, 1969. *The Dawn Warriors*

Bragging Rights

Yesterday's newspaper mentioned mass graves just discovered
where the other side was doing its thing not so long ago.
The next page bore a photograph of an old Afghan man,

arms outstretched, eyes imploring: eighteen, he said,
were sleeping together in his house,
all of his family, when our side dropped those bombs

aimed at enemies departed weeks before.
Two, the old man said, sobbing,
had emerged from the rubble,

two: he and his small daughter,
all that remained.
Today's newspaper indicated that overall our side

is proud and happy because its bombs
have been 70% accurate.

Alexander, *2011, p. 20*

Introduction

Thinking about Human Aggression, Past and Present: Alexander and Tinkle's (1968) Review of Lorenz and Ardrey
Bobbi S. Low

The altruistic tendencies of man most likely arose directly out of the interplay between increasingly elaborate intergroup aggressiveness and intragroup cooperativeness originating in parental behavior; the same process was more than likely fundamental in the rapid evolutionary increase in man's brain size.

—ALEXANDER AND TINKLE 1968: 2470

Writing (and reading) book reviews can be a real bore: the main types include the "just the facts, ma'am" versions ("in chapter 3, evolution is covered…"); the rant by a dear enemy; the rave by an acolyte. But this is not your daughter's book report; written more than 40 years ago, this short piece foreshadows Alexander's subtle and encompassing theory of cooperation, and the interplay between cooperation and aggression, as well as laying out, with remarkable sophistication for the time, the interactions likely between natural selection—cultural selection—topics of deep current interest.

Only two years after Williams's 1966 *Adaptation and Natural Selection*, Alexander and Tinkle made a clear argument for individual costs and benefits driving cooperation, at a time when others (like my major professor) had not yet thought through the difficulties in generating, maintaining, and spreading "species adaptations" of the naïve group-selection sort. And they made the clear, novel, argument that the driving force behind cooperation among males was not escaping predators or finding food, but conspecific intergroup aggression; that is, that other humans were (as they probably remain) the strongest selective forces in the evolution of *Homo*. The force of conspecific arms races drove both aggression and cooperation, and a rapidly enlarging brain (see "A New Theory of Cooperation" introduced by Frank, this volume).

Alexander and Tinkle walked a fine line to delineate, as well as could be done at the time, the fact that natural selection had surely shaped human aggression and cooperation in context, despite a widespread then-current assumption that there was no connection between any part of the human genotype and aggression. Today we know about some genetic influences (as is usual, from uncovering cases that are so far from normal as to be noticeable); but Alexander and Tinkle were effective long before this knowledge, by pointing out that costs and benefits of aggression and cooperativeness differed for various individuals with the context, and that open aggression was more often profitable for males rather than females.

So, before the term "behavioral ecology" really existed, Alexander and Tinkle laid out the argument that the particular structure of human groups (in time, space, composition), combined with sociality and intelligence, led to the development of a delicate and complex within-group balance between cooperation and competition, and increasingly intense between-group competition. Alexander continued to develop theoretical aspects of the intelligence-cooperation-competition tension (e.g., 1979, 1986, 1987), and Flinn and colleagues (2005) have fleshed out Alexander's argument further.

Alexander was amazingly prescient; even today there are papers asking whether human group structure might influence behavioral ecology (e.g., Bowles 2009)— yet Alexander and Tinkle raised this issue more than 40 years ago, and set it in the broadest (and most relevant) context: that of reproductively important resources. Others followed (Chagnon 1997; Wrangham 1985; Manson and Wrangham 1991; Low 1993, 2000; ch. 13, this volume) but we were merely filling in the details. The main outlines of the argument were set out in this short piece. In traditional societies, a man's skill in war tended to be closely tied to his lifetime reproductive success. As a Yanomamö man made clear to Napoleon Chagnon (1997: 191) in discussing raids, it is really all about women and the resources to get them, whether those are cattle, sheep, goats, or something else.

Perhaps the historical oddity of a grand-scale Cold War in the twentieth century obscured the importance of individual and coalitional striving—striving that evolved because it influenced reproductive success. That is the core; in proximate terms, one can list endless "causes" of warring. Even scholars of military history tended to miss the importance of intergroup conflict as reproductive strategy, though the historian John Hale (1985:22) called warfare through the Middle Ages "violent housekeeping," which does suggest individual offense and defense focused on resources. From very small tribal conflicts on, leaders and skillful fighters traditionally fought in the middle of the fray; since the Middle ages, as John Keegan (1987) noted, commanders have moved farther from danger and the "front" so that risks and benefits to fighting grew ever more separate. But the *functional* significance of the evolution of coalitions to fight other groups' coalitions has not changed.

Not every observation Alexander and Tinkle made turns out to be strictly true, but their argument is solid. They named humans as the only group other than social insects that made war; we know now that, for example, vervet monkeys have intergroup aggression over range boundaries (Cheney 1986); and chimpanzee males go on raids much like the Yanomamö (e.g., Manson and Wrangham

1991, Mitani et al. 2010, Watts et al. 2006), though neither of these cases is entirely comparable (nor are ants). Human warfare may, however, be unique among intra-specific intergroup conflict, in having "disinterested" third-party involvement (e.g., within-nation legal powers, United Nations peacekeeping forces), as well as having a separation across individuals in who pays the costs and who reaps the benefits of warfare, from "taking the King's shilling" to the US draft, and today's US recruitment primarily of young men who have few other prospects. It is true that the institutions don't always work, as in the Rwandan genocide, but they are unique in existing at all. (This is another phenomenon on which Alexander later expanded: e.g., Alexander 1986, 1987.)

Alexander and Tinkle suggested that small group size may have mandated some rather close inbreeding, resulting in the favoring of intensive outbreeding in some groups. Perhaps, but first-cousin marriages (prohibited in 25 US states, allowed in another six only if the individuals cannot reproduce, or in one case [Maine], have a certificate of genetic counseling) are in fact, preferred marriage forms across traditional societies. The commonest *types* of cousin marriage across traditional societies reinforce the general argument Alexander and Tinkle make (and which Alexander expanded explicitly in 1979: 15): Most of these societies are patrilocal, and "father's brother's daughter" is the associated preference. Avunculocal and matrilocal residences favor other arrangements (father's sister's daughter, and mother's brother's daughter). Both resources and male-male cooperation within group are at stake (e.g., Flinn and Low 1986). Interestingly, the one type that would round out our survey—mother's sister's daughter—is virtually unknown, probably because women in most traditional societies control few resources, and do not, as do vervet monkey females, fight openly in intergroup conflicts.

So even particular "not-quite-facts" of their time, when we find out more, con-tribute to Alexander and Tinkle's general argument. For me, the saddest statement that turned out not to be true was: *It is a significant step forward that the questions receiving attention today are not* whether *man evolved, but* how *he evolved.* I think that statement was made more in hope than conviction. During the late 1960s and the 1970s, Alexander publicly debated noted creationists, hoping, perhaps, to make that statement come true. Nonetheless, today we have thinly disguised creationism in the "intelligent design" group, and 44% of Americans are sure that humans did *not* evolve, but were created (presumably with all other life forms) by God in the last 10,000 years (Dawkins 2009, citing a 2008 Gallup poll). If you include those who think humans did evolve, but under God's guidance, the total is 80%.

One aspect peculiar to Dick was that whenever he wrote anything, he ended up trying to explain the world. In longer pieces, this could be frustrating. (Once, when I was working on historical demography, I thought I remembered that he had made a prescient comment in print about demographics. I asked him, and he said yes, he had—but he couldn't remember where, and I never did find it again.) But in short reviews, this tendency has a mind-expanding impact; tak-ing on human evolution, the interplay between aggression and cooperation, gets readers thinking about more than what Lorenz and Ardrey had written, and that is crucially important. In another review on a book on biological control, Alexander (1975) raised, far before others were thinking about it, what is now called the "pesticide treadmill" and the evolutionary problem of how, by relying on pesticides and herbicides in agriculture, we simply select for ever-more-resistant pests, a problem that still plagues us today.

Alexander and Tinkle ended with a crucial caveat worth remembering:

> Finally, we do not believe, as Ardrey and Lorenz both imply, that knowledge concerning man's evolutionary history, regardless of the revelations it may involve, can in any way restrict what man is able to accomplish in manipulating his own behavior toward any desired end. Knowledge of our evolutionary background cannot close doors; it can only open them.

May it be so.

References

Alexander, R.D. 1975. Natural enemies in place of poisons. *Nat. Hist.* 84(1): 92–95. (Book review of Paul DeBach's *Biological Control by Natural Enemies.*)

Alexander, R.D. 1979. *Darwinism and Human Affairs.* Seattle: University of Washington Press.

Alexander, R.D. 1986. Biology and law. *Ethol. Sociobiol.* 7: 167–173.

Alexander, R. D. 1987. *The Biology of Moral Systems.* Hawthorne, NY: Aldine DeGruyter.

Bowles, S. 2009. Did warfare among ancestral hunter-gatherers affect the evolution of human social behaviours? *Science* 324:1293–1298.

Chagnon, N. 1997. *Yanomamö.* 5th ed. Ft. Worth, TX: Harcourt Brace.

Cheney, D. 1986. Interactions and relationships between groups. In: B. Smuts, D, Cheney, R. Seyfarth, R. Wrangham, and T. Struhsaker (eds.), *Primate Societies.* Chicago: University of Chicago Press, pp 267–281.

Dawkins, R. 2009. *The Greatest Show on Earth: The Evidence for Evolution.* New York: Free Press.

Flinn, M., Geary, D., and Ward, C. 2005. Ecological dominance, social competition, and coalitionary arms races: Why humans evolved extraordinary intelligence. *Evol. Hum. Behav.* 26: 10–46.

Flinn, M., and Low, B. 1986. Resource distribution, social competition, and mating patterns in human societies. In: D. I. Rubenstein and R. W. Wrangham (eds.), *Ecological Aspects of Social Evolution: Birds and Mammals.* Princeton, NJ: Princeton University Press, pp. 217–243.

Hale, J.R. 1985. *War and Society in Renaissance Europe.* New York: St. Martin's Press.

Keegan, J. 1987. *The Mask of Command.* London: Jonathan Cape.

Low, B. 1993. An evolutionary perspective on war. In: H. Jacobson and W. Zimmerman (eds.), *Behavior, Culture, and Conflict in World Politics.* Ann Arbor: University of Michigan Press, pp. 13–56.

Low, B. 2000. *Why Sex Matters.* Princeton, NJ: Princeton University Press.

Manson, J., and Wrangham,V. 1991. Intergroup aggression in chimpanzees and humans. *Curr. Anthro.* 32(4):369–390.

Mitani, J., Watts, D.P., and Amsler, S.J. 2010. Lethal intergroup aggression leads to territorial expansion in wild chimpanzees. *Curr. Biol.* 20(12): R507–R508.

Watts, D., Muller, M., Amsler, S., Mbabazi, G., and Mitani, J.C. 2006. Lethal intergroup aggression by chimpanzees in the Kibale National Park, Uganda. *Am. J. Primatol.* 68: 161–180.

Williams, G.C. 1966. *Adaptation and Natural Selection*. Princeton, NJ: Princeton University Press.

Wrangham, R. 1985. War in evolutionary perspective. In: D. Pines (ed.) *Emerging Synthesis in Science*. Santa Fe, NM: Santa Fe Institute, pp. 123–132.

A COMPARATIVE REVIEW

Excerpt from Alexander, R.D., and Tinkle, D.W. 1968. Review of *On Aggression* by Konrad Lorenz and *The Territorial Imperative* by Robert Ardrey. *Bioscience* 18:245–248.

On Aggression, by Konrad Lorenz, Harcourt, Brace & World, New York, 1966.
The Territorial Imperative, by Robert Ardrey, Atheneum, New York, 1966.

Few recent books have been reviewed as many times in rapid succession, or with as much vehemence in both defense and derogation, as *On Aggression* by Konrad Lorenz (1966, Harcourt, Brace and World) and *The Territorial Imperative* by Robert Ardrey (1966, Atheneum). The principal reason for this attention—and for the disagreements—is that Lorenz and Ardrey have tried to write about one of the most sensitive and important questions facing man: his nature as determined by and determinable from his evolutionary history. The two books have often been reviewed together because they share the basic theme that man is an aggressive animal and that this aggressiveness is in some way a product of the evolutionary process. *On Aggression* is a personal commentary from a professional zoologist with an extensive background of training, thought, and investigation in behavioral biology. Ardrey, on the other hand, is no biologist, but he has produced a fascinating narrative that is remarkably well-documented. Unfortunately, one of its fascinating aspects is the disarming ease with which it travels back and forth between major insights and ridiculous oversimplifications. Both men write in ways tending to rekindle old, pointless arguments of the instinct vs. learning variety. Although they profess to be presenting evolutionary arguments, both men have mixed into their discussions some peculiarly nonevolutionary or antievolutionary themes.

Man is indeed an elaborately aggressive organism, and the nature of the evolutionary background for this aggressiveness is a legitimate problem. He is also probably the most extensively altruistic of all organisms. Ardrey and Lorenz take the evolutionary basis of his aggressive tendencies as their major themes, but, in general, seem to muddle the problem of evolving his altruistic tendencies. Critics argue that the prominence of these seemingly opposed tendencies in man's behavior indicates that the characteristic that really evolved was merely the capacity for either behavior, as the situation demanded. They contend that the developmental basis of such behaviors in man is too complex, and that they are too indirectly related to the genotype, for selection to accumulate genes directly correlated with either aggression or altruism as such. We believe that this view, too, is an oversimplification. It would be naive to try to explain man's preoccupation with sex without reference to natural selection; perhaps it is only a little less naive to do so with aggression.

The nature of the evidence on which reasonable answers about man's evolutionary history can be based and the problem of what it means for man to make discoveries about himself regarding such attributes as aggression and altruism are the important questions that arise from reading Lorenz's and Ardrey's books. Several previous reviewers have concentrated on detailed criticisms; we would like instead to consider the general question of the role of aggressiveness and territoriality in man's evolution and build an hypothesis on what we think are appropriate attitudes toward these human attributes. We will also try to identify what we believe are flaws in Lorenz's and Ardrey's arguments.

It is a significant step forward that the questions receiving attention today are not whether man evolved but how he evolved. Doubt seems no longer to exist in the minds of reasonable and knowledgeable persons that man is a product of evolution—a result of the same basic process that has produced all life. A major consequence of this realization is that whatever characteristics may be construed to be uniquely or most decidedly human are thereby automatically categorized as producible through natural selection.

If the size of his brain is used as the chief index to man's evolutionary divergence (and this seems reasonable not only because of the importance of brain function in specifying man but also because brain size increases correlate with paleontological and archeological evidence of increasing complexity of social organization and various cultural phenomena), then there seem to be at least three major puzzles concerning man's evolution from a nonhuman primate:

1) How could his brain increase in size so rapidly from australopithecine to modern man (50–150,000 generations)?
2) What caused the increase in brain size to go so far beyond that of all other primates?
3) What caused the brain apparently to stop increasing in size some 50–100,000 years ago?

Regardless of the indirectness of the relationship between man's genotype and those aspects of his phenotype that we generally refer to as "intellect," we must conclude that variations in intellect were subjected to unusually intense selective action, that this selection was consistent across a long period, and that it carried man's intellectual capabilities right up to their present condition.

Let us consider the basic process by which natural selection operates. First, it always involves competition between alternate genetic elements within species. Even in interspecific competition, evolution occurs as a result of some variants within one or both species outreproducing the other variants. Although selection actually works through favoring certain individual organisms, the result is change in gene frequencies in populations.

There seem to be three possible kinds of intraspecific competition or three different levels of intensity at which selection can operate on alternative genetic elements:

1) Differential reproduction without direct interaction, and no confrontation between competitors.
2) Partial or complete exclusion of competitors from the best (or only) sources of food, mates, and shelter through aggressiveness and territoriality.
3) Elimination of competitors or potential competitors by killing them; this could include cannibalism, or the elimination of competitors with food being obtained without additional risk or energy expenditures.

Of these kinds of intraspecific competition, the first would usually result in the slowest evolutionary change, the others, in order, in increasingly rapid change. The questions we would ask about man's evolution are (1) which kinds of competition were involved, (2) which were most likely predominant, and (3) what were the sizes and compositions of the units among which each kind of competition operated? In other words, which operated only among individuals and which among social groups, such as families, of different sizes and complexities?

Differential reproduction without direct competition occurs in every species of organism, whether or not the other forms of competition also occur. Exclusion of competitors through aggression or some form of territoriality is widespread among animals with complex behavior - such as vertebrates, arthropods, and cephalopods and may be universal among such organisms during times when food, shelter, or mates are in short supply. Nearly all modern primates seem to be territorial.

Killing of competitors and cannibalism are rarely observed, and it is usually difficult to obtain evidence whether observed cases represent evolved functions or incidental effects resulting from some other kind of selective action. Few animals seem to be cannibalistic—none as much as man's fossil record suggests was the case during his evolution. On the other hand, reviewers of these books who emphasize that little intraspecific violence occurs in most animals in the wild are simply reminding us that responses to aggression also evolve. That ritualization and threat can be effective in establishing dominance without injuries or death is only evidence that inhibition of aggression as well as aggression is subject to natural selection. Aggressive interactions are also crucial when they are conducted solely by threats; such ritualization, on the other hand, is only effective when, on the average, the subordinate gains by giving in. When commodities (food, shelter, mates) are in sufficiently short supply, no advantage is to be gained by giving up. Unlike Lorenz and many other behavioral biologists, we do not find it reasonable or necessary to consider either inhibition of aggression or altruistic behavior as "species" adaptations, evolved to assist the species at the expense of the reproduction of the individuals showing such responses.

The chances seem remote that man evolved without a significant amount of intraspecific aggression occurring continuously and, in fact, guiding his evolution to some extent. We would go further and agree with Lorenz and Ardrey that a more elaborate and extensive array of intraspecific aggressiveness may have been involved in man's evolution than in that of any other animal. This is not to say that any particular kind or instance of human aggression at present may not have grown out of a purely cultural context. We are simply agreeing that, during a long period—perhaps all—of man's evolution, aggressive behavior was directly favored by selection. Under these circumstances there must have been increases in the frequency of many genes that increased the effectiveness of aggression. As with most other human traits, and all human behavior, it is difficult to understand the developmental and hereditary basis of aggressive behavior in any individual or any particular instance; selective action on such a trait must operate in exceedingly indirect fashions. Aggressiveness may easily be modified by culture, and discernible variations in aggressiveness based on genetic differences may be rare or absent among men today. These facts, however, cannot be used to deny the possibility of a genetic background for either the general intensity and quality or the prevalence of aggressiveness in humans. Neither do such conclusions lead us to the remarkable parallels Ardrey draws when he supposes, for example, that a scientist who "place[s] at the disposal of the machinery of war the most sophisticated attainments of his discipline" is "fill[ing] out from the particularity of his learning the generality of that open instinct, the territorial imperative, and, having done so ... [acting] according to the finished pattern with the predictability of a capricorn beetle."

Excluding certain social insects, man is the only warring species, and one of the few that commonly engages in interindividual death battles. Perhaps only in man are all of the necessary abilities for such behavior combined. Other animals never developed necessary equipment for killing conspecifics, lack ability to organize for group warfare, lack the ability for recognizing and sparing near relatives, or have been unable simultaneously to resolve the conflicting necessities of intragroup (or family) tolerance and intergroup hostility. In any species engaging in the more violent kinds of intraspecific competition, ability to recognize and spare close relatives would be highly favored. Such an effect, moreover, would have been facilitated by man's tendency to live in small bands or family groups. Members of one's own band could automatically be treated as relatives, or tolerated and even assisted; those of other bands could equally automatically be treated as competitors or the enemy.

Let us take a closer look at what early man was presumably like in order to understand better the significance of the above suggestions. Sometime during his early evolution man became more carnivorous than any modern primate. He hunted his food, and this would have placed a selective premium on individuals capable of improving their weapons, their bipedal locomotion, and their ability to hurl weapons at elusive prey.

Up to this point, there may have been relatively mild selection favoring larger brains (by which is implied—properly, we believe—more complex brain function). Cooperation among individuals of a family in hunting could have favored effective communication systems which would have, in turn, allowed for passing on more cultural information to offspring. Such families, with the favorable genetic endowment of larger brains and thus better ability to absorb and remember past experiences and to associate cause and effect relationships, must have been better hunters and also better at transmitting to offspring the benefits of experience. There must also have been sexual selection in the same contexts, for it would certainly have been to the advantage of females to choose among potential mates those whose intelligence and hunting prowess would cause the maximum survivorship of their offspring.

One way or another, family groups evidently increased in size, consisting of more than a pair of adults, and perhaps in some cases three generations of individuals, all of which had more in common, both genetically and culturally, than they had with members of other such groups. The degree of inbreeding may have been rather high within such groups. This could have resulted in the rapid fixation of certain genotypes and favored intensive outbreeding within groups, owing to prevalence of genes deleterious in the homozygous condition, perhaps leaving an effect in incest taboos of modern man.

As males in family groups aged, they would be unable to maintain dominant positions. However, it may have been of advantage to younger members of the group to tolerate such individuals, thereby benefiting from their experience and wisdom. Such behavior would not only select for long adult life but make for greater cohesiveness between generations and cause groups to increase in size without fragmentation and to persist longer. Cooperation between parents and grandparents might allow surer recognition and encouragement of offspring in culturally transmissible skills such as tool making and hunting. It could free younger adults for hunting and other essential activities, and it would allow a longer period for passing on the accumulated culture to each successive generation. Such processes as these should rapidly incorporate into a stable and long-persisting group not only genes for greater intelligence but also any useful cultural attributes introduced into the group. Under such conditions, "postreproductive" becomes a difficult term to define.

The social structure of early man was also probably conducive to the development of elaborate intraspecific aggression. Each family group would have differed from every other one in cultural as well as genetic traits, to a degree depending upon its stability and cohesiveness. The individuals of such groups were surely able to recognize members of their own group, and, further, to recognize some of their closer relatives (at least their own offspring) within the group. Direct aggression between family groups could have resulted in rapid shifts in gene frequencies in the population as a whole. On the other hand, altruistic behavior toward other individuals within groups would also have been favored by selection, both because of the necessity of belonging to a group and because it would result in the favoring of genetically related individuals. Intragroup cooperativeness does not preclude intragroup competition, as, for example, in baboon colonies today.

Elaborate parental behavior, which includes both recognition of relatives and a kind of altruism (toward one's offspring), and elaborate aggressive and territorial behavior go hand-in-hand in a wide array of animals. They are almost universally linked. It seems to us that man's altruistic tendencies, as well as his aggressiveness, could have been favored by ordinary natural selection. There is no need to involve the supernatural or to speak of "species" adaptations. We do not understand Ardrey's tendency to divorce aggressive and reproductive behavior; aggressive and territorial behavior cannot evolve unless it enhances reproduction, and there is no evidence making this argument problematic in any way. We certainly disagree with Lorenz's conclusions that man failed to develop inhibitions to aggression and that this was because for a long period of his history he was unable to kill his fellow man.

Let us consider in more detail the extent and nature of intergroup aggression in early man. As a result of spatial isolation of family groups and an exclusive kind of social organization such as occurs in many primates (and man) today, each family group would have been to a large extent a gene pool and micro-culture of its own. Different groups might be expected to have varied in average intelligence, in the degree of intragroup cooperation, and in the nature of weapons, hunting ability, and experience.

If shortages of essential commodities such as food and shelter were the rule, then when groups contacted one another, we suppose that one usually attacked the other, killing the males and possibly the young, and appropriating the females. The successful band in these battles could accumulate experiences increasing the probability of success in subsequent encounters. Repetition of intergroup interactions should select for greater intelligence, increasing aggressiveness between groups, and, simultaneously, increasing cooperativeness and altruism within each group.

In short, we visualize a situation in man's early hunting ancestry in which reproductive individuals characteristically lived in groups, and in which some groups, possessing higher frequency of individuals of greater intelligence, were able by intragroup cooperation and communication to exterminate and replace adjacent groups. Such a process could bring about increasing uniformity among surviving groups, by assimilating intergroup genetic and cultural variation faster than it could be produced, and ultimately decrease the profit to be gained from direct intergroup strife. As this condition was approached, more cohesiveness among splintering bands might have led to sizeable tribes and nations with a corresponding extension of the allegiances of individuals.

What forces could have promoted cohesiveness in band structure? Large predators eliminate lone individuals or small groups among modern primates that live in tightly organized bands. This may be the only kind of selective action that has produced large bands in primates. Intraband competition, promoted by recognition of near relatives deriving from complex parental activities, would run counter to increases in band size and cohesiveness. Advantages deriving from cooperation in killing large game have often been used to explain development of large bands of primitive men. This not only presumes a dependence upon large game, but it does not seem likely to explain groups of more than a dozen or so able-bodied hunters. Furthermore, as man's weaponry improved and his behavioral complexity increased, the minimal size of groups effective in this context would decrease, not increase.

For long periods during man's evolution organized bands may have served as protection against, not other species of predators, but other bands of humans. In such case there would be no theoretical upper limit on band size. If intraband selection worked continually against increasing band size and cohesiveness, then we would expect that reduction of interband friction resulting from lowered population density (for example, from disease), tendencies toward agriculture, and, eventually, the development of means of protecting large groups within which each family had its own territory (armies and farms) would result in shifts toward increased isolation of small family groups and coincidentally toward monogamy.

To return now to the three questions given at the outset, we believe that man's brain size increased so rapidly and diverged so far from the brains of other primates (1) because man's chief competitors all during his evolution were other men and (2) because the competition was of a most direct and extensively aggressive sort, an increasing amount of it operating between family groups of growing size and complexity and with increasingly effective cultural transmission. We believe that brain size stopped increasing when culture became so elaborate and social groupings so large and complex that recognition of allies largely lost its association with degree of genetic relatedness. The result would have been a re-direction of altruistic behavior previously directed toward genetic relatives until its selective advantages were reduced or nullified. As a result, genetic variations reflected in variations in brain size or complexity would largely lose their selective advantage, and evolutionary increases in brain size would level off.

From the preceding discussion, several useful questions arise for the continued investigation of man's evolutionary history: What was the breeding and social structure of known groups of early man and how much group interchange occurred? How were new bands formed? What was the extent of intergroup aggression? How was intragroup competition expressed? When did cultural means of recognition of fellow band members or tribesmen arise? Were males and young and old members of competing groups ordinarily killed by victorious bands? Were males more often eaten in cannibalistic ceremonies? Was the degree of cooperation between bands a function of the degree of their genealogical relationship? Not all of these questions can be answered in regard to primitive man, but they provide a frame of reference. They are the same questions that should be asked, but often are not, with regard to the behavior of other primates and so-called primitive men still in existence today.

To conclude, the story of man's evolution seems to have been that of individuals becoming able to recognize themselves as members of larger and larger groups of increasing complexity of social organization. The altruistic tendencies of man most likely arose directly out of the interplay between increasingly elaborate intergroup aggressiveness and intragroup cooperativeness originating in parental behavior; the same process was more than likely fundamental in the rapid evolutionary increase in man's brain size. Man's tendency to become involved in wars was almost surely directly favored by selection for a long period of his evolution and, therefore, in some important sense, is not a kind of degenerate or degraded behavior resulting from civilization.

Finally, we do not believe, as Ardrey and Lorenz both imply, that knowledge concerning man's evolutionary history, regardless of the revelations it may involve, can in any way restrict what man is able to accomplish in manipulating his own behavior toward any desired end. Knowledge of our evolutionary background cannot close doors; it can only open them. If man's history did involve "nature red in tooth and claw," it is no less to our advantage to comprehend where we have been, and possibly of very great benefit in insuring that we realize whither it may be that we wish to go and the best way of moving in that direction.

Richard D. Alexander and Donald W. Tinkle
Museum of Zoology and Department of Zoology
University of Michigan, Ann Arbor

Kinship, Parental Care, and Human Societies

To a Life Mate

should we be primitives you and I
could lie together naked between bearskins

staring at the night-time sky
listening to the forest

and dreaming sweet sadnesses
about all that must have gone before us

and all that must come after

Alexander 2011, p. vii

Concealed Ovulation in Humans: Further Evidence
Beverly I. Strassmann

Introduction to Alexander, R.D. and Noonan, K.M. 1979. Concealment of Ovulation, parental care, and human social evolution. In N.A. Chagnon and W.G. Irons (eds.). *Evolutionary Biology and Human Social Behavior: An Anthropological Perspective.* 436-453. North Scituate, MA: Duxbury Press.

I first met Richard Alexander in 1976 when he directed my summer efforts to help my sister Joan with her research on paper wasps (*Polistes annularis*). This study entailed standing on a ladder painting dots on the wasps, sticking it out as long as possible while the wasps grew increasingly agitated. In the Texas heat, my sister wisely wore a bee veil and painter's coveralls while I wore skimpier clothing and got lots of wasp stings. At the end of the summer, I met with Alexander in his office in the Museum of Zoology and he asked me: "What is kin selection?" Alexander was unique among professors in his enthusiasm for discussing science one-on-one with undergraduates and his wonderful courses (Evolution and Human Behavior and Evolutionary Ecology) were in a league of their own. His legacy includes the many evolutionary biologists and anthropologists who got their scientific start in these courses.

I was fortunate that during my undergraduate years Alexander was writing *Darwinism and Human Affairs (D&HA)*—the book that developed his theoretical insights about evolution and culture, as well as nepotism and reciprocity. I feel that I was witness to an important time in the history of science. More than 30 years later, D&HA is still being translated into additional languages. During my first year (1986–1987) of Ph.D. fieldwork among the Dogon of Mali, I used D&HA to teach evolutionary theory to my research assistant, Sylvie Moulin. We had a small house in a remote village with a shade shelter next to an ancient baobab. Sylvie would read D&HA amid the cacophony of weaver finches and ask me questions about difficult paragraphs. In this African field setting where most human interactions take place in a web of kinship, D&HA came dramatically to life. Sylvie abandoned her competing reading material, Clifford Geertz's *Understanding Culture*, to the cobwebs in a corner of her room.

Alexander sparked fire in the minds of his undergraduate students by giving them the opportunity to figure out answers to problems in biology that hadn't already been fully solved. We were expected to come up with something original, and he challenged us to write a publishable essay—possibly one that would identify an error he had made. I wrote my essay, "Sexual selection, paternal care, and concealed ovulation in humans," on an idea that sprang from Alexander's article with Katie Noonan (1979). They began their article with a discussion of distinctively human attributes—attributes not expressed in other species to the degree that they are in humans, including consciousness, foresight, tool use, language, and culture. To this list they added 25 additional traits, many of which are "sexually asymmetrical," reflecting in particular "interactions of the sexes in connection with parental care." The crux of their argument is that as intergroup competition became a principal guiding force in human evolution, the complexity of social competition and cooperation within groups increased, selecting for intelligence, consciousness, and foresight, as well as for increased parental care to impart social skills to offspring.

Intergroup competition had not previously been invoked to explain either the distinguishing human attributes or the unusual extent and duration of parental care in humans relative to other primates. Looking back, it is clear that the Alexander and Noonan article was a scientific watershed because it stimulated a vast literature and has stood the test of time. For example, Bowles and Gintis (2011) recently emphasized intergroup competition as the driving force for intragroup cooperation in humans and their arguments trace directly back to Richard Alexander.

Testing the hypothesis that paternal care is crucial for offspring survival and social success is currently an active area of research in human behavioral ecology. Strongly supportive evidence for the hypothesis has been found in the Ache of Paraguay where children without fathers were 3.9 times more likely to be killed in each year of childhood and children of divorced parents were 2.8 times more likely to be killed (Hill and Hurtado 1996). Many studies in other societies found no association between paternal presence and juvenile survival—perhaps because mothers, grandparents, and other individuals took up the slack when the father was absent (see Winking et al. 2011). Although fathers in many societies engage in little direct care of infants, they have an important effect on the social competitiveness of offspring—the paternal role emphasized by Alexander and Noonan. Among the Martu aborigines of Australia, for example, the presence of fathers is associated with an earlier age at initiation, which is the gateway to reproduction for sons (Scelza 2010). Comparative data on humans and other primates link the evolution of paternal care to the development of pair bonds, as suggested by Alexander and Noonan's scenario for the evolution of concealed ovulation (Fernandez-Duque et al. 2009). Recent research on neuroendocrine mechanisms has shown that lower testosterone and higher prolactin levels are markers of fatherhood and pair-bonding (Gray et al. 2002, Gray et al. 2007, Alvergne et al. 2009). Paternal care is not merely a cultural overlay; instead it is supported by a suite of evolved proximate mechanisms.

CONCEALED OVULATION

Alexander and Noonan's pivotal contribution was to link increased paternal investment in humans to paternity certainty and the evolution of concealed ovulation. They argued that the advertisement of ovulation might cost a female her mate's parental investment in two ways: "(1) by attracting competing males who threaten his confidence of paternity, and (2) by freeing him, after a brief consort period, to seek copulations with other fertile females who would compete later with her for his parental care" (Alexander and Noonan 1979, p. 443). In my paper for Alexander's course, I added a further argument that stemmed from the tradeoff between mating effort and parental effort in males (Strassmann 1981). I suggested that subordinate males who are the least successful at getting multiple mates would have the most to gain by paternal behavior, while the males successful as polygynists would gain least. Further, I suggested that the concealment of ovulation was favored by selection because it enabled females to garner the paternal investment of subordinate males with whom they had a confluence of interest. Both subordinate males and females benefited from increased paternal investment (Strassmann 1981). Alexander and Noonan had suggested that the dominant males would be the first to enter into pair bonds with females concealing ovulation. In primate species, dominance rank and reproductive success are usually correlated, but the evidence has supported the view that females sometimes prefer middle or low-ranking males (Wroblewski et al. 2009).

In his 1990 paper "How did humans evolve?," Alexander used strong logical arguments to reject the other hypotheses (Symons 1979, Benshoof and Thornhill 1979, Burley 1979, Hrdy 1979) that appeared the same year (1979) that he and Katherine Noonan published their article. His discussion of these hypotheses (ch. 7, this volume) illustrates the technique of strong inference (Platt 1964) and shows Alexander's uncanny ability to test alternative hypotheses using qualitative information. His 1990 paper explains why females evolved to conceal ovulation not only from their mates but also from themselves. Females who were conscious of their own ovulation would be in an excellent position to obtain good genes outside the pair bond but would assume the risk associated with continually deceiving an "intimate associate" about the timing of ovulation. Moreover, if females exploited their knowledge of the timing of ovulation, then selection would favor males who withheld paternal investment. Female deceit, in conjunction with continued male paternal investment, would not be evolutionarily stable.

RECENT CLAIMS FOR "HUMAN OESTRUS" AND "OVULATORY CUES"

Alexander and Noonan's hypothesis on concealed ovulation has been scrutinized by countless biologists, anthropologists, and evolutionary psychologists. One recent argument is that ovulation in women is not concealed after all (e.g., Haselton and Gildersleeve 2011, Gangestad and Thornhill 2008). This reaction is due to a misperception that ovulation is concealed only if there is a *total absence* of detectable phenotypic changes during ovulation. Instead, what Alexander and Noonan (p. 442) actually sought to explain were the reasons why natural selection *reduced* the conspicuousness of ovulation in the human lineage:

> In nonhuman primates the general period of ovulation always appears to be more or less dramatically signaled to males (even if only by pheromones or other means not obvious to human observers). All of the nonhuman primates in which females are known to show "pseudo-estrus" are group-living species, while the least obvious signs of ovulation seem to occur in monogamous species like gibbons, or polygynous species, like gorillas, which tend to live in single-male bands. *Human females are thus unique in that they give little or no evidence of ovulation and may be receptive during any part of the ovulatory cycle. Although some women have discovered ways to determine the time of their ovulation, it is clear that selection has reduced the obviousness of ovulation during human evolution, apparently to women themselves as well as to others (emphasis added).*

The hypothesis that human females experience estrus (Gangestad and Thornhill 2008) is at odds with the demonstrable difficulty of detecting ovulation in women. Until the 1920s, even medical experts believed that ovulation occurred during menstruation rather than at midcycle (Strassmann 1996). In a cross-cultural sample of 186 preindustrial societies, I confirmed an earlier report by Paige and Paige (1981) that the most prevalent belief was that conception occurs immediately after menstruation—a view also supported by in-depth field studies of the Dogon of Mali (Strassmann 1996) and Hadza foragers of Tanzania (Marlowe 2004). Needless to say, no man can point to the ovulating women in the room with any appreciable success.

A careful study of American women by Sievert and Dubois (2005) asked: Do women who think they know when they ovulate truly make accurate assessments? Women collected daily urine samples from day 5 of the menstrual cycle through the day they believed they ovulated. The last three samples were tested for a luteinizing hormone (LH) surge and if at least one sample tested positive for an LH surge, then that woman was scored as correct in her assessment for that cycle. To detect ovulation, the women most frequently relied upon changes in cervical mucus ("spinnbarkeit"), abdominal pain ("mittelshmertz"), and the expectation that ovulation occurs at midcycle—such information derives from modern medical research to which ancestral women would not have been privy. Using these signals, only 28% of women who believed that they knew when they ovulated actually gave accurate assessments. When the analysis included women who used basal body temperature as an assessment technique, the accuracy rate went up to 36.1%. These results suggest that even in the presence of contemporary medical information, ovulation is concealed for most women (Sievert and Dubois 2005).

The inaccuracy of self-reports of ovulation does not preclude the possibility of subtle changes across the menstrual cycle in women's sexuality, and identifying such changes has become a particularly active area of research in evolutionary psychology. A recent review concludes that ovulation cues can be found in women's social behaviors, body scents, voices, and possibly, aspects of physical beauty with effect sizes ranging from small (d = 0.12) to large (d = 1.20) (Haselton and Gildersleeve 2011, see also Alvergne and Lummaa 2009).

Most studies in this arena are based on the hypothesis that psychological adaptations should differ when a woman is in the ovulatory phase of the menstrual cycle versus when she is not. The investigators then proceed to document differences in lap dance tips, vocal attractiveness, and other outcome variables at "fertile" versus "infertile" cycle phases. Upon finding differences, the investigators then assume that they have found support for their adaptive hypotheses about male and female sexual strategies. Although this is a burgeoning literature, few studies (Puts 2006, Welling et al. 2007, Roney and Simmons 2008) concern themselves with examining the underlying mechanisms. As Gangestad and Thornhill (2008, p. 997) noted: "The precise endocrine mechanisms that regulate these changes remain largely unknown."

At present, there are no hormones that are strong candidates for explaining the reported ovulation cues. Testosterone contributes to libido in women (Wylie et al. 2010) and therefore might be a hormone of interest. In a recent study (Rothman et al. 2011), however, the difference in serum testosterone and free testosterone between the midcycle and midluteal phases of the menstrual cycle was small and not statistically significant. Estradiol and estrone are other hormones of possible interest, but it is androgens and not estrogens that are prescribed to increase libido in women (Schwenkhagen and Studd 2009). Like the androgens, the estrogens are elevated at both midcycle and the mid-luteal phase (Rothman et al. 2011). A hormone that displays a sharp midcycle peak and is not elevated during the luteal phase would be a stronger candidate for explaining behavioral changes that occur only during or immediately prior to ovulation. Although luteinizing hormone might qualify, it has not (to my knowledge) been reported to influence behavior in humans. If a hormonal basis for ovulatory cues is found, it will be necessary to rule out the possibility that the cues are merely sideeffects rather than evolved mechanisms that promote adaptive strategies (see Haselton and Gildersleeve 2011). Researchers have not set forth their standards of evidence for detecting adaptive design in the observed menstrual cycle correlates (see Reeve and Sherman 1993). Instead they assume that if changes are detected in outcome variables at "fertile" and "infertile" phases of the menstrual cycle, that this evidence alone is sufficient to document psychological adaptations.

When studies of ovulatory cues involve surveys, exceptional diligence is needed to prevent subjects from intuiting that the researcher's focus is on the menstrual cycle, as that knowledge could influence women's responses. Such cueing from the researcher can be extremely subtle, as demonstrated in a study in which subjects read a newspaper article about a mass murder and were asked about the precipitating causes. The respondents' answers were subconsciously influenced by the questionnaire's fictitious letterhead. When the letterhead said "Institute for Personality Research" respondents emphasized personality variables, whereas they identified social-contextual variables when it said "Institute for Social Research" (Norenzayan and Schwartz 1999). In a classic study, women who were manipulated by experimenters into believing that they were premenstrual reported experiencing a significantly higher degree of physical symptoms, such as water retention, than did women who were led to believe that they were intermenstrual (Ruble 1977). In any research involving self-reports, it is challenging to prevent the questions from shaping the answers (Schwartz 2010); researchers of ovulatory cues should address this problem up front.

Laeng and Falkenberg's (2007) study of female sexual response at three phases of the menstrual cycle is noteworthy because it employed pupillary size as an index of "interest." Changes in pupil size are not under conscious influence and do not rely on responses to questionnaires, diaries, or ratings that might reflect participants' beliefs. In this study, women's pupils got larger during the ovulatory phase of the menstrual cycle when they viewed photographs of their boyfriends but not when they viewed the boyfriends of other subjects, or if they used oral contraceptives. I wonder what results might be obtained if women were asked to view photos of dogs (own dog, random dog) to further control for recognition effects and non-sexual responses. In the sex research literature, female sexual response did not change over the menstrual cycle as measured by vaginal blood volume, vaginal pulse amplitude, labial temperature, and so forth (summarized in Laeng and Falkenberg 2007). International data from the Demographic and Health Surveys showed no evidence for a peak in coitus at midcycle for couples in stable unions (Brewis and Meyer 2005). At best, the data are conflicting.

If there are indeed "ovulatory cues" in humans, then it should be possible to document the mechanistic basis for the cues. Given the absence of evidence for the underlying mechanisms, it is premature to claim that women possess *psychological adaptations* surrounding mating that take account of whether they are ovulating or that men shift their behavior in response to subtle ovulatory cues emitted by women (Haselton and Gildersleeve 2011). Not only is the mechanistic basis for the notion of "human estrus" or "ovulation cues" insubstantial, the theoretical basis is also highly problematic. To make this argument, I turn to evidence from the Dogon of Mali.

MENSTRUAL CYCLES IN THE DOGON OF MALI

Having experienced the difficulty of testing hypotheses on the evolution of concealed ovulation, I was motivated to do my Ph.D. on questions about human reproductive behavior that could be addressed empirically. I entered a Ph.D. program in biology at Cornell University where key faculty disallowed research on humans and where one professor informed me that I was, in his view, incapable of doing fieldwork in Africa. To solve this problem, I returned to Michigan, where I knew that I could count on Richard Alexander's unflinching support. A great aspect of Alexander's mentoring is that he gave students free reign to develop their own projects—those who did not like this system called it "sink or swim." For several months I sifted through library books about traditional peoples, eventually choosing the Dogon after learning about them in the Time-Life series *Peoples of the Wild* (Pern 1982). The Dogon had all the features I was looking for: menstrual huts, no contraception, polygyny, nucleated villages, and an indigenous religion that had not been fully supplanted by a world religion. My dissertation focused on: (1) the biology of menstruation, (2) the function of menstrual taboos, and (3) variation in female fecundability (Strassmann 1990).

In the mid-eighties, there was no previous, long-term, prospective study of menstrual cycles in a population that was experiencing natural fertility. Whereas North American women have about 400 menses during their lifetimes, Dogon women in my data set had about 100 menses, most of which occurred in women too young or too old to become pregnant. Repeated menstrual cycles were characteristic of subfecund and infertile women. By contrast, women between the ages of 20 and 35 years spent most of their time pregnant or in lactational amenorrhea—for these women, menstruation and ovulation were extremely rare events (Strassmann 1997, Strassmann & Warner 1998). To take women's evolved reproductive biology into account, evolutionary psychologists should sample women during all phases of the interbirth interval (pregnancy, lactational amenorrhea, and menstrual cycling). Studies that are restricted to undergraduate women who report regular menstrual cycles are not well suited for testing adaptive hypotheses.

Evidence from the Dogon is also pertinent to the theoretical expectation that females engage in a dual mating strategy, seeking good genes from extra pair copulation (EPC) during ovulation while limiting threats to paternal care by remaining faithful to their partners at other stages of the menstrual cycle (Symons 1979, Benshoof and Thornhill 1979, Gangestad et al. 2002). If Dogon females were prone to a dual mating strategy in which they sought EPCs during ovulation, then their rate of extra pair paternity (EPP) should be much higher than it is. The rate of EPP in the Dogon is 1.8% (N = 1704 father-offspring pairs), suggesting that in this traditional society EPCs are rare—at least during the fertile period. Menstrual huts are a cultural feature of Dogon society that helps to prevent cuckoldry by forcing females to disclose the onset of pregnancy and the resumption of fertility after lactational amenorrhea (Strassmann 1992, 1996). Extra pair paternity was more than two-fold higher when the menstrual taboos were not enforced versus when they had been abandoned, a situation associated with religious conversion (Strassmann et al. 2012). Menstrual taboos were found in most pre-industrial societies and are not unique to the Dogon (Strassmann 1992). They are a cultural tactic enforced by males that helps to circumvent the concealment of ovulation (Strassmann 1992, 1996).

With the possible exception of the Jewish Halakha laws (Boster et al. 1998), males do not use menstrual huts or other menstrual taboos to count to day 14 or to identify the precise timing of the fertile period of the menstrual cycle. When lactational amenorrhea lasts about 20 months—as in the Dogon—knowing that a woman has resumed menstrual cycling provides strong evidence that she will soon be fertilizable and must be mate-guarded (Strassmann 1996). Hormonal data show that the postpartum resumption of the menses is closely tied to the resumption of ovulation (Howie et al. 1982). Sometimes ovulation occurs first, other times menstruation is first—either way the two events are usually only about two weeks apart, which is minimal compared with the previous two-and-a-half years when the woman was not fertilizable. The evidence for widespread menstrual taboos in conjunction with the pervasive misimpression that the fertile period occurs immediately after menstruation, suggests that: (1) human males are greatly concerned about their risk for cuckoldry, and (2) their strategies usually do not precisely pinpoint the fertile period within the menstrual cycle.

Notwithstanding some fanciful accounts, it was males who imposed the taboos on females in every society in which menstrual taboos were directly observed by the ethnographer (Strassmann 1992). When females can successfully hide their menses, then they can avoid more intensive mate guarding when they are cycling, which may free them to copulate with extra-pair partners. They can also keep greater control over knowledge of the genetic father's possible identity. In societies as diverse as the Inuit of the Arctic (Balikci 1970) and the Dogon of Mali (Strassmann 1992, 1996), fear that females might hide their menses was a major male concern. When a woman is forced to obey menstrual taboos, then her reproductive strategies are constrained—the husband and his family have the same knowledge that she has about the timing of the menses and in the absence of genetic data, this knowledge is important for paternity assessments (Strassmann 1992, 1996). Hormonal data show that menstrual hut visitation is an honest signal of menstruation in Dogon women of reproductive age (Strassmann 1996). If women want paternal care for their offspring, and to bear sons who stand a chance of being accepted into their social father's patrilineage, then they must signal menstruation honestly. In the Dogon, there is usually one menstrual hut and one shade shelter for each patrilineage—the two structures are placed in close proximity to each other so that the women at the menstrual hut can be monitored. Menstrual taboos are embedded in religion because in all societies religions play a major role in enforcing sexual morality (Strassmann et al. 2012). The addition of supernatural threats to social norms is aimed at increasing compliance with the taboos (Strassmann 1992, 1996).

In sum, females may benefit from EPCs, but they cannot engage in them without risking the loss of paternal investment. When Dogon women divorce, they do so immediately after leaving the menstrual hut because at that time they are demonstrably nonpregnant (Strassmann 1992, 1996). In contrast with the report that American undergraduate men are more proprietary toward partners who are near ovulation (Gangestad et al. 2002), mate guarding in the Dogon is said to be more intensive immediately after the menses—the time when the Dogon believe that females are most fertile and are most prone to deserting their mates (Strassmann 1992, 1996). I would expect that due to the riskiness of EPCs, women usually seek to package "good genes" and paternal investment together in one man at a time—as best they can. When the man proves deficient in either regard, then women can use EPCs as a strategy for securing another combination package with a different (hopefully better) man. I present this "combo hypothesis" as an alternative to the "dual mating hypothesis" which holds that EPCs are timed to occur during ovulation. The "combo hypothesis" unlike the "dual mating hypothesis" predicts relatively high paternity certainty in humans and low levels of sperm competition.

Aside from the Dogon, the only genetic data on paternity certainty in a traditional, small-scale society come from a study of the Yanomamo of Brazil and Venezuela. In this study, the EPP rate was 9.1% in a sample of 132 offspring (Neel and Weiss 1975). A survey of 67 genetic studies reported that the median EPP rate was 1.7% (range 0.4-11.8) for men who were not sampled at paternity testing laboratories (Anderson 2006). Another review of the literature concluded that EPP rates are around 2% in Europe and North America (Simmons et al. 2004). Together with the Dogon result, there is emerging evidence for high paternity certainty in many human populations.

The diverse morphology of spermatozoa in human semen samples and the low quality of human semen samples (Cooper et al. 2010) are indicative of a species in which postcopulatory sexual selection was minor or even trivial. Species with a high degree of postcopulatory sexual selection have reduced variation in sperm morphology (Calhim et al. 2007, Kleven et al. 2008) and a high percentage of motile sperm per ejaculate (Møller 1988, Pizzari and Parker 2009). Human semen samples do have some features in common with those of chimpanzees (*Pan troglodytes*), but the evidence points to far less sperm competition in humans (Anderson et al. 2007, Simmons et al. 2004). Chimpanzees have higher sperm numbers, a reduced duration of epididymal transit, and the ability to maintain high sperm counts in successive ejaculates (Marson et al. 1991, Anderson et al. 2007). Chimpanzee spermatozoa also have a higher mitochondrial membrane potential which may improve sperm swimming speed or longevity (Anderson et al. 2007).

For humans, the ratio of testes mass to body mass (0.06) is similar to that of orangutans (*Pongo pygmaeus*) (0.05), three times that of gorillas (*Gorilla gorilla*) (0.02), and 22% that of chimpanzees (*Pan troglodytes*) (0.27) (Harcourt et al. 1981). Gorillas usually (but not always) live in single-male groups with only one sexually active male (Harrison and Chivers 2007) whereas chimpanzees live in multimale groups and have a promiscuous breeding system (Wroblewski et al. 2009). Orangutans have a dispersed harem polygynous social system in which there are two adult male morphs (flanged and unflanged); females prefer to mate with the flanged males during the periovulatory period and flanged males sire most (but not all) of the offspring (Harrison and Chivers 2007, Stumpf et al. 2008). Although the ejaculates of chimps and orangutans are similar in volume, orangutans have only one-tenth the sperm concentration (Graham 1988). The ratio of seminiferous tubules to connective tissue in human testes is 1.3—similar to that of monogamous gibbons (1.1), but far smaller than that of the primates with multi-male breeding systems: chimpanzees (2.4), baboons (*Papio*) (2.8), and macaques (*Macaca*) (2.2) (Schultz 1938, Harcourt et al. 1981). Thus, several lines of evidence suggest that postcopulatory sexual selection has been relatively weak in humans, casting doubt on the hypothesis that women have an evolved tendency to seek EPCs during ovulation.

Alexander and Noonan's hypothesis on concealed ovulation has been challenged by studies that report that the sexual strategies of men and women are contingent on phase of the menstrual cycle—fertile or infertile (Haselton and Gildersleeve 2011, Gangestad and Thornhill 2008). I have outlined several reasons for being skeptical of these reports: (1) the high level of paternal investment that characterizes most human populations is incompatible with high levels of cuckoldry, (2) genetic studies show relatively high paternity certainty compared with the levels predicted by the "dual mating strategy" hypothesis, (3) morphological and physiological comparisons of humans with other primates do not provide evidence for postcopulatory sexual selection, (4) ethnographic and demographic studies show that ovulation is indeed concealed, and (5) the mechanistic basis for ovulation cues has not been identified. It is an honor to introduce Alexander and Noonan's chapter on concealed ovulation, as it sparked in me a lifelong interest in the divergent strategies of males and females, and in the reproductive events of menstruation, ovulation, fertility, and cuckoldry. It is a classic article, and with each reading I notice something new.

References

Alexander, R.D. 1979. *Darwinism and Human Affairs*. London: Pitman Publishing Limited.

Alexander, R.D., and Noonan K.M. 1979. Concealment of ovulation, paternal care, and human social evolution. In: N.A. Chagnon and W. Irons (eds.), *Evolutionary Biology and Human Social Behavior: An Anthropological Perspective*, pp. 436–453. Belmont, CA: Duxbury Press.

Alexander, R.D. 1990. *How Did Humans Evolve? Reflections on the uniquely unique species*. University of Michigan Museum of Zoology Special Publication No. 1. 1–38.

Alvergne, A., Faurie, C., and Raymond, M. 2009. Variation in testosterone levels and male reproductive effort: insight from a polygynous human population. *Horm. Behav.* 56:491–497.

Alvergne, A., and Lummaa, V. 2009. Does the contraceptive pill alter mate choice in humans? *Trends Ecol. Evol.* 25(3):171–179.

Anderson, K.G. 2006. How well does paternity confidence match actual paternity? Evidence from worldwide nonpaternity rates. *Curr. Anthropol.* 47(3):513–520.

Anderson M.J., Chapman, S.J., Videan, E.N., Evans, E., Fritz, J., Stoinski, T.S., Dixson, A.F., and Gagneux, P. 2007. Functional evidence for differences in sperm competition in humans and chimpanzees. *Am. J. Phys Anthro.* 134:274–280.

Balikci, A. 1970. *The Netsilik Eskimo.* Garden City, NY: Natural History Press.

Benshoof, L., and Thornhill, R. 1979. The evolution of monogamy and loss of estrus in humans. *J. Social Biol. Struct.* 2:95–106.

Boster, J.S., Hudson, R.R., and Gaulin, S.J.C. 1998. High paternity certainties of Jewish priests. *Am. Anthro.* 100(4):967–971.

Bowles, S., and Gintis, H. 2011. *A Cooperative Species: Human Reciprocity and Its Evolution.* Princeton, NJ: Princeton University Press.

Brewis, A., and Meyer, M. 2005. Demographic evidence that human ovulation is undetectable (at least in pair bonds). *Curr. Anthro.* 46:465–471.

Burley, N. 1979. The evolution of concealed ovulation. *Am. Natural.* 114:835–858.

Calhim, S., Immler, S., and Birkhead, T.R. 2007. Postcopulatory sexual selection is associated with reduced variation in sperm morphology. *PLoS ONE* 2(5):e413. doi:10.1371/journal.pone.0000413.

Cooper T.G., Noonan, E., von Eckardstein, S., Auger, J., Baker H.W.G., Behre, H.M., Haugen, T.B., Kruger, T., Wang, C., Mbizvo, M.T., and Volgelsong, K.M. 2010. World Health Organization reference values for human semen characteristics. *Hum. Repro. Upd.* 16(3):231–245

Fernandez-Duque, E., Valeggia, C.R., and Mendoza, S.P. 2009. The biology of paternal care in human and nonhuman primates. *Annu. Rev. Anthro.* 38:115–130.

Gangestad, S.W., Thornhill, R., and Garver, C.E 2002. Changes in women's sexual interests and their partners' mate retention tactics across the menstrual cycle: evidence for shifting conflicts of interest. *Proc. R. Soc. B.* 269:975–998.

Gangestad, S.W., and Thornhill, R. 2008. Human oestrus. *Proc. R. Soc. B.* 275:991–1000.

Graham, C.E. 1988. Reproductive physiology. In: Schwartz, J.H. (ed.), *Orangutan Biology.* New York: Oxford University Press, pp. 91–116.

Gray P.B., Kahlenberg, S.M., Barrett, E.S., Lipson, S.F., and Ellison, P.T. 2002. Marriage and fatherhood are associated with lower testosterone in males. *Evol. Hum. Behav.* 23:193–201.

Gray, P.B., Parkin, J.C., and Samms-Vaughan, M.E. 2007. Hormonal correlates of human paternal interactions: A hospital-based investigation in urban Jamaica. *Horm. Behav.* 52:499–507.

Harcourt, A.H., Harvey, P.H., Larson, S.G., and Short, R.V. 1981. Testis weight, body weight, and breeding system in primates. *Nature.* 293:55–57.

Harrison M.E. and Chivers, D.J. 2007. The orangutan mating system and the unflanged male: A product of increased food stress during the late Miocene and Pliocene? *J. Hum. Evol.* 52:275–293.

Haselton, M.G., and Gildersleeve, K. 2011. Can men detect ovulation? *Curr. Direct. Psychol. Sci.* 20(2):87–92.

Hill, K., and Hurtado, A.M. 1996. *Ache Life History: The Ecology and Demography of a Foraging People.* New York: Aldine de Gruyter.

Howie, P.W., McNeilly, A.S., Houston, M.J., Cook A., and Boyle H. 1982. Fertility after childbirth: Postpartum ovulation and menstruation in bottle and breast feeding mothers. *Clin. Endocrin.* 17:323–332.

Hrdy, S.B. 1979. Infanticide among animals: A review, classification, and examination of the implications for the reproductive strategies of females. *Ethol. Sociobiol.* 1:13–40.

Kleven, O., Laskemoen, T., Fossøy, F., Robertson, R.J., and Lifjeld, J.T. 2008. Intraspecific variation in sperm length in negatively related to sperm competition in passerine birds. *Evolution.* 62(2): 494–499.

Laeng, B., and Falkenberg, L. 2007. Women's papillary responses to sexually significant others during the hormonal cycle. *Horm. Behav.* 52:520–530.

Marlowe, F.W. 2004. Is human ovulation concealed? Evidence from conception beliefs in a hunter-gatherer society. *Arch. Sex. Behav.* 33(5):427–432

Marson, J., Meuris S., Cooper, R.W., Jouannet, P. 1991. Puberty in the male chimpanzee: progressive maturation of semen characteristics. *Biol. Repro.* 44:448–455.

Moller, A.P. 1988. Ejaculate quality, testes size and sperm competition in primates. *J Hum. Evol.* 17:479–488.

Neel, J.V., and Weiss, K.M. 1975. The genetic structure of a tribal population, the Yanomama Indians. 12. Biodemographic studies. *Am. J. Phys. Anthro.* 42:25–52.

Norenzayan, A., and Schwarz, N. 1999. Telling what they want to know: participants tailor causal attributions to researchers' interests. *Euro. J. Soc. Psych.* 29:1011–1020.

Paige, K.E., and Paige, J.M. 1981. *The Politics of Reproductive Ritual.* Berkeley: University of California Press.

Pern, S. 1982. *Masked Dancers of West Africa: The Dogon.* Peoples of the Wild Series. Amsterdam: Time-Life Books.

Pizzari, T. and Parker, G.A. 2009. Sperm competition and sperm phenotype. In *Sperm Biology, An evolutionary perspective* (eds Birkhead TR, Hosken DJ, and Pitnick S), pp 205–244. London: Academic Press.

Platt, J.R. 1964. Strong Interference. *Science.* 146:347–353.

Puts, D.A. 2006. Cyclic variation in women's preferences for masculine traits: potential hormonal causes. *Hum. Nat.* 17:114–127.

Reeve, H.K., and Sherman, P.W. 1993. Adaptation and the goals of evolutionary success. *Q. Rev. Biol.* 68:1–32.

Roney, J.R., and Simmons, Z.L. 2008. Women's estradiol predicts preference for facial cues of men's testosterone. *Horm. Behav.* 53:14–19.

Rothman, M.S., Carlson N.E., Xu M., Wang C., Swerdloff R., Lee P., Goh V.H.H., Ridgway E.C., and Wierman M.E. 2011. Reexamination of testosterone, dihydrotestosterone, estradiol and estrone levels across the menstrual cycle and in postmenopausal women measured by liquid chromatography-tandem mass spectrometry. *Steroids.* 76:177–182.

Ruble, D.N. 1977. Premenstrual symptoms: a reinterpretation. *Science.* 197:291–292.

Scezla, B. 2010. Father's presence speeds the social and reproductive careers of sons. *Curr. Anthro.* 51(2):295–303.

Schultz, A.H. 1938. The relative weight of the testes in primates. *Anatom. Rec.* 72:387–394.

Schwarz, N. 2010. Measurement as cooperative communication: what research participants learn from questionnaires. In: G. Walford, E. Tucker, and M. Viswanathan (eds.), *The SAGE Handbook of Measurement.* London: Sage Publications, pp. 43–61.

Schwenkhagen, A. and Studd, J. 2009. Role of testosterone in the treatment of hypoactive sexual desire disorder. *Maturitas.* 63:152–159.

Sievert, L.L. and Dubois, C.A. 2005. Validating signals of ovulation: do women who think they know, really know? *Am. J. Hum. Biol.* 17(3):310–320.

Simmons, L.W., Firman R.C., Rhodes G., and Peters M. 2004. Human sperm competition: testis size, sperm production and rates of extrapair copulations. *Anim. Behav.* 68:297–302.

Strassmann, B.I. 1981. Sexual selection, paternal care, and concealed ovulation in humans. *Ethol. Sociobiol.* 2:31–40.

Strassmann, B.I. 1990. *Reproductive Ecology of the Dogon of Mali.* PhD. dissertation, University of Michigan, Ann Arbor.

Strassmann, B.I. 1992. The function of menstrual taboos among the Dogon: defense against cuckoldry? *Hum. Nat.* 3:89–131.

Strassmann, B.I. 1996. Menstrual hut visits by Dogon women: a hormonal test distinguishes deceit from honest signaling. *Behav. Ecol.* 7:304–315.

Strassmann, B.I. 1997. The biology of menstruation in *Homo sapiens*: total lifetime menses, fecundity, and nonsynchrony in a natural fertility population. *Curr. Anthro.* 38:123–129.

Strassmann, B.I., Kurapati, N.T., Hug, B.F., Burke, E.E., Gillespie, B.W., Karafet, T.M., and Hammer, M.F. 2012. Religion as a means to assure paternity. *PNAS* 109:9781–9785.

Strassmann, B.I., and Warner, J.H. 1998. Predictors of fecundability and conception waits among the Dogon of Mali. *Am. J. Phys. Anthro.* 105:167–184.

Stumpf, R.M., Emery Thompson, M., Knott, C.D. 2008. A comparison of female mating strategies in *Pan troglodytes* and *Pongo* spp. *Int. J. Primatol.* 29:865–884.

Symons, D. 1979. *The Evolution of Human Sexuality.* New York: Oxford University Press.

Welling L.L., Jones, B.C., DeBruine, L.M., Conway, C.A., Law Smith, M.J., Little A.C., Feinberg D.R., Sharp M.A., and Al-Dujaili, E.A. 2007. Raised salivary testosterone in women is associated with increased attraction to masculine faces. *Horm. Behav.* 52(2): 156–161.

Winking, J., Gurven, M., and Kaplan, H. 2011. Father death and adult success among the Tsimane: implications for marriage and divorce. *Evol. Hum. Behav.* 32:79–89.

Wroblewski, E.E., Murray, C.M., Keele, B.F., Schumacher-Stankey, J.C., Hahn, B.H., and Pusey, A.E. 2009. Male dominance rank and reproductive success in chimpanzees, *Pan troglodytes schweinfurthii. Anim. Behav.* 77:873–885.

Wylie, K., Rees, M., Hackett, G., Anderson, R., Bouloux, P.M., Cust, M., Goldmeier, D., Kell, P., Terry, T., Trinick, T., and Wu, F. 2010. Androgens, health and sexuality in women and men. *Maturitas.* 67:275–289.

CONCEALMENT OF OVULATION, PARENTAL CARE, AND HUMAN SOCIAL EVOLUTION

Excerpt from Alexander, R.D. and Noonan, K.M. 1979. Concealment of ovulation, parental care, and human social evolution. In N.A. Chagnon and W.G. Irons (eds.), *Evolutionary Biology and Human Social Behavior: An Anthropological Perspective.* North Scituate, MA: Duxbury Press.

Alexander et al. (this volume, chapter 15) describe reasons for assuming, from its current attributes, that the human species has been polygynous during much of its recent evolutionary history (i.e., that, generally speaking, fewer males than females have contributed genetically to each generation, although not necessarily that harems have been involved). Considerable evidence already indicates that humans have essentially always lived in bands of close kin, probably containing more than a single adult male (e.g., Lee and DeVore 1968). These two characteristics, however, fit a large number of nonhuman primate species. Alone they tell us nothing about how the human species came to possess its numerous distinctive and social attributes.

Here we approach this question by first listing and discussing a number of distinctive human attributes. Surprisingly, most of these attributes are sexually asymmetrical or involve the interactions of the sexes. They suggest that the human male is not particularly unusual among primate males, except that he is generally more parental than the males of other group-living species. On the other hand, the human female is distinctive in several regards, most dramatically in undergoing menopause and in the concealment of ovulation. Menopause has been associated with parental care by the female (Williams 1957; Alexander 1974; Dawkins 1976), and we shall argue that concealment of ovulation is associated with the unusual amount of parental care by the male. Several other distinctively human attributes, such as length of juvenile life and helplessness of young juveniles, indicate that an increase in the prominence of parental care was one of the most dramatic changes during evolution of the human line.

Concealment of ovulation, as a strategy for obtaining parental care, seems likely to evolve only in certain kinds of social situations. We believe that by considering the nature of these situations, together with circumstances that could lead to the evolution of increased parental care, it is possible to gain insights into the very general question of how humans evolved their distinctive sexual and social attributes.

Most lists of distinctively human attributes include, in some form, the following and little else:

1. Consciousness (self-awareness)
2. Foresight (deliberate planning, hope, purpose, death-awareness)
3. Facility in the development and use of tools (implying consciousness and foresight)
4. Facility in the use of language and symbols in communication (implying consciousness and foresight)
5. Culture (a cumulative body of traditionally transmitted learning— including language and tools, and involving the use of consciousness and foresight)

As suggested by the parenthetical comments, these five attributes are closely related to one another, perhaps inseparable. They are also not strictly comparable to one another: thus, consciousness and foresight are aspects of the human *capacity* for culture, while language and tools are simultaneously *vehicles* and *aspects* of culture. Culture in turn, by its existence and nature, and through its changes, becomes a central aspect of the environment in which the capacities of individuals to acquire and use consciousness, foresight, and facility in the development and use of language and tools have been selected.

Although these five attributes may once have been regarded as uniquely human, it now seems likely that all occur in other primate species, and chimpanzees alone may possess all five, though not in the form or to the degree that they are expressed in humans (Lawick-Goodall 1967; Gallup 1970; Premack 1971; Fouts 1973; Rumbaugh et al. 1973; Gardner and Gardner 1969, 1971; Mason 1976). The problem in understanding this set of related attributes, and why humans possess them, is to determine their relationship to the reproductive success of individuals during human history, when the capacities and tendencies to express them were originating and being elaborated. Despite the attention paid to them, these attributes have not been extensively analyzed as contributors to reproductive success.

Numerous other traits are also distinctive to humans (d, in the list below) or distinctively expressed in humans, as compared to their primate relatives (de). They may be either cultural (c) or noncultural (nc) in origin, and universal (u) or not universal (nu) among humans. We first list these additional attributes, then discuss their significance in trying to reconstruct the evolutionary background of human sociality. Our reason for presenting this list, in developing an argument about the relationship of parental care and the concealment of ovulation during human social history, is to emphasize how many distinctive human attributes are somehow sexually asymmetrical (as) rather than symmetrical (s) in their expression (probably all but 6 and 30 in the following list), and how many involve: (1) interactions of the sexes in connection with parental care (especially 9–16) and (2) group-living (especially 17–30). We shall argue that concealment

of ovulation could only evolve in a group-living situation in which the importance of parental care in offspring reproductive success was increasing, and that
these two circumstances together describe a large part of the uniqueness of the
social environment of humans during their divergence from other primates. We
make no pretense that the following list is a complete set of attributes unique to
humans.

6. Upright locomotion usual (de, nc, u, s)
7. Frontal copulation usual (de, c & nc, u, as)
8. Relative hairlessness (de, nc, u, as)
9. Longer juvenile life (d, nc, u, as)
10. Greater infantile helplessness (de, nc, u, as)
11. Parental care frequently extending into and even across the offspring's adult life (d, c, u?, as)
12. Unusually extensive paternal care for a group-living primate (de, c?, nu?, as)
13. Concealed ovulation in females (sometimes described as continuous sexual receptivity, continuous estrus, "sham" estrus, or lack of estrus (d, nc, u, as)
14. Greater prominence of female orgasm (perhaps - but see Lancaster, in press) (d, nc?, u, as)
15. Unusually copious menstrual discharge (d, nc, u, as)
16. Menopause (d, nc, u, as)
17. Close association of close kin of both sexes, sometimes throughout adulthood (de, c, u?, as)
18. Extensive extrafamilial nepotism (de, c, u?, as)
19. Extensive extrafamilial mating restrictions (de, c, u?, as)
20. Socially imposed monogamy (d. c, nu, as)
21. Extreme flexibility in rates of forming and dissolving coalitions (d, c, u, as)
22. Systems of laws imposed by the many (or powerful) against the few (or weak) (d, c, nu, as)
23. Extensive, organized, intergroup aggression; war (d, c, u?, as)
24. Group-against-group competition in play (d, c, u?, as)
25. Ancestor worship (d, c, nu, as)
26. Political and other kinds of appointed, elected, or hereditarily succeeding leaders (d, c, nu, as)
27. The concepts of gods and life-after-death (d, c, nu, as)
28. Organized religion (d, c, nu, as)
29. Nationalism; patriotism (d, c, nu, as)
30. Polities of thousands or millions of nuclear families (d, c, nu, s)

What challenges, in the form of differential reproduction, caused the emphasis
on (1) consciousness and foresight, (2) social living, and (3) parental activities,

revealed in the above list of human attributes (cultural and physiological)? If we assume that the above combination of attributes arose as a result of physical or nonhuman biotic selective forces, such as climate, predators, or food shortages, then an evolutionary sequence diverging humans so far from other species in the particular directions they have taken, with all the intermediate stages becoming extinct, appears difficult to reconstruct (Alexander 1971). In the absence of any clear evidence of a massively unique selective environment in these respects for humans, we would have to postulate that our uniqueness as a prehuman primate preadapted us to respond uniquely to some not-so-unique concatenation of environmental conditions, thereby evolving humanness. We regard this hypothesis as unlikely.

The alternative is that something about the evolving human species itself explains the differential reproduction that led to the divergence of the human line, and the extinction of close relatives along the way. This possibility is immediately tantalizing, since it suggests a solution to the problem of human uniqueness. Once set in motion, such a selective force could be self-propelling and, simultaneously, capable of suppressing similar trends in closely related lines. It further implies a reason for the continuing elaboration of cultural patterns, outracing genetic changes so dramatically as to render them trivial in regard to rates of behavior change.

The attribute that could cause differential reproduction leading to human uniqueness, we believe, is an increasing prominence of direct intergroup competition, leading to an overriding significance in balances of power among competing social groups, in which social cooperativeness and eventually culture became the chief vehicle of competition. The probable relevance of complex social competition to human intelligence, consciousness, and foresight—and elaborate social tendencies—has been emphasized before (Darwin 1871; Keith 1949; Fisher 1958; Alexander 1969, 1971, 1974, 1975a, 1975b, 1977a, 1978a, in press a; Bigelow 1969; Alexander and Tinkle 1968; Carneiro 1970; Flannery 1972; Wilson 1973). Intergroup competition has not, however, been linked to the human emphasis on parental care or the unusual sexual attributes of humans.

If intergroup competition was a principal guiding force in human evolution and if groups as a result grew in size and in social unity among the individuals comprising them, then parental care would have increased in value in two general ways. First, larger group sizes inevitably meant intensified competition for resources within groups. Juveniles, lacking the strength and sophistication to compete successfully for themselves, would have benefited increasingly from parental protection and assistance in securing resources for growth and future reproduction.

Second, the intensification of both within- and between-group competition implies a growing conflict in individual reproductive strategies by pitting the value of direct competition with other group members against potential gain from their cooperation in inter-group competition (or within-group alliances). Individual reproductive success would depend increasingly on making the right decisions in complex social situations involving self, relatives, friends, and enemies. Critical choices would be aided by experience, and an intimate knowledge of the particular social environment. Parents who could impart to their offspring this information and the social skills for using and expanding it, while providing guidance during the vulnerable years of learning, would have realized increasing reproductive advantages over parents who failed to so equip their offspring. Buffered against physical and social disaster by parental protection, the human juvenile may have evolved to abandon efforts at serious direct competition, becoming increasingly helpless over longer periods, while evolving extraordinary abilities to absorb and retain information and develop skills through attachment, identification, imitation, and more formal learning in early years. In other words, because of the existence of groups intensely competitive against one another, and because of the complexity of social competition within groups evolving to be effective in intergroup competition, the human species in some sense became its own most important selective environment, and the pressure of evolutionary change focused increasingly on parental care.

Tendencies to infanticide or enforced desertion of infants would benefit males to the extent that such practices hastened ovulation in females and preserved female reproductive effort for the male's own offspring. The increased parental care of human mothers, its greater duration, and the wider spacing of babies associated with more intensive early parental care and infant helplessness would enhance the benefits of infanticide and enforced desertion of children to males acquiring females from other males. Thus, an important aspect of male parental care may have been protection of the child against other group males competing for the female as a reproductive resource. Observations by Bygott (1972) suggest that if a chimpanzee mother with an infant joins a new group, the infant is vulnerable to infanticide by group males. Accounts of men (or women) killing children made fatherless by inter-community warfare and exchange of women, such as among the Yanomarno (Biocca 1969), imply that this may have been an important selective context for paternal care in human history as well, and another indirect consequence of the extension of juvenile dependence.

In ancestral humans, then, an orphan, even at an advanced juvenile stage, was probably doomed to social impotence and reproductive failure, if not pre-reproductive death, unless it was a female old enough to interest a mature male; the extremely derogatory connotation of words for "fatherless" juveniles in nontechnological societies supports this inference (see Alexander 1977a). On the other hand, juveniles with powerful parents and other relatives must have been essentially certain of high success. Indeed, the unstratified or egalitarian bands presumed to represent the ancestral kind of human sociality can almost be defined by saying that in them the major resource by which reproductive competition could be maximized is kinspeople (see also Chagnon chapter 14).

To clarify these circumstances we first examine the social situations in which concealment of ovulation might evolve and compare the resulting model with extant nonhuman primates, then consider the situations in which increases in parental care leading to the human condition might be favored.

WHY CONCEAL OVULATION?

The human female has commonly been described as "continually sexually receptive" because she may willingly mate at any time during the menstrual cycle (James 1971). Most other female mammals mate only during a brief estrus period occurring around the ovulatory period. To refer to the human female's sexual behavior simply as "continuous receptivity," however, seems a gross oversimplification. First, this "receptivity" is unlike the relatively uninhibited receptivity of some estrous female mammals, which may accept essentially any male. By comparison, the human female's behavior might best be described as a kind of selective or low-key receptivity, commonly tuned to a single male, or at least to one male at a time. From the point of view of males not bonded to a particular female, it might just as well be termed "continuous nonreceptivity." It is a truly remarkable attribute of human females that their ovulation is often essentially impossible to detect, even, in some cases, through medical technology (Sturgis and Pommerenke 1960; Behrman 1960; Cohen and Hanken 1960).

Conditions in nonhuman primates that seem to approach those of the human female are: (1) sham estrus in langurs (Blaffer Hrdy 1974, 1977), (2) sexual receptivity outside the ovulatory period, especially in rhesus (Loy 1970) and chimpanzees (van Lawick-Goodall, 1971; Lancaster, in press, reviews other primate cases of mating outside the usual estrus period around ovulation), and (3) relatively few external signs of ovulation in gibbons (Carpenter 1941), orangutans (Rijksen 1975), gorillas (Schaller 1963; Hess 1973; Nadler 1975), and possibly bonnet macaques (Simonds 1965; Rahaman and Parthasarathy 1969; MacArthur et al. 1972). In the last case we are assuming that advertisement of estrus by pheromones is not unusually exaggerated in species with few visual signs; any efforts to quantify advertisement of estrus among species are necessarily restricted to visual signs because no effort has been made to accomplish this with pheromones.

In nonhuman primates the general period of ovulation always appears to be more or less dramatically signaled to males (even if only by pheromones or other means not obvious to human observers). All of the nonhuman primates in which females are known to show "pseudo-estrus" are group-living species, while the least obvious signs of ovulation seem to occur in monogamous species like gibbons, or polygynous species, like gorillas, which tend to live in single-male bands. Human females are thus unique in that they give little or no evidence of ovulation and may be receptive during any part of the ovulatory cycle. Although some women have discovered ways to determine the time of their own ovulation, it is clear that selection has reduced the obviousness of ovulation during human evolution, apparently to women themselves as well as to others.

Emphasis on the so-called continuous responsiveness of the human female, in trying to model its history, has focused attention on the proximate effect that the male is able to enjoy copulation more or less whenever he desires it. Thus, it has been argued that sexual competition is reduced among human males, allowing larger group sizes and more extensive cooperation (Etkin 1963; Washburn and Lancaster 1968; Pfeiffer 1969), or, that females keep their males at home by supplying constant sex (Washburn and DeVore 1961b; Campbell 1966; Morris 1967; Crook 1972). Those approaches have not provoked explanations for the male's retaining his interest in frequent copulation with his mate when fertilization of her ovum is unlikely—that is, when she is not ovulating or when she is pregnant. Focusing on concealment of ovulation within the essentially continuous receptivity of the female, on the other hand, raises the question of the value to the female of this concealment. Evidently it deceives all males in the female's vicinity, both those with whom she is copulating and those with whom she is not copulating. Such deceit would be valuable to the female only if recognition of the ovulatory period somehow caused males to gain at her expense. In the absence of parental investment by males, advertisement of ovulation would both assure the female of copulation at times appropriate to fertilization and increase the likelihood of competition among males, thus increasing her likelihood of securing a male of unusual competitive ability (e.g., Cox and Le Boeuf 1977). Concealment of ovulation and more or less continuous receptivity would have little value in this circumstance, although some tolerance of mating may be expected to appear if male insistence was quite deleterious to resisting females (e.g., McKinney 1975, on ducks). Conversely, if males invest parentally, sexual advertisement of ovulation might cost a female her mate's parental care in two ways: (1) by attracting competing males who threaten his confidence of paternity, and (2) by freeing him, after a brief consort period, to seek copulations with other fertile females who would compete later with her for his parental care.

We suggest that concealment of ovulation evolved in humans because it enabled females to force desirable males into consort relationships long enough to reduce their likelihood of success in seeking other matings, and simultaneously raised the male's confidence of paternity by failing to inform other, potentially competing males of the timing of ovulation. If these events occurred in a situation in which paternal care was valuable, but not sufficiently valuable to males to offset philandering (Trivers 1972), and in which desertion was frequent when confidence of paternity was low, they could tip the balance, making increased paternal investment profitable to males. This strategy would be most likely to be successful for females perceived by males to be of unusual value as mates. Thus, its effectiveness would have been magnified during human history first by abilities and tendencies of males to kill or enforce abandonment of fatherless offspring, and, second, by the rise of differences in heritable wealth and status.

To explore the significance of concealed ovulation, we shall now consider variations in the advertisement of ovulation among female primates living in different kinds of social groups, which might be considered to parallel certain aspects of the social environment in which human sexuality evolved. We shall begin with the most monogamous forms, such as the white-handed gibbon. The male and female form an isolated, territorial pair and raise their offspring alone. There are indications that adults are hostile to strangers of the same sex, and their territorial tendencies would seem to minimize direct sexual competition (Ellefson 1974). Males are virtually certain of their paternity because, usually, no other males are around to compromise it, and females are relatively free of competition from other females for their mate's parental care. It might be supposed that the human species lived in this fashion during much of its evolutionary history, and that the so-called continuous receptivity of the human female evolved because such females tended to keep their husbands home while other husbands went roving, presumably to satisfy persistent sexual cravings. Such an argument, however, does not describe the behavior of modern gibbons, which mate only when the female ovulates, and it is difficult to defend on a logical basis. If males that stayed at home, tending their mates and their offspring, outreproduced faithless males, their doing so would surely not depend on whether or not their mates provided them with constant sex. Indeed, in such circumstances, constant sex would be no more than a useless and potentially dangerous distraction from the business of staying alive and healthy and rearing one's offspring. Rather than being continuously receptive, females in such families should evolve to be receptive only during a brief period associated with ovulation—just long enough, in other words, to maximize the likelihood of fertilization at the appropriate times. They should not be gaudy or extravagant about their responsiveness, communicating no further than the mate. The male, likewise, should evolve to be interested in sex only during the same period and only with his mate. Gibbons, which appear to behave as these arguments suggest they should, do not provide a suitable model for the history which determined the nature of present human sexuality. Only if continuous sexuality in the male consistently yielded opportunities for philandering should it be maintained by selection and defenses against philandering be evolved by females because of the value of male parental effort; these conditions are likely only when families are either polygynous or in relatively close proximity.

In orangutans one male often controls a territory containing more than one female (Rijksen 1975). It is not clear how effectively such males exclude other males, and males move about, sometimes in groups, and apparently rape females (Rijksen 1975; MacKinnon 1971). Females sometimes approach males for copulation and are said to prefer "high-ranking" males (Rijksen 1975). Males evidently provide no parental care. From these reports, we might predict that orangutan females advertise ovulation more than gibbon females, and are sexually responsive only at ovulation time. Females may also be receptive when approached by strange males, if such males might benefit from killing an offspring sired by another male or when resistance to rape would be detrimental (infanticide, however, has not been observed in orangutans). To the extent that orangutan females select males, they might profit from concealing ovulation altogether and ovulating reflexively. Males would then gain by copulating with any available female whenever they could. But the orangutan situation, which is still known only very sketchily, appears not to have led to extensive male parental care, nor does it involve extensive group-living, both of which seem essential to a model of human social and sexual history.

Considering single-male primate bands next, such as gorillas[1], we might again predict a relatively low rate of sexual activity, and a relatively nonobvious estrus period. Although females in a band must compete with one another to some extent for the single dominant male's attention, male competition apparently consists largely in securing and maintaining a harem of females. To the extent that this is true, females would gain little by advertising estrus more than is necessary to secure the male's attention at ovulation time, unless other potentially better males are constantly in range of their signals, and able to use them to usurp ownership of the band. As with gibbons, the male's confidence of paternity is probably high—at least in species in which harems are held for long periods—and some paternal care should thus be evident. Because males are essentially certain of their paternity, females would gain little in terms of a larger share of paternal care by concealing ovulation and prolonging receptivity. Sexual behavior in gorillas is indeed infrequent, estrus is not sharply advertised, and the silverback male's evident willingness to repel potential predators could be interpreted as paternal effort (Schaller 1963; Fossey 1970, 1971).

> 1. Although blackback, younger males occur with gorilla bands including but one silverback male, and some bands have two silverback males, we believe gorilla bands are appropriately termed single-male, or at least are different from such species as chimpanzees, cynocephalus baboons, and rhesus macaques, because of the evident dominance of one male and the fact that one silverback male generally determines band movements (Schaller 1963: Fossey 1970, 1971). Harcourt, Stewart, and Fossey (1976) argue that the dominant silverback gorilla male inhibits other males from mating more effectively than do dominant baboon or chimpanzee males and suggest that some blackbacks are offspring of the silverback male.

In multi-male bands of nonhuman primates one finds the most dramatic advertisements of sexual receptivity, the most obvious and intense sexual competitiveness, and—aside from single-male harems for which ownership changes frequently (e.g., langurs, Blaffer Hrdy 1974, 1977)[2]—the most striking cases of receptivity outside the ovulation period. In baboons, macaques, chimpanzees, and a few other forms that live in multi-male troops, the females develop bright-colored swollen rumps during estrus. It is difficult to explain these gaudy swellings except on the assumption that females in such groups gain by competing for the attention of the dominant males. Paternal care is evidently minimal in such groups, compared to one-male bands and isolated monogamous pairs, and one apparent correlate is a low confidence of paternity. The gaudy females appear to be competing for the attention of the males most capable of physically monopolizing them at ovulation. There is no necessary correlation between a male's ability to monopolize a female and his ability to provide paternal care; such a correlation might evolve, for example, if females consistently gained from protection against predators (or other males) during the period of sexual receptivity. If monopoly gives a male high confidence of paternity, his willingness to invest parentally may be raised; in contrast, if a female copulates with a subordinate male, and he is quickly supplanted after copulation, neither male is likely to be willing to provide paternal care.

The trend in the three kinds of social situations just discussed—isolated, territorial pairs, single-male bands, and multi-male bands—seems opposite to that which gave rise to the human condition, at least in regard to advertisement of ovulation by the females and the extent of paternal care. Yet it is commonly accepted that humans evolved in multi-male bands. On the other hand, it is difficult to see how concealment of ovulation could evolve in either isolated monogamous families or single-male harem groups, since there would be no "extramarital" males from which to conceal ovulation. Nor does the evolution of tendencies toward regular and continual sexuality in mated human males and females seem likely in either single-male groups or monogamous families unless takeovers by new males occur frequently enough (Blaffer Hrdy 1974, 1977).

> 2. Langurs are probably most appropriately compared to multi-male troops because their rapid changes of ownership by males, and the sham estrus of females by which even pregnant females mate with the new owner of a troop, suggest that rapid shifts of ownership are not a novel or recent phenomenon, as has heen argued in criticisms of Blalfer Hrdy's (1977) conclusions, presaged by Mohnot (1971) and Sugiyama (1967)—see Alexander (1974).

It might be argued that humans were extensively group-living, or multi-male in social structure, when the distinctive attributes of the human female evolved. Perhaps our ancestors lived in isolated monogamous families or in single-male harem-polygynous groups, with initially extensive male parental care followed by expanded group size in response to predator pressure. For two reasons, this alternative seems a less likely route to human sociality than the prevalent view that pre-humans lived in relatively large multi-male bands. First, humans are thought to have evolved in open woodland or savannah (e.g., Lee and DeVore 1968; Kolata 1977) where, today, there are no monogamous primates and few single-male, harem-polygynous ones. Moreover, primate species in such habitats today (e.g., langurs, vervets, paras) show anti-predator behavior which seems counter to the trend in human evolution-hiding, secretiveness, and little direct confrontation of predators. Man's most similar relatives, chimpanzees, which inhabit woodland and savannah, have a multi-male, polygynous social structure. Second, the selective pressures generally thought to have been important in human social evolution—cooperative defense, hunting, and inter-group competition—favor group sizes probably greater than one male and his harem. So, even if humans began in small harem groups, they obviously (and probably early) achieved multi-male situations which tend to threaten pair bonds. Cooperation by males against predators occurs both in multi-male bands with harems (hamadryas and gelada baboons) and in so-called "promiscuous" multi-male bands (cynocephalus baboons). In both types of social structure a male's confidence of paternity is threatened to some degree, probably more in the latter case. In summary, we believe that two coincident circumstances can explain the evolution of concealment of ovulation. The first is a social situation in which females of reproductive age are not completely inaccessible to males other than their mates or consorts (e.g., multi-male groups or defensible multi-female territories). The second is a growing importance of parental care such that the value to a female of a male's prowess in monopolizing her at ovulation time would be overshadowed by the value of male prowess and willingness as a providing or protective parent. Gradual evolution of concealment of ovulation by females behaving so as to maximize their mate's confidence of paternity—hence his likelihood of behaving paternally—would with each step toward concealment improve the female's ability to secure her mate's parental care. Because no male could tell when a female was ovulating, only a male who tended her more or less continuously could be sure of the paternity of her offspring. Occasional forced or clandestine matings outside the pair bond, in the absence of information about ovulation, would have a very low likelihood of resulting in pregnancy (e.g., Tietze 1960).

According to what sequence of changes might the human female have evolved her current uniqueness in regard to the cycling of sexual receptivity and external signs of ovulation? One might consider three possibilities: either (1) external evidence of estrus diminished first, with receptivity later increasing in duration; (2) receptivity became more or less continuous, with external evidence of ovulation later diminishing; or (3) these two changes occurred together. Since we favor the hypothesis of a human ancestor living in multi-male groups we tend initially to eliminate the first of these three possibilities.

Presumably sham estrus, or receptivity outside the ovulatory period, is effective only if it actually mimics estrus. It can only do this by becoming elaborate like true estrus, or by a reduction in the elaboration of true estrus. One might imagine that females gained directly from damping signs of ovulation, if pairing and some male parental behavior preceded concealment of ovulation, as they must have. In an extensively group-living species, however, at this stage, opportunities for males to be polygynous must have been numerous, detracting from the value of the pair bond to males, but not to females. It is difficult to see how a female could keep her mate by damping sexual signals. Rather, a gradual extension of signals beyond the ovulatory period seems a more likely way to lure the male into giving more parental care than he would give if he based his effort on a correct determination of the time of ovulation.

Steep differentials in female quality would enhance the effectiveness of this deception. The highest-quality females (in terms of ability to bear and rear offspring, and, perhaps, to enhance their status or other correlates of reproductive success), would be in demand by the highest-quality males most prone to polygyny; such females could afford to prolong estrus longer than others without risk of desertion by the male. Low-quality females would be at least partially excluded from mating with the highest quality males by this ploy. Competition for matings by lower-quality males would be reduced (because top-quality males are already committed), making concealment of ovulation advantageous to low-ranking females in securing at least some parental care for their offspring. Once sham estrus became prevalent, reduction of external signs of ovulation seems inevitable if only on the basis of cost reduction. So we postulate that sham estrus evolved for a time to mimic true estrus, and subsequently the elaborateness of both true and sham estrus was reduced.

Bright sexual skins in hamadryas and gelada baboon females at first seem to contradict our hypothesis because these species have harem-polygynous breeding systems. Paternity is more certain here than in the promiscuous polygynous bands of savannah baboons, and one might therefore expect the evolution of paternal care and concealed ovulation in females. However, observations by Kummer (1968b) suggest that hamadryas females, too, may be competing for the attention of high-quality males. Although harem-masters do not fight over estrous females or mate extra-maritally they do battle over anestrous females so that females occasionally pass between harems. Females are more "spatially independent" while in estrus, perhaps inviting competition between their harem-masters and extra-marital males, primarily bachelors. Fighting takes place frequently between harem-masters and bachelors, and occasionally between two or more harem-masters. Mating occurs in crowded areas near sleeping cliffs, and adulterous matings with bachelors are common, reducing the confidence of parenthood of harem masters. Males in this system provide little direct parental care to their offspring, and the willingness of females to stray and mate with other males suggests that securing the best possible mating is more central to their reproductive strategy than maintaining a high confidence of paternity.

CONCEALED OVULATION AND RAPE

When females do not reveal their ovulations, unmated males no longer have the opportunity to compete for ovulating females. If our interpretation that human females are better viewed as continuously *nonreceptive* or continuously *selectively receptive* is correct, then males not bonded to a female have no way of obtaining reproductively effective matings except by increasing their numbers of matings, more or less randomly with respect to ovulation. Moreover, since female receptivity does not correlate with ovulation, there is no reason, except convenience and lowered risk, for males to seek only receptive females. A forced mating might even be as likely to lead to actual success in reproduction, since unwillingness sometimes would imply a pair bond, hence the possibility of paternal care for the resulting offspring from a cuckolded male. In this sense, then, concealed ovulation and some aspects of rape in humans may be historically related. As females evolved to deny males the opportunity to compete at ovulation time, copulation with unwilling females became a feasible strategy for achieving some reproduction. A raped female, moreover, might sometimes lose too much by revealing the event to her mate, and this would increase the likelihood of rapists' going unpunished. Compared to other primates, then, a mating with a willing human female is less likely to lead to reproduction (because she is less likely than a willing nonhuman primate female to be ovulating). A mating with an unwilling female is more likely to do so (because she is more likely than an unwilling nonhuman primate female to be ovulating, also more likely to have a male who gives paternal care and whom, on that account, she is unlikely to inform of the rape).

In many societies in which rape occurs rather frequently, women submit to avoid being hurt and usually do not complain later (e.g., New Guinea: Matthiessen 1962; Kenya: LeVine 1977). Such rapes are not necessarily associated with psychological pathology in the males or murder, characteristic of a significant proportion of rapes in the U.S. (but see Amir 1971). The association of rape with murder and psychological pathology in males in the United States may reflect the severe penalties traditionally incurred for violating socially imposed monogamy. Only the most deprived males (in actuality or by delusion) would be inclined to behave as though viewing rape as a viable reproductive strategy in relation to other reproductive alternatives; having committed such a crime, fear of discovery might lead such males to murder their victims—canceling any reproductive gain, but escaping certain death if caught.

CONCEALED OVULATION AND FEMALE ORGASM

Female orgasm was once regarded as unique to humans. Although recent studies suggest that this is untrue (see review by Lancaster, in press) the frequency of orgasm in the human female, and perhaps its intensity or outward signs, may still be unique. To examine the significance of this situation we may note first that orgasms in other primate females have been described in species such as rhesus macaques which live in multi-male groups and rather consistently show sexual receptivity outside the ovulatory period (Lindburg 1971).

Orgasm in the human female, and perhaps other primates as well, may increase the likelihood of fertilization (Fox et al. 1970; but see Masters and Johnson 1966: 122-124). But when should there be external signs of orgasm? We suggest two possibilities: (1) orgasms may sometimes increase the likelihood of abortion, thus decreasing the likelihood of paternity mistakes in some circumstances, and (2) external signs of orgasm may communicate the female's sexual satisfaction to the male. In the latter case the apparent tendency of female orgasm, in humans at least, to resemble male orgasm, in the apparent absence of a correlation with an event paralleling release of gametes (Masters and Johnson 1966), may suggest (l) that the correlation is to some extent with the pleasure or satisfaction of male orgasm to the male (so that the external signs of a female's orgasm suggest to the male pleasure on her part similar to his own), and (2) that this aspect of the female orgasm may be a communicative device which tends to raise her male's confidence that the female is disinclined to seek sexual satisfaction with other males. To the extent that this interpretation is correct, the obvious outward signs of female orgasms should (1) mimic male orgasms and (2) frequently involve deception, with females pretending to have orgasms when they do not. Moreover, both this function and effects on likelihood of fertilization suggest that orgasms should (1) occur most frequently in (a) deeply satisfying or long-term interactions with males committed to the female and her offspring (Gebhard 1966), and (b) with dominant males or males with obviously superior ability to deliver parental benefits, and (2) occur least frequently in brief or casual encounters, or in copulation with a partner unsatisfactory in the above regards. All of these contingencies seem compatible with what is known about orgasm in human females, while few of them seem to characterize male orgasm. All appear to be consistent with the information reported by Lancaster, except possibly for the suggestion that female macaques in the laboratory achieve orgasms sooner if they have been deprived of sexual activity. According to our speculation, female orgasms would perhaps be more likely in females trying to obtain or keep a "good" male, identified as a male with much parental investment to offer. Thus, we believe that this feature of human female sexuality too may be linked to male parental care and the threat of desertion.

PARENTAL CARE AND HAIRLESSNESS

A possible relationship exists between the relative hairlessness of humans and their emphasis on parental care. In the several published arguments on this question it seems to have been overlooked that the least hairy of all humans are their juveniles. Hairlessness in young mammals otherwise seems to correlate with multiple births and a helpless period in the nest. Humans may be the only mammal giving birth to a naked single offspring. We suggest that the selective value of being a juvenile, or of giving that impression, should be investigated in efforts to explain the gradual evolution of hairlessness in humans, and its present distribution among humans of different ages and sexes.

UPRIGHT LOCOMOTION AND PARENTAL CARE

Upright locomotion has been associated with the evolution of hunting and tool use (Eirnerl and DeVore 1965; Morris 1967). Washburn and DeVore (1961b) suggest that parental care may be implicated in the evolution of bipedalism, as well. They argue that, as the human infant became more and more helpless, and the human mother more hairless, the baby was no longer able to cling unassisted, as nonhuman primate babies do from a few days after birth. Upright posture may, in fact, have tended to evolve first in females in the context of carrying infants. If so, then with the possible exception of frontal copulation (7) and unusually copious menstrual discharge (25), all of the attributes which we have identified as uniquely expressed in humans appear to be related in some rather direct fashion to parental care and other forms of nepotism.

SELF-AWARENESS AND CONCEALED OVULATION

Whether or not the above hypotheses about concealed ovulation are precisely correct, any argument from adaptiveness must take into account that the human female has evolved to conceal her ovulation not only from others in her vicinity but evidently from herself as well. The timing of her own ovulation is not commonly a part of a woman's conscious knowledge. Unless one assumes that other primate females, such as chimpanzees, are also not aware of their extensive behavioral and physiological changes at ovulation, this fact suggests that it has somehow been reproductive for human females actually to lose awareness of their own ovulation.

If disappearance of outward signs of ovulation was favored because they deceived other individuals, suppression of self-awareness of ovulation, unless merely incidental, may appropriately be hypothesized to have furthered the deception. What is implied is that social deception may sometimes be more successful if the deceiver is unaware of it (Alexander 1975b). Evolutionists can scarcely fail to be puzzled and intrigued that with all the complexity of human consciousness, and its apparent pervasion of all of our social interactions, humans nevertheless express surprise, disbelief, or even outright anger at the suggestion that their behavior has evolved to maximize their genetic reproduction. Unless one rejects the primacy of differential reproduction this response indicates that it has been advantageous for us to think that we have other goals. Perhaps what has actually been advantageous is for us to be able continually to deceive others in regard to the nature and extents of our likely gains as a result of particular activities with effects on them. That is, except for avowed enemies such as members of alien or competitive groups, it may have been more detrimental, during human history, to us as individuals to have others construe continually that our immediate motivations for each social act were personal gains relative to those with whom we are interacting or for them to know exactly how much we are likely to gain from the interaction. Yet these are the precise criteria of success in differential reproduction.

It is difficult to avoid the conclusion that we are better at convincing others that our acts are truly beneficial to them and theirs, as opposed to ourselves, if we believe it ourselves. Righteousness is a valuable aid and effective deceiver. Deliberate lying is notoriously difficult as a social strategy; we are extraordinarily clever at detecting it, vindictive and grudge-holding in our response to it, and much more likely to be fooled by the selfish person with a sincere belief in his own altruism and integrity.

The ability to deceive, partly by self-deception as to motives, we here suggest to be a central part of human sociality, and of consciousness and self-awareness in the human individual (see also Alexander 1975a). Concealed ovulation we view as a particularly powerful and instructive case of deception of others, linked with self-deception and made more effective by it.*

Literature Cited

Alexander, R.D. 1969. Comparative animal behavior and systematics. In *National Academy of Systematics*. Publication 1692. Pp. 484–520.

Alexander, R.D. 1971. The search for an evolutionary philosophy of man. *Proceedings of the Royal Society of Victoria* 84:99–120.

Alexander, R.D. 1974. The Evolution of Social Behavior. *Annual Review of Ecology and Systematics* 5:325–383.

Alexander, R.D. 1975a. The search for a general theory of behavior. *Behavioral Science* 10:77–100.

Alexander, R.D. 1975b. Natural selection and chorusing behavior in Acoustical Insects. In *Insects, Science and Society*. D. Pimentel, ed. Pp. 35–77. New York, Academic Press.

Alexander, R.D. 1977. Natural selection and the analysis of human social behavior. In *Changing Scenes in the Natural Sciences*, 1776–1976. C.E. Goulden, ed. Pp. 283–337. *Academy of Natural Sciences*, Special Publications 12, Philadelphia: the Academy.

Alexander, R.D. 1978. Natural selection and societal laws. *In Science and the Foundations of Ethics*. IV. T. Engelhardt and D. Callahan, eds. Hastings-on-Hudson, New York: Hastings Institute of Society, Ethics and the Life Sciences.

Alexander, R.D. In press a. Evolution, social behavior, and Ethics. *In Science and the Foundations of Ethics*. III. *Morals, Science and Society*. T. Engelhardt and D. Callahan, eds. Hastings-on-Hudson, New York: Hastings Institute of Society, Ethics and the Life Sciences.

Alexander, R.D. In press b. Natural selection and social exchange. In *Social Exchange and Developing Relationships*. R.L. Burgess and T.L. Huston, eds. New York: Academic Press.

Alexander, R.D. and Tinkle, D.W. 1968. Review of On Aggression by K. Lorenz and The Territorial Imperative by Robert Ardrey. *Bioscience* 18:245–248.

Amir, M. 1971. *Patterns of Forcible Rape*. Chicago: University of Chicago Press.

Behrman, S.J. 1960. Detection and ovulation. *Postgraduate Graduate* 27:12–17.

Bigelow, R.S. 1969. *The Dawn Warriors*. Boston: Little, Brown.

Biocca, E. 1969. *Yanomama: the Narrative of a White Girl Kidnapped by Amazonian Indians*. New York: Dutton.

Blaffer Hrdy, S. 1974. Male-male competition and infanticide among the Langurs (*Presbytis entellus*) of Abu, Rajasthan. *Folia Primatologica* 22:19–58.

---

* Beverly Strassman has remarked to us that nowhere in this chapter have we emphasized that concealment of ovulation would tend to favor subordinate males, inferior in direct physical competition for oestrus females but usually willing and capable in regard to parental effort. We think her point is well taken.

Blaffer Hrdy, S. 1977. Infanticide as a primate reproductive strategy. *American Scientist* 65:40–49.

Bygott, J.D. 1972. Cannibalism among wild chimpanzees. *Nature* 238:410–411.

Campbell, B.G. 1966. *Human Evolution*. Chicago: Aldine.

Carneiro, R.L. 1970. A theory of the origin of the state. *Science* 169:733–738.

Carpenter, C.R. 1941. The menstrual cycle and body temperature in two gibbons (*Hylobates lar*). *Anatomical Record* 79:291–296.

Cohen, M.R. and Hanken, H. 1960. Detecting Ovulation. *Fertility and Sterility* 11:497–507.

Cox, C.R. and LeBoeuf, B.J. 1977. Female incitation of male competition: a mechanism in sexual selection. *American Naturalist* 111:317–335.

Crook, J.H. 1972. Gelada baboon herd structure and movement: a comparative report. *Symposia of the Zoological Society of London* 18:237–258.

Darwin, C. 1871. *The Descent of Man, and Selection in Relation to Sex*. 2 Vols. London, John Murray.

Dawkins, R. 1976. *The Selfish Gene*. Oxford: Oxford University Press.

Eimerl, S. and DeVor, I. 1965. *The Primates*. New York: Time-Life Books.

Ellefson, J.O. 1974. A natural history of the white-handed gibbons in the Malayan Penninsula. *In Gibbon and Siamang. 3. Natural History, Social Behavior, Reproduction, Vocalizations, Prehension*. D.M. Rumbaugh, ed. Pp. 2–134. Basel: Karger.

Etkin, W. 1963. Social behavior factors in the emergence of man. *Human Biology* 35:299–310.

Fisher, R.A. 1958. *The Genetical Theory of Natural Selection*. New York: Dover Press (Originally published in 1930).

Flannery, R. 1972. The cultural evolution of civilizations. *Annual Review of Ecology and Systematics* 3:399–426.

Fouts, R.S. 1973. Acquisition and testing of gestural signals in four young chimpanzees. *Science* 180:978–980.

Fossey, D. 1970. Making friends with mountain gorillas. *National Geographic Magazine* 137:48–67.

Fossey, D. 1971. More years with mountain gorillas. *National Geographic Magazine* 140:574–585.

Fox, C.A., Wolff, H.S., and Baker, J.A. 1970. Measurement of intra-vaginal and intra-uterine pressures from human coitus by radiotelemetry. *Journal of Reproduction and Fertility* 22:243–251.

Gallup, G.G. 1970. Chimpanzees: self-recognition. *Science* 167:86–87.

Gardner, B.T. and Gardner, R.A. 1969. Teaching sign language to a chimpanzee. *Science* 165:664–672.

Harcourt, A.H., Stewart, K.S., and Fossey, D. 1976. Male emigration and female transfer in wild mountain gorilla. *Nature* 263:226–227.

Hess, J.P. 1973. Some observations on the sexual behavior of captive lowland gorillas, *Gorilla gorilla gorilla*. *In Comparative Ecology and Behavior of Primates*. R.P. Michael and J.P. Crook, eds. Pp. 508–581. New York, Academic Press.

James, W.H. 1971. The distribution of coitus with the human inter-menstruum. *Journal of Biosocial Science* 3:159–171.

Keith, S. 1949. *A New Theory of Human Evolution*. New York: Philosophy Library Inc.

Kolata, G.B. 1977. Human Evolution. Hominids of the Miocene. *Science* 197:244–245, 294.

Kummer, H. 1968b. *Social organization of Hamadryas Baboons. A field study. Bibliotheca Primatologica*, No. 6, Pp. 1–189. Basel: S. Karger.

Lancaster, J.B. In press. Sex and Gender in Evolutionary Perspective. In *Sex and Its Psychosocial Derivatives*. H. Katchadourian and J. Martin, eds.

Lawick-Goodall, J. van, 1967. Mother-offspring relationships in free-ranging chimpanzees. *In Primate Ethnology*. D. Morris, Ed. Pp. 287–346. Chicago, Aldine.

Lee, R.B. and DeVore, I. eds. 1968. *Man the Hunter*. Chicago, Aldine.

LeVine, R.A. 1977. Gusii sex offenses. In *Forcible Rape: The Crime, the Victim, the Offender*. D. Chappell, R. Geis and G. Geis, eds. Pp. 189–226. New York: Columbia University Press.

Lindburg, D.G. 1971. The Rhesus Monkey in North India: An ecological and behavioral study. *In Primate Behavior. Developments in Field and Laboratory Research*. L.A. Rosenblum, ed. Pp. 1–106. New York: Academic Press.

Loy, J. 1970. Perimenstrual sexual behavior among Rhesus monkeys. *Folia Primatologica* 13:286–297.

MacArthur, J.W., Ovadia, J., Smith, O.W., and Bashir-Farahmond, J. 1972. The menstrual cycle of the bonnet monkey (*Macaca radiata*). Changes in secretion of cervical mucus, vaginal cytology, sex skin and urinary estrogen secretion. *Folia Primatologica* 17:107–121.

MacKinnon, J.R. 1971. The Orang-utans in Sabah today. *Oryx* 11:141–191.

Mason, W.A. 1976. Environmental models and mental modes. Representational processes in the great apes and man. *American Psychologist* 31:284–294.

Masters, W.H. and Johnson, V.E. 1966. *Human Sexual Response*. Boston: Little, Brown.

Matthiessen, P. 1962. *Under the Mountain Wall: A Chronicle of Two Seasons in the Stone Age*. New York: Viking.

McKinney, F. 1975. The evolution of duck displays. *In Function and Evolution in Behaviour*. G. Baerends, C. Beer, and A. Manning, eds. Pp. 331–357.

Mohnot, S.M. 1971. Some aspects of social change and infant-killing in the Hanuman Langur, *Presbytis entellus* (Primates: Cercopithecidae) in Western India. *Mammalia* 35:175–198.

Morris, D. 1967. *The Naked Ape*. New York, Dell Publishing Co.

Nadler, R.D. 1975. Cyclicity in tumescence of the perineal labia of female lowland gorillas. *Anatomical Record* 18:791–797.

Pfeiffer, J.E. 1969. *The Emergence of Man*. New York: Harper and Row.

Premack, D. 1971. On the assessment of language competence in the chimpanzee. *In Behavior of Nonhuman Primates*. A.M. Schrier and F. Stollnitz, eds. Pp. 185–228. New York, Academic Press.

Rahaman, H. and Parthasarathy, M.D. 1969. Studies on the social behavior of the bonnet monkeys. *Primates* 10:149–162.

Rijksen, H.D. 1975. Social structure in a wild orangutan population in Sumatra. *In Comtemporary Primatology: Fifth International Congress of Primatology, Nagoya, Japan*. S. Kondo, M. Kawai and A. Hara, eds. Pp. 373–379. Basel: S. Karger.

Rumbaugh, D.M., Gill, T.V., and Glaserfeld, E.C. von, 1973. Reading and sentence completion by a chimpanzee (Pan). *Science* 182:731–733.

Schaller, G.B. 1963. *The Mountain Gorilla: Ecology and Behavior*. Chicago: University of Chicago Press.

Simonds, P.E. 1965. The Bonnet Macaque in South India. *In Primate Behavior: Field Studies of Monkeys and Apes*. I. DeVore, ed. Pp. 175–196. New York: Holt, Rinehart and Winston.

Sturgis, S.H. and Pommerenke, W.T. 1960. The clinical signs of ovulation—a survey of opinion. *Fertility and Sterility* 1:112–132.

Sugiyama, Y. 1967. Social organization of Hanuman Langurs. *In Social Communication Among Primates*, S.A. Altmann, ed. pp. 221–236. Chicago: University of Chicago Press.

Tietze, C. 1960. Probability of pregnancy resulting from a single unprotected coitus. *Fertility and Sterility* 11:485–488.

Trivers, R.L. 1972. Parental investment and sexual selection. *In Sexual Selection and the Descent of Man, 1871–1971*. B.H. Campbell, ed. Pp. 136–179. Chicago: Aldine.

Washburn, S.L. and DeVore, I. 1961b. Social behavior of baboons and early man. *In The Social Life of Early Man*. S.L. Washburn, ed. Pp. 91–105. Chicago: Aldine.

Washburn, S.L. and Lancaster, C.S. 1968. The evolution of hunting. In *Man, the Hunter*. R.B. Lee and I. DeVore, eds. Pp. 293–303. Chicago, Aldine.

Williams, G.C. 1957. Pleiotropy, natural selection and the evolution of senescence. *Evolution* 11:398–411.

Wilson, E.O. 1973. On the queerness of social evolution. *Bulletin of the Entomological Society* 19:20–22.

Human Childhood

The Ones Who Continue

to dream expect to be a part
of children growing up,
of savoring comfort and pain,
the sociality of love,
of meeting all those challenges of life,
that we would like to go on and on
because, of course, they
are, somehow, everything.

Alexander, 2011, p. 254

INTRODUCTION

Altriciality, Neoteny, and Pleiotropy
Paul W. Turke

BROKEN RULES AND DASHED HOPES

It is a perilous game for a cricket biologist to write about human evolution. Anthropologists have claimed this topic, and we have rules.

The first rule is, if you must mention another animal make it a primate, preferably a nice, sexy, matriarchal one like the bonobo. The second rule is, "the standard model of evolution"—the one proposing that adaptive design is organized primarily at the level of the individual, with the effect of producing self-interested individuals—is surely inadequate for explaining ourselves.

Dick Alexander repeatedly breaks both rules in his "Uniquely Unique" paper. Fly larvae (a.k.a. maggots) are brought into the mix, for example, and there is a conspicuous absence of models (of the sort which are popular once again) claiming that we've evolved to be selfless group benefactors via implausible forms of group selection or fanciful cultural co-evolutionary effects. Instead, nepotism (Hamilton, 1964) and social reciprocity (Trivers, 1971; Alexander 1974; 1987) are said to be the key drivers of human within-group cooperation, and even more perilously Alexander suggests that within-group cooperation facilitates out-group competition of the nastiest sort.

The "Uniquely Unique" paper has additional problems. It is sweeping in scope, which is not conducive to neat data tables and statistically significant p-values. This leaves it open to supercilious sniping, which it has gotten, but, as Alexander is aware, the price one pays for addressing a large and important topic is often just that. He states up front that his goal is not to have the last word. Rather, he strives to pull together a grand account of human evolution that passes the initial test of having all of its component parts fit together; and he hopes aloud that it will stimulate others to formulate additions and improvements.

In the first respect, I see great success. I see a roadmap for understanding human evolution that is far superior to anything that came before. In the second respect, the rest of us have largely let him down. There are probably many reasons for this (including offense taken over the aforementioned broken rules), but the lack of a suitable target cannot be one of them. After all, he tells us pointedly that his human hairlessness argument (pp. 26–29) is low-hanging fruit—ripe for improvement, so to speak. Accordingly, I will devote my remaining space to this most glaring problem, and to the related problem of altriciality.

ALTRICIALITY

Altriciality is a little known term outside of evolutionary biology. It refers to physical helplessness. Precociality is its antonym. Alexander points out that human infants are extremely altricial—like maggots—and he carefully reviews the explanations previously given for the evolution of altriciality in the various species in which it is present. He finds much to agree with in these explanations, but he then does what he has been very good at throughout his career: he generalizes.

> The general adaptive explanation for the evolution of altriciality seems to be this: to the extent that a juvenile is relieved of the necessity to protect itself from extrinsic hostile forces of nature (such as predation), it is freed to devote a greater proportion of its calories to improving its performance at some later stages of juvenile life or to becoming a better adult. (p.24)

Alexander goes into considerable detail describing the circumstances that might diminish the value of precociality in certain species, which in turn potentially opens the door to altriciality. For example, since it is very difficult to make a maggot nimble or fierce, why try? Resources are likely better spent on improving feeding efficiency, he argues, because feeding is more in keeping with the general character of maggots than nimbleness or fierceness. For hominins the general argument is the same, but the specific argument is very different. He proposes that early physical precociality is sacrificed for the sake of mental precociality, which I believe makes good sense under the assumption that mental precociality is a crucial primer for successful reproduction in the milieu of an ever-escalating social competition which came to characterize hominins at some early point, and which still characterizes us today.

In short, under Alexander's model altriciality can replace precociality if it frees up calories that can be used to sufficiently improve some later occurring function. Tradeoffs of this sort bring to mind pleiotropy.

LIFE'S "HAIRIEST" PROBLEM: FUNCTION PORTENDS DYSFUNCTION

I learned most of what I know about pleiotropy from Alexander. All of his many graduate students and postdocs would have heard him recite Dobzhansky's dictum, "heredity is particulate, but development is unitary" (cited in Wigglesworth, 1961:111); and they would have heard him interpret it to mean that all genes therefore are both pleiotropic and epistatic. It is difficult to know whether most biologists have embraced this conclusion (I suspect that many haven't), but it is clear that Alexander is not alone in recognizing that pleiotropic genes must at least be very common. Williams (1957), for example, made genes that have advantageous effects early but deleterious effects late (i.e., antagonistic pleiotropic genes) the centerpiece of his theory of the evolution of senescence, and he felt confident in doing so not because he could identify even one such gene but because his understanding of how organisms allocate effort to different functions as they develop convinced him of their ubiquity (cf. Williams, 1966). Shoring up this conviction, Kirkwood and Rose (1991) have pointed out that any gene that diverts resources from repair functions to immediately improve some other function (e.g., making gametes, mating, caring for offspring) increases the risk of downstream dysfunction. Similarly, I have proposed that development itself is always antagonistically pleiotropic, which if correct forces the conclusion that all of the specialized structures and functions that comprise all organisms—from bacteria to people—are built from antagonistic pleiotropic genes (Turke, 2008; 2013).

Development is always antagonistically pleiotropic, I believe, because it is always driven by differentiation, and differentiation always diminishes totipotency. Thus, while damaged tissues or organs can be repaired to a point, or regenerated *in toto* in some instances with considerable effort and risk (by keeping pluripotent cells on standby, or by dedifferentiating and then redifferentiating, as plants sometimes do), there are limits to repair and regeneration, and the limits are proportional to the degree of differentiation. I cannot, for example, regrow a limb or a kidney because too many long and tortuous differentiation pathways were traveled the first time I grew them. The only possibility I'm left with is to regenerate them secondarily by combining one of my germ cells with someone else's to recreate the totipotency that I started with. This is the very same conundrum that Williams (1957) identifies at the beginning of his famous article on senescence in which he points out that a capacity for complete, primary morphogenesis is inexorably transformed into an inability to merely maintain what is already formed. A fuller discussion cannot be given here, but this claim—that it is more difficult to indefinitely maintain a differentiated structure than an undifferentiated one—contributes to the decision to repair damage, or not to, which in turn is the basis for the evolution of Weismann's (1893) germ-soma distinction (Turke, 2008; Chao, 2010). Germ of course is undifferentiated and maintained indefinitely, whereas soma is differentiated and disposable (see, e.g., Kirkwood and Cremer, 1982).

Thus, if development (via differentiation) does indeed portend dysfunction for the reasons I have just outlined, we can conclude that the specialized adaptations that produce precociality are, in every instance, built from antagonistic pleiotropic genes. Furthermore, the spread, persistence, and continued expression of such genes depends on the persistence of circumstances that allow the positive side of antagonistic pleiotropy to outweigh the negative, which in turn depends on both the timing and magnitude of expressed effects (e.g. Hamilton, 1966). From Alexander's hypothesis, the absolute magnitude of the benefit that early hominins derived from physical precociality began to decrease and continued to do so because of technological advances, such as fire and improved weapons, and because of changes in social organization, such as more committed fathers and more alloparents (e.g., Hrdy, 2009). Simultaneously these technological and social upgrades were increasing the proportion of individuals living to older ages, which would have meant that adult hominins increasingly experienced the negative side of antagonistic pleiotropy, and because the usefulness of older adults was increasing (see "Uniquely Unique," p.14), the force of selection against the dysfunction of old age was almost certainly on the rise. Furthermore, as Alexander emphasizes, changes in the social milieu of our ancestors were creating more profitable uses for the calories that physical precociality requires. Thus, due to this combination of changing circumstances a hypothetical tipping point was reached in which the net effect of many of the genes that build physical precociality became negative.

Nevertheless, since extant humans and apes are genetically very similar, it seems unlikely that hominin evolution led to the wholesale elimination of genes for physical precociality. Rather, I suspect that we retain most such genes, but that their expression has been silenced or at least dampened by a relatively small number of regulatory genes. Hormones are key mediators of such regulation, which brings us to hairlessness.

Hormones are among the most important signaling agents that turn soft, pudgy, unaggressive, hairless human infants into NFL linebackers, but they do so rather slowly and with dampened effects in modern humans, relative to in extant apes and presumably in apelike early hominins. To be relatively soft, pudgy, and hairless—which even adult linebackers are compared to adult apes—is to be, in a word, neotenic. I suggest, therefore, that the evolution of human neoteny (and hairlessness, which is a key feature of our neoteny) can be explained as follows. Neoteny is the proximate basis for altriciality—that is, it exists as a collection of physical traits that result in physical helplessness. Neotenic features, including hairlessness, are exaggerated in young humans and persist into adulthood because regulatory genes slow the production and effects of maturation hormones, this in turn dampens the early expressions of the many antagonistic pleiotropic genes that contribute to physical precociality, and also dampens and delays the late deleterious effects of these same antagonistic pleiotropic genes (examples and additional explanation, below). Therefore, altriciality and neoteny manifest, saving calories, which, as Alexander suggests, are then diverted to cognitive development, and delayed senescence also manifests, which is directly advantageous in individuals that increasingly achieve reproductive success through lifelong learning and age-related social connectedness. Thus, lifespan extension is a direct adaptive consequence of neoteny/altriciality. It is also worth noting that essentially the same argument (only some of the details differ) can be used to explain the neoteny/altriciality and extended lifespan of some social insect queens (cf. Alexander et al., 1991).

Alexander anticipates part of my explanation for neoteny and hairlessness by recognizing that both are part of a developmental process that results in altriciality (p.28). However, he believes that this proximate-level explanation is inadequate or at least incomplete because it does not identify any directly adaptive functions of neoteny and hairlessness. He is led to this conclusion, though, only because he does not connect the argument to antagonistic pleiotropy, which means that he also does not connect it to the directly adaptive outcome that neoteny confers by delaying senescence. Without these connections, neoteny and the hairlessness that it entails might indeed be seen as non-adaptive incidental effects.

Alexander mentions senescence theory, and its pleiotropic underpinnings (p. 28), just as I do here, perhaps hinting that he has a sense of the importance of these concepts to his overall hypothesis; but he does not, at least not explicitly, draw from them the same implications that I do. Rather, he seems simply to be illustrating awareness of the fact that non-adaptive traits can evolve and persist due the inability of selection to remove them. He then moves on without further discussion to propose that hairlessness is possibly adaptive by virtue of promoting heat transfer from parents to young offspring. Alexander is not entirely convinced by his own proposal, however, foremost because it leaves unanswered "why older juveniles and adult humans are relatively hairless" (p.29). My extension of his primary argument potentially solves this problem by suggesting that a neotenic phenotype that extends into adulthood provides a "somatic environment" conducive to diminished and delayed expression of negative (senescence-causing) pleiotropic effects.

The previous sentence succinctly expresses the crux of my argument. Let me be less succinct. According to Williams (1957:400), many genes will express "opposite effects on fitness at different ages, or more accurately, in different *somatic environments*" (emphasis his). Two hypothetical examples illustrate how fitness effects might flip from positive to negative in organisms such as apes and humans due to changes in somatic environment that are relatable to neoteny and altriciality.

> (1)   Consider a gene for an enzyme which folds into a specific configuration when it finds itself in a pristine cytoplasmic environment. As such, it efficiently catalyzes a biochemical reaction that is required at all stages of the life cycle. However, in the course of an individual's physical maturation (with its many metabolic requirements), cell cytoplasm inevitably changes—water content often decreases, various metabolites build up or decline, glucose concentrations rise or fall, pH changes, oxidative and other damage occurs, etc. These changes in somatic environment alter the enzyme's folding pattern and hence its final shape, which diminishes its function; as a result, an original positive fitness effect eventually becomes negative.
>
> The scenario just given is hypothetical insofar as it does not identify a known gene or enzyme, but Shi et al. (2008) have identified actual enzymes that change as described for the general reason I have suggested. For my hypothetical enzyme, as well as Shi et al.'s real ones, what might slow the change from positive to negative fitness effect? One obvious possibility

would be to slow the rate of physical maturation, which would extend the duration of a neotenic-appearing phenotype and thus help to maintain a pristine cytoplasmic environment; and as suggested earlier this could be accomplished via mutation in relatively small numbers of regulatory genes.

(2) Consider a suite of hormones that increase leanness, muscle strength, hair growth, and aggressiveness. These proximal effects give an advantage in most types of physical competition, but they also lead to more fighting, more wounds, and more infections. The net effect of this suite of hormones is nonetheless positive in up-and-coming adolescents and young adults, given their fully functional immune systems and peak ability to regenerate damaged tissues, but becomes negative once tissue regeneration and immune function begin to falter, as they inevitably do over time. What might slow this transition? Down regulation of hormone production.

In the first example, the change from positive to negative fitness effect occurs because of changes in somatic environment occurring at the level of cytoplasm biochemistry. In the second example, more distal environmental factors come into play—factors such as the likely outcome of fighting when the entire soma is in its prime versus when it is in decline. Both examples illustrate Alexander's claim that all genes are both pleiotropic and epistatic (see above), and at the same time they show how changes in somatic environment—that is, changes that increase and extend neoteny—can alter gene expression in a manner that results in altriciality *and* an extended lifespan. Such changes of course would likely be disfavored in apes (because, as Alexander suggest, apes benefit greatly from physical precociality) and favored in hominins (because, as he also suggests, hominins don't).

Some might wonder at this point whether hairlessness, specifically, could be decoupled from neoteny? I suspect that it could be, but I cannot think of a reason that would justify the effort, especially if hairlessness is even slightly beneficial to infants for the reason Alexander has suggested. One might similarly ask whether the lifespan extension that has occurred in hominins could have been achieved without extending neoteny into adulthood? I cannot envision how, without changes so extreme as to render us entirely different from what we are.

A (MOSTLY) FALSE RUMOR AND A FINAL SPECULATION

Many of the leading evolutionary biologists of the past half-century have from time to time been rumored to be too gene-centered in their approach. To paraphrase Mark Twain, such rumors are greatly exaggerated. George Williams, for example, who is surely among this era's most influential evolutionary biologists, and surely a target of such rumors, was acutely aware of the role that environment plays in gene expression as evidenced by the central importance he gave to what he called the "somatic environment" in determining the fitness effects of pleiotropic genes (see Williams, 1957, and above). Williams' theory of senescence, as well as the whole of his work, greatly influenced Alexander and it is this particular perspective—that phenotypic effects of a given gene manifest from interaction with an environment that is multilayered, and in which the layers of importance may include everything from other co-resident genes to the social environment—that has made Alexander particularly well suited to unraveling the most unique of human traits, including our extreme physical altriciality, mental precociality, and of course lifelong learning. It is this nuanced view of development, a view that sees plasticity of response as the raison-d'être of phenotype, that allowed Alexander to see as perhaps no one did before him that culture is nothing other than the "cumulative effect of the inclusive fitness maximizing behavior…of all humans who have ever lived" (Alexander, 1979:68); and it is also a view, imparted to his students, with considerable back and forth, as I recall, that is yielding creative solutions to a wide variety of extraordinarily difficult evolutionary problems (see, e.g., West-Eberhard, 2003; Crespi and Summers, 2005; Crespi, 2008; Sherman et al., 2008; Betzig, 2009, 2010; Flinn et al., 2009; Frank and Crespi, 2011).

I'll close with a speculation, which I think is appropriate in a volume discussing the collected works of Dick Alexander. A central tenet of the Medawar-Williams theory of the evolution of senescence is that organ systems should senesce in tandem. Thus, the human immune system, for example, should not regularly fail ahead of cardiovascular, renal, respiratory, digestive, and neurological systems. However, equality in the timing of system failure does not imply equality in the amount of effort that has been put into achieving in-tandem senescence of vital systems. It is possible, in other words, that one system more than others has been a limiting factor in the extension of the human lifespan to a maximum of a little more than 100 years. In particular, from Alexander's altriciality hypothesis there is reason to suspect that maintaining cognitive function to advanced ages has been an especially difficult feat to achieve. This follows because although silencing or removing the genes for physical precociality became an option during hominin evolution, quite the opposite is predicted for genes contributing to enhanced cognitive ability in children and young adults. These genes are likely to have been added and/or up-regulated during hominin evolution, and if their late-life effects are detrimental to the central nervous system—which is expected if diminished totipotency/regenerative capacity is, as argued, the price individuals pay for enhanced early function—then the loss of cognitive function will have been a long-running constraint on hominin lifespan extension. Sadly, as health care, sanitation, and other modern inventions keep more people alive longer, the devastating nature of this constraint is becoming more and more apparent.

Acknowledgments

I thank Dick Alexander, Laura Betzig, Bernie Crespi, Bev Strassmann, and Kyle Summers for a number of thoughtful suggestions.

References

Alexander, R.D. 1974. The evolution of social behavior. *Ann. Rev. Ecol. Syst.* 5: 325–83.

Alexander, R.D. 1979. *Darwinism and Human Affairs.* Seattle: University of Washington Press.

Alexander, R.D. 1987. *The Biology of Moral Systems.* New York: Aldine-de Gruyter.

Alexander, R.D. 1990. How did humans evolve? Reflections on the uniquely unique species. *Univ. of Mich. Spec. Pub.* 1:1–38.

Alexander, R.D., Noonan, K. M., and Crespi, B. J. 1991. The evolution of eusociality. In: Sherman P.W. et al. (eds.), *The Biology of the Naked Mole-Rat.* PrincetonUniversity Press, Princeton, NJ, pp. 3–44.

Betzig, L. 2009. But what is government itself but the greatest of all reflections on human nature? *Politics and the Life Sciences,* 28:102–105.

Betzig, L. 2010. The end of the republic. In P. Kappeler and J. Silk, eds. *Mind the Gap: Primate Behavior and Human Universals,* pp. 153–68. Berlin: Springer Verlag.

Chao, L. 2010. A model for damage load and its implications for the evolution of aging. *PLoS Genet.* 6: e1011076.

Crespi, B. 2008. Genomic imprinting in the development and evolution of psychotic spectrum conditions. *Biol. Rev. Cambr. Phil. Soc.* 83:441–493.

Crespi, B. and Summers, K. 2005. Evolutionary biology of cancer. *Trends Ecol. Evol.* 20:545–552.

Flinn, M.V., Muehlenbein, M.P., and Ponzi, D. 2009. Evolution of neuroendocrine mechanisms linking attachment and life history: the social neuroendocrinology of middle childhood. *Behav. Brain Sci.* 32:27–28.

Frank, S.A. and Crepsi, B. J. 2011. Pathology from evolutionary conflict, with a theory of X chromosome versus autosome conflict over sexually antagonistic traits. *Proc. Nat. Acad. Sci. USA* 108 (Suppl. 2):10886–10893.

Hamilton, W.D. 1964. The genetical evolution of social behavior. *J. Theoret. Biol.* 7: 1–52.

Hamilton, W.D. 1966. The moulding of senescence by natural selection. *J. Theoret. Biol.* 12(1):12–45.

Hrdy, S.B. 2009. *Mothers and Others.* Cambridge: HarvardUniversity Press.

Kirkwood, T.B.L. and Cremer, T. 1982. Cytogerontology since 1881: A reappraisal of August Weismann and a review of modern progress. *Hum. Genet.* 60: 101–121.

Kirkwood, T.B.L. and Rose, M.R. 1991. Evolution of senescence: Late survival sacrificed for reproduction. *Phil. Trans. R. Soc. B,* 332: 15–24.

Sherman, P.W., Holland, E. and Shellman Sherman, J. 2008. Allergies: Their role in cancer prevention. *Q. Rev. Biol.* 83: 339–362.

Shi, J. Dertouzos, J., Gafni, A., Steel, D.G., and Palfey, B. A. 2006. Single-molecule kinetics reveals signatures of half-sites reactivity in dihydroorotate dehydrogenase A catalysis. *Proc. Nat. Acad. Sci. USA,* 103:5775–5780. PMID: 16585513.

Trivers, R.L. 1971. The evolution of reciprocal altruism. *Q. Rev. Biol.* 46: 35–57.

Turke, P.W. 2008. Williams' theory of the evolution of senescence: Still useful at fifty. *Q. Rev. Biol.* 83: 243–256.

Turke, P.W. 2013. Making young from old: how is sex designed to help? *Evolutionary Biology,* (in press). DOI 10.1007/s11692-0139236-5.

Weismann, A. 1893. *The Germ-Plasm: A Theory of Heredity,* translated by W. N. Parker and H. Rönnfeldt. London: Walter Scott.

West-Eberhard, M.J. 2003. *Developmental Plasticity and Evolution*. New York: Oxford University Press.

Wigglesworth, V.B. 1961. Insect polymorphism—a tentative hypothesis. In: J. S. Kennedy (ed.), *Insect Polymorphism*. London: Royal Entomological Society.

Williams, G.C. 1957. Pleiotropy, natural selection, and the evolution of senescence. *Evolution* 11: 398–411.

Williams, G.C. 1966. Natural selection, the costs of reproduction, and a refinement of Lack's principle. *Am. Natural.* 100:687–690.

ALTRICIALLTY

Why Are Human Babies Helpless?

Excerpt from Alexander, R.D. How Did Humans Evolve? Reflections on a Uniquely Unique Species. *University of Michigan Museum of Zoology Special Publications* 1:1-38.

Altriciality, or physical helplessness, is widespread among juvenile animals, but the human neonate, which is distinctly more helpless than any of its primate relatives, is probably the most famous of all altricial juveniles (Zeveloff and Boyce, 1980 and Dienske, 1986, survey an extensive literature). Although I concentrate here on the human juvenile, I have tried to consider how to account for altriciality wherever it occurs.

The words "altricial" and "precocial" are used primarily in the ornithological literature and are defined in most dictionaries in terms of their application to newly hatched birds (see also Gill, 1990). Altricial hatchlings, as with sparrows, starlings, and pigeons, are more or less naked and helpless; they may be blind and are usually ectothermic. Food is brought to the nest for them by their parents. In contrast, precocial hatchlings—as with chicks, ducklings, pheasants, and quail—typically are covered with down, agile, homeothermic, have their eyes open, and are more or less ready to move out alongside their mother and pick up their food themselves. There are degrees of intermediacy (for example, goslings are somewhat more helpless than ducklings, the eyes of owl hatchlings are closed but not those of hawks, etc.). Gill (1990, pp. 369-70) exemplifies and illustrates eight different categories of hatchlings originally established by Nice (1962) based on "primary criteria of mobility, open or closed eyes, presence or absence of down, and the nature of parental care...." The extreme differences between (1) most songbirds and (2) mound-builders and most ducks, shorebirds, and "fowl-like" birds is probably the reason why the terms altricial and precocial were applied so readily to birds, as well as studied there more extensively than in other organisms.

As Nice (1962) and Case (1978) noted, other animals also display variations paralleling those found in birds. Newborn mice and rats are naked, blind, and more or less helpless. They are born in a nest where they remain for some time. In other mammals, such as some ungulates, on the other hand, newborn often are able to stand alone within a few minutes, and some, such as horses, are able to gallop alongside their mother in less than an hour. Newborn ungulates may travel considerable distances with their mothers, who are following a herd in more or less normal movements. Again, there are intermediates: canine and feline babies are blind when born but not naked, and not as helpless as most newborn rodents; and some ungulate newborns are physically less capable than others.

It is useful to apply the concepts of altricial and precocial even more widely, for example to insect juveniles. Maggots, and the maggot-like larvae of some insects with complete metamorphosis (e.g., honeybees), can be regarded as altricial (in a broader application of the term, so can all insect larvae). In contrast, the nymphs of insects with incomplete metamorphosis, such as grasshoppers and crickets, are precocial in the same sense as some baby mammals and birds. Again, there are intermediates. For example, within the Family Gryllidae (Order Orthoptera), including all crickets, most juveniles would be seen as precocial. Their exoskeletons are hard, and they are agile, quick, and seek out their own food right from hatching; there is no parent alive to assist them. But in genera such as *Anurogryllus,* in which the female cricket prepares a closed burrow with a food cache before she lays her eggs, and then tends her babies until she dies—feeding them small, apparently unfertilized trophic eggs—the hatchlings are soft and fat, resembling termite juveniles (West and Alexander, 1963). Many other examples could be given: thus, caterpillars may be soft and helpless or quick-moving and covered with urticaceous hairs or other defenses. Many internal parasites, especially those living in the alimentary tracts of their hosts, have some of the features of altricial juveniles. Certain adult insects, such as queens in large-colony eusocial forms, possess some of the characteristics of altricial juveniles (Alexander et al., 1991).

Ricklefs (1974, 1975, 1979a, 1979b, 1983) has contributed extensively to the development of theory that helps explain altricial and precocial juveniles. Initially he showed that altricial nestlings of birds grow faster than the more precocial nestlings of related species, and he eventually concluded (1983, pp. 11-12) that "The overwhelming advantage to altricial development seems to be rapid growth.... Although adoption of the altricial condition may increase vulnerability to predation and enhance the effects of exposure to bad weather, these are presumably more than compensated for by the brevity of the development period." Faster growth may actually be the adaptive function of altriciality in a wide variety of species. The human embryo and neonate both grow faster than do the embryos and neonates of their primate relatives (Sacher and Staffeldt, 1974). It is difficult to believe, however, that this is the full explanation for the altriciality of the human infant. Unlike altricial birds, for example, the human juvenile has a juvenile life as long as or longer than those of the closest relatives of humans whose juveniles are all less altricial (Smith, 1989). As Montagu (1961, p. 56) notes, "man is born and remains more immature for a longer period than any other animal."

Dienske (1986) has reviewed and criticized previous theories about human altriciality, giving good reasons for doubting that human newborns are altricial simply because more advanced neonates could not pass the pelvic passage, and pointing out that although apes have pelvic passages that are larger in relation to their babies' heads than are those of rhesus macaques, the apes' babies are more altricial. It does not seem likely either that the human baby is simply born at an earlier stage through shortening of the gestation period, since the great apes have about the same gestation periods as humans, or slightly shorter (Schultz, 1956; Sacher and Staffeldt, 1974), and the 12-month gestation briefly postulated for Neanderthals because of a presumed larger pelvic opening has since been discounted (Rosenberg, 1986; Greene and Sibley, 1986; Trevathan, 1987; Trinkaus, 1987). Dienske also doubted that altriciality occurs because the human baby's brain is small in size, since it is comparable in size, in relation to the body weight, to that of other primates. Dienske summarized the evidence that in humans the adult brain is much larger, in relation to its size at birth, than those of other primates, and he wondered if this might not have something to do with altriciality, supposing that this difference might mean that the neonate's brain is less developed. As he put it: a neonate brain that is smaller in relation to the adult brain "...implies a greater immaturity if many parts of the neonatal brain are still in a (rudimentary) stage of functioning." Although this hypothesis is probably correct, it need not imply that this aspect of altriciality is explainable simply as a result of a physiological or developmental constraint or that human neonates are "embryos," simply born at an earlier stage of development (e.g., Kuttner, 1960). That view would not account for the early mental precociality of the human juvenile compared to ape juveniles or engage the question of the pattern of development of function in the human brain.

Zeveloff and Boyce (1982) seek an adaptive hypothesis for human altriciality. Concordant with the arguments developed here, they suggest (p. 540) that "...monogamous pair-bonds and concomitant opportunities for paternal investment may contribute to the evolution of human altriciality," and that monogamy and paternal care were made more likely by increased confidence of paternity. They also argued that increased time for learning, from an increased length of the juvenile period, is the main benefit of altriciality. My arguments here differ from theirs in that (1) they sometimes seem to be suggesting that monogamy and paternal care evolved because of altriciality ("an altricial neonate will offer greater potential for male parental care "- p. 537) rather than vice versa and (2) they seem to assume that altriciality (a) depends on a shorter gestation period and (b) is responsible for the longer learning period (see also Case, 1978; Zeveloff and Boyce, 1980). The extensiveness and profundity of learning and maturational changes in human juveniles following ages 11–13 indicates that it is appropriate to refer to the human juvenile period as lengthened in comparison to those of related primates even if the earliest time of possible reproduction is about the same for chimpanzees and humans (Smith, 1989). This extended juvenile period may have evolved for the same reasons as human neonate altriciality and not simply as a necessary or incidental result of it—that is, because it contributed to the long learning period, or period of plasticity, that enables human juveniles to absorb and cope with the complexities of culture and human sociality.

In general, we are not hard-pressed to provide adaptive hypotheses for pre-cocial organisms having the attributes that cause us to label them as precocial. It seems obvious why a baby ungulate would gain from being able to run alongside its mother soon after birth. Precocial birds are often tended only by their mothers, are usually hatched in vulnerable nests on the ground, and typically eat the kinds of food that can be captured by moving about on the ground or in the water (Nice, 1962). Juvenile insects in species with incomplete or gradual metamorphosis (that is, the precocial sort) live without parents in dangerous locations, and they are usually able either to run or leap, or else they produce various kinds of poisons or other deterrents to predators. Their abilities to do these things are what causes us to see them as precocial. The same is true, more or less, of relatively precocial lar-vae in insects with complete metamorphosis (that is, larvae with urticaceous hairs, unusual locomotory abilities, or any other special defensive features), even though the larva itself, and the form of metamorphosis ("complete") that it typifies, can be seen as an evolutionary trend toward an altricial feeding stage.

The question that remains is: Why, when extrinsic sources of mortality are removed, do juveniles become soft, fat, helpless, and maggot-like? Is there a gen-eral answer, other than the unsatisfying or incomplete one that particular selective pressures are relieved or removed, or that there is some advantage to the parents rather than to the juvenile itself? Case (1978), for example, offers several possibili-ties for the latter, but all seem to depend on shorter gestation periods and young being smaller in altricial forms. Although some cases of altriciality may fit one or more of the arguments invoking advantages to parents, neither shorter gesta-tion periods nor smaller young is always associated with altriciality; thus, altricial neonates of humans are larger than less altricial neonates of their relatives (Sacher and Staffeldt, 1974). Because some altricial organisms seem to have evolved shorter juvenile periods (e.g., songbirds), and others longer juvenile periods (e.g., humans), any general explanation will have to take both conditions into account.

In seeking a broad explanation, we may first note that juvenile life has two main functions: to get to the adult stage without dying and to become the best possible adult. For our purposes the latter need have no more precise definition than to be maximally capable of doing whatever an adult has to do in one's own species to reproduce as well as or better than anyone else. Presumably, the only selective reason for the existence of a juvenile stage is that it gives rise to an adult that is sufficiently better at reproducing to more than offset the disadvantages of juvenile life such as longer generation time and investment of calories in other than repro-duction *per se.*

The traits and tendencies that make one a better adult are not likely to be syn-onymous with traits that enable one to bypass or deal successfully with particular hazards along the pathway to adulthood. In other words, the two functions of juvenile life are not likely to be achieved by precisely the same directions of selec-tion. What we call precociality evidently represents expensive ways of dealing with hazards that may terminate juvenile life—expensive in the sense that they interfere with selection that otherwise could cause the juvenile to use more of its life effort in preparing to be a more reproductive adult.

In the course of becoming satisfactory adults, juveniles must do two things: grow and develop. Growth enables the juvenile to reach an appropriate adult size at the appropriate time or season. Development, which can be defined as differential growth or change in different tissues or organs, involves changing from the form or function that best serves the juvenile to that which best serves the adult. But this description is still far too simple. Juvenile life is not necessarily unitary: in different forms it can be subdivided into multiple stages, each of which takes its own form, growth rate, developmental rate, and way of functioning. Obviously, all of these things may be affected by changes in the nature or emphasis of the forces that affect the juvenile's success, such as sources of mortality.

Complex metamorphoses—as illustrated by parasites (especially those with multiple hosts), anuran amphibians, and insects—presumably evolve when appropriate forms and functions for different stages of the life cycle vary widely. The larval stage of insects with complete metamorphosis lives in habitats that are suited to feeding and growth. Development is primarily restricted to the pupal stage, which follows the larval stage. The adult does not resemble either of the two juvenile stages or, in general, live in the same habitat. A similar pattern is exhibited by anuran amphibians, with the feeding, aquatic tadpole eventually transforming during a relatively short period into an adult that is dramatically different in form and function and lives in a different range of habitats.

Such patterning during the juvenile life may be considerably more subtle, yet requires understanding if we are to explain the nature of the human juvenile and the patterning of its life. Thus, the altricial juveniles of songbirds for the most part live at first in a nest hidden from predators or inaccessible to them but shortly become capable of flight and leave the nest (songbird nests must often become increasingly vulnerable to predation as the juveniles grow and the parents visit the nest increasingly frequently to feed them). Following fledging, the juvenile songbird's life soon becomes that of an independent flying bird which lives more or less in the adult habitat. As Ricklefs (1983) specified, his description of altriciality as a way of providing more calories for the growth process thus applies only to the earliest part of juvenile life—the time spent in the nest. That period, moreover, necessarily includes not only rapid growth but the development required to transform an altricial hatchling into a feathered, coordinated fledgling capable of flight and with keen sensory apparatus enabling it to avoid predators and locate its own food. In precocial birds these parts of development largely precede hatching. We are required to assume that altriciality, involving only a brief initial part of a songbird's juvenile life, provides sufficient advantage in growth rate to more than compensate for the delay in initiating the dramatic developmental changes necessary for transformation into a suitable fledgling, which in at least some cases might appropriately be described as having achieved a certain precociality.

In some senses songbird hatchlings parallel certain altricial juvenile insects, such as the fly larvae called maggots that occur in dung, carrion, fungi, and other short-lived and vulnerable habitats. These larvae do not seem well protected from predation, weather, or deterioration of their microhabitats. Why, then, do they take on the aspect of being altricial? I would guess because they cannot do anything about the serious threats in their habitat, so that their best strategy is to get through the dangerous feeding stage and out of the larval habitat as fast as possible. As with internal parasites, who may be protected, their particular form of altriciality has caused them to evolve to become mere "sacks" of efficient nutrition-grabbing ability. In their temporary and dangerous hatchling environments they load themselves with nutrients as fast as possible and drop off or crawl out of the dangerous place where they have secured their food to grow and develop in safer locations. One might suspect that they are highly precocial in terms of their ability to ingest their medium rapidly.

Songbird and insect patterns of altriciality are not easily compared to those of humans. First, neither songbird hatchlings nor maggots have evolved coincidentally altriciality *and* a longer juvenile life as have humans. Second, unlike songbirds, in humans parental care remains extensive not only throughout the juvenile's life but well into its adult life.

The general adaptive explanation for the evolution of altriciality seems to be this: to the extent that a juvenile is relieved of the necessity, or any importance, of evolving to protect itself from extrinsic hostile forces of nature (such as predation), it is freed to devote a greater proportion of its calories to improving its performance at some later stages of juvenile life or to becoming a better adult. This will be true, regardless of the means by which the relief is effected, whether by direct or continual parental solicitude or by having been placed in a safe location by a now deceased or departed parent. It will also be true regardless of the means by which the improved later performance is effected: whether primarily by growth or development. Protected juveniles are also free to bring back into their juvenile life—further in time and to a greater extent—traits, tendencies, and events (learning or practicing) that are devoted to enabling their survival or competition as older juveniles or as adults.

On this hypothesis allowing more calories to be devoted to growth is, as Ricklefs (1983) maintained, surely a widespread advantage of altriciality. In songbirds this is possible because the nest is either hidden or off the ground and inaccessible to predators, and because in general both parents provide food. In some rodents and subterranean crickets, as well as larvae injected by their mothers into safe locations such as inside wood, parallels in degrees of security from predation and the physical environment have evolved.

Prior to parturition the human embryo grows faster than do the embryos of other primates (Sacher, 1975), and the neonate is considerably larger in both body weight and brain weight (Sacher and Staffeldt, 1974). Accordingly, it seems possible that the altricial nature of the human neonate yields a significant part of its advantage as an increase in growth rate during the embryonic stage. The human brain, however, also changes in mass during postnatal juvenile life several times as much as the brains of other primates (Sacher and Staffeldt, 1974). Postnatal body size changes in primates are considerably more variable, with gorillas adding much more to their mass than humans or other apes and humans adding more than chimpanzees (Sacher and Staffeldt, 1974). Effects of early altriciality on growth rates during juvenile life—even as a consequence of the unusual intensity and duration of parental solicitude—thus do not seem likely to explain the distinctiveness of the entire pattern of human development.

The overall problem in understanding the adaptiveness of altriciality and precociality is thus one of trade-offs between different life stages, whether similar or different activities or structures are being compared. Precocial birds have larger brains than altricial birds (Gill, 1990), but the altricial human infant's brain is larger than those of its less altricial relatives (Sacher and Staffeldt, 1974; Dienske, 1986). In birds, the larger brain of the precocial bird is presumably used to protect it from predators and other more or less immediate threats—that is, to provide it with skills that increase the likelihood it will reach the adult stage. In humans, however, the size and construction of the neonate's brain has likely evolved for a different reason—in a way that enabled it eventually to develop into an extraordinarily large and complex brain that functions primarily in the complex social activities required for reproductive success in the adult stage.

I hypothesize, then, that early physical precociality has been sacrificed in human juveniles partly in favor of later mental precociality. I suggest that early physical precociality was expendable because human parents became almost entirely responsible for the survival of the juvenile to adulthood, and for its failure to survive when this outcome occurs. This responsibility could only evolve, of course, if the parents were capable of giving sufficient parental care of the appropriate type and if their interests very broadly overlapped those of the offspring. On this theory, the characteristics of the brain of the human infant are investments toward the development of a better adult (or late juvenile) brain and are less involved in the survival of the (early) juvenile than the enlarged brain of a precocial bird or other animal.

If the general idea about altriciality presented here is correct, then to understand the altriciality of human babies thoroughly we will need to understand what kinds of attributes make the best possible adult (or late juvenile) human. I think the answer is, generally speaking, intelligence and social capability. If Humphrey's (1976) argument about the evolution of the human intellect is correct, we should expect the physically altricial human juvenile to become, at some point, intellectually and socially precocial, as suggested by Alexander and Noonan (1979) in their discussion of parental care and the concealment of ovulation. We should expect that the juvenile human begins practicing to be socially successful much earlier in life and on a much more massive scale early in life than is possible for less altricial primate juveniles. I think that this prediction is consistent with the seemingly inordinate attention given by humans to the concept of so-called intelligence quotients, with their connotation of intellectual precociality, attempting to measure mental development, or "age," in relation to physical or chronological age. It is also consistent with the efforts of many human parents to parade their offspring as socially and intellectually precocial. Many aspects of human juvenile life also give the appearance of social-intellectual precociality—such as early smiling, humor, language acquisition, and engagement in reciprocity. Throughout human history even young adult humans probably depended on their parents and other relatives for success, emphasizing the importance of considering that even aspects of early human juvenile life may take their form and function because of events that will transpire many years later in late juvenile or adult life.

I also think of humans as having brought far back into their juvenile life many kinds or instances of social-intellectual-physical play, as practice, and of them having done so as part of what we usually have termed evolving increasing degrees of altriciality. I think as well that they have been able to do this because of the dramatic increase in the amount and effectiveness of parental care during human evolution. For humans, explicitly, it seems to me that the question is moot as to what extent the general increase in parental care took place because it allowed human juveniles to devote more calories to becoming better adults, and to what extent it occurred because it literally saved juveniles from death, thereby incidentally allowing the juveniles to devote more of their calories to becoming better adults. In either case, my hypothesis requires that in young juvenile humans more calories are now devoted to better performances later in life, including adult life, and I think this means being socially more effectively cooperative as a way of being reproductively more competitive, than was the case in the days of, say, *Homo erectus,* or of species ancestral to *Homo.*

The intellectual-social precociality of the human juvenile was probably responsible for two prominent evolutionary theorists, G. Evelyn Hutchinson (1965) and George Williams (1966), making independently the intriguing suggestion that the creative intellect of adult humans is an incidental effect of selection for high intellectual capacity in juvenile humans. The only way this argument seems to me to have credibility is if it is applied only to the late juvenile (or early adult) stages, when juveniles are actually striving to enter the breeding population. Even with this qualification, it seems to me that the length of adult human life and the evidence of rather dramatic increases in control of resources during adulthood, together with the fact that the juvenile survives and succeeds primarily at the behest of its parents and other adult close relatives, tends to deny the argument that human brain complexity functions primarily during juvenile life.

Discussion.—Altriciality apparently evolves, then, when an infant gives up on protecting itself and turns its protection over more or less entirely to extrinsic forces. This protection may come about through continued attention by the parent or because the juvenile lives in a protected situation, where it may have been placed by its parent. If the extrinsic force is long-term care by parents or a permanently safe location, the offspring is also free to evolve an extended juvenile life, if by this it improves its adult performance sufficiently to compensate the added time and expense involved. Altriciality in some attributes may even evolve in an unsafe situation if thereby the juvenile can grow at the fastest possible rate and escape the unsafe situation. In such cases, obviously, juvenile life will decrease in duration. Or, as in the case of internal parasites protected by being inside their host or eusocial queens protected by their workers and soldiers, physical "altriciality" may continue for the entire life of the organism, in the interests of turning all effort to activities more directly reproductive than (useless) protection. Although two parents (or any tending parents) are not required for the evolution of altriciality, biparental care is a common situation, explaining the association with monogamy in birds and mammals.

HAIRLESSNESS IN LATE JUVENILE AND ADULT HUMANS

I have not taken up the question of why humans remain relatively hairless as juveniles and adults, and I confess that I do not have a convincing hypothesis. Nor, evidently, does anyone else. Some comparative discussion around this question, however, may be useful.

The argument has been made repeatedly that ". .. the evolution of human hairlessness was somehow associated with temperature regulation in a tropical environment" (Kushlan, 1980, p. 727: see also Campbell, 1966; Leakey and Lewin, 1977). Morris (1967) suggested that hairlessness enabled humans to lose heat more efficiently, as when chasing large game. Schwartz and Rosenblum (1981, p. 10) noted that "As the ratio of surface/volume decreases in a static series of adult primates, the advantage of an insulating coat diminishes as well." The human baby, however, is both the smallest human and the least hairy, and the human female, smaller than the male, is next. Neither juveniles nor females do much chasing of large game. As Darwin (1871) noted " ... other members of the order of primates, ... although inhabiting various hot regions, are well clothed with hair." No other predator that captures its prey after chases in the hot areas of the world has lost its hair. Nor has any other species at all taken up the particular pattern of hairlessness that the human species exhibits, in which there is (1) a single naked infant rather than a litter of naked infants and (2) a naked infant that is carried as opposed to being hidden in a nest or a marsupial pouch. It is at least possible that in the hominid line a hairless baby evolved first, followed by older juveniles, the adult female, and finally the adult male.

It has also been argued that humans lost their hair because parasites were such a problem, but then we have to ask why parasites were more of a problem with humans than with other species. None of these or other hypotheses given so far for human hairlessness (clothing, fire, and shelter made hair superfluous: Glass, 1966; early humans were aquatic: Napier, 1970; sexual selection, species identification, social appeasement, neoteny: Darwin, 1871; Keith, 1912; Guthrie, 1970; see also Kushlan, 1980) seems to have gained much momentum.

It appears that other relatively hairless mammals have lost their hair for reasons that cannot be used to explain human hairlessness. The following arguments on this topic are modified from Alexander (1991b), a chapter written for a volume on naked mole-rats. The prevalence of ectotherrny among altricial mammal and bird juveniles, and the ability of human babies to survive extreme lowering of the body temperature for long periods, suggest a general connection between ectothermy and altriciality, and raise questions about the apparent connections between ecto- thermy and nudity in some cases.

Mammalian Variations in Hairlessness.—Mammals are the organisms that have hair and produce milk. There are analogues for both traits in other organisms (pigeons produce a milk-like food for their young—see discussion in Gill, 1990— and many organisms have hair-like structures), but no homologues.

The amount and kind of hair varies extensively among different mammals. Relative hairlessness occurs in a variety of mammals, rarely for reasons that are entirely obvious (Lyne and Short, 1965; W. J. Hamilton, 1973; Jarvis, 1981). For some aquatic mammals, both marine and freshwater (e.g., cetaceans, sirenids), a layer of fat beneath the skin seems to have proved a more appropriate correlate of homeothermy than a coat of hair, partly because hair causes drag in an aquatic environment, reducing the efficiency of locomotion, and partly because trapped air in pelage or plumage can be lost through compression, for example as a result of diving. Although a variety of aquatic mammals have retained a hair coat (seals, walruses, polar bears, otters, mink, beaver), these seem invariably to be either spe- cies that live in cold climates or species that spend a significant amount of time out of the water. Some mammals have replaced part or all of the hair coat with armor of one sort or another (armadillos, pangolins, ant-eaters—in some armadillos abundant ventral hair is retained); such forms also live in mild or tropical climates (Walker, 1975). Several large, entirely terrestrial (elephant, rhinocerus) or primar- ily terrestrial (hippopotamus) mammals have lost a hair coat in favor of a thick, leathery skin. It has been postulated that these tropical forms have a low body surface area in relation to their body mass and therefore have gained by increasing their ability to lose heat through the skin. As predicted from this hypothesis, tem- perate zone, montane, and rain forest-dwelling relatives of these forms have more hair, and juveniles of these forms have more hair than adults (Walker, 1975). A few mammal species (suids and some primates) are somewhat intermediate, having lost much of their hair (Lyne and Short, 1965). Many mammals that bed down or nest in contact with their young, or carry infants on their venters, have lost much of the hair on their venters and around the mammary glands (e.g., suids, rodents, some primates). In such cases the young juveniles are also either virtually hair- less (rodents), relatively so (suids), or only lightly haired on the particular parts of their anatomy that regularly contact the mother (primates) (birds that brood altricial young also sometimes have bare patches on their venters).

Only two mammal species additional to the above groups have virtually hairless adults and older juveniles: naked mole-rats and humans. Each of these species appears to have evolved nudity independently of any other mammalian forms, since their close relatives are all relatively hairy. The exception to this statement is that nearly all rodent newborns are naked, so that in fact only the older juveniles and adults of naked mole-rats have diverged in this regard from other rodents. The hairlessness of non-newborns in naked mole-rats and humans is also similar in that in neither case is there either a dramatically thickened skin or armor (although relatively more in naked mole-rats); it differs, of course, in that adult humans have retained abundant hair on the head, in the pelvic region, and in the armpits. The nudity of non-newborns in these two species, representing two of seven or more independent origins of relative hairlessness, seems more reminiscent of the kind of hairlessness of newborn altricial mammals; these two species may also be the only mammals in which nudity in newborns (probably) preceded nudity in older juveniles and adults.

Hairlessness in newborns is widespread in mammals, as is absence or near absence of feathers in newly hatched birds (see discussion of altriciality above). This is probably the reason hairlessness has been regarded as part of a neotenic trend, which may be a correct view in terms of developmental processes but does not provide an explanation in evolutionary or selective terms. Phenomena such as neoteny and allometry may represent inertial or constraining forces, in the sense that natural selection must always operate on "last year's model," but in the same sense all genetic, developmental, physiological, and morphological attributes of organisms represent inertial elements for selection. Unless one assumes that natural selection is helpless in the face of such inertias, the search for evolutionary (selective) explanations continues in approximately the same fashion as in the absence of information about such inertias. The general assumption of such searches is that selection is the *principal* (not the sole) guiding force of evolution.

Hairlessness and Ectothermy.—Newborn mammals that are both naked and sometimes left by the mother in a nest also tend to be ectothermic, as do altricial vertebrates in general, and this implies that there is merit in attempting to relate the evolution of hairlessness to that of altriciality (see also Case, 1978). The human baby is not ectothermic, but it is often said to be unusually capable of surviving periods of lowered body temperature, and this feature may not be entirely independent of its extreme altriciality. An ectothermic organism is one that relies for its body temperature largely or entirely on external sources. Such organisms are often described as having "poor" or "inadequate" means of thermoregulation. This view is not productive of hypotheses as to the origin and basis of the trait, unless one imagines, again, that selection has somehow been ineffective and a trait that is disadvantageous has evolved. Such traits do evolve, as in senescence (Williams, 1957), but only under special conditions such as pleiotropy, with beneficial and deleterious gene effects continuing in concert whenever they derive from the same indivisible chunk of genetic material. Such deleterious traits are saved only because their inevitable companion traits are sufficiently beneficial to overcome the deleterious effects and no way of divorcing the two effects has yet appeared. In no organism, apparently, has a reason been generated for regarding ectothermy as a deleterious pleiotropic effect, and, contrary to the situation with senescence, no circumstances seem to exist that make such an explanation likely. Accordingly, it seems parsimonious to assume that ectothermy evolved in altricial juveniles and naked mole-rats because it is somehow directly advantageous, or because it allowed some other change that was advantageous.

One correlate of ectothermy is a rather low metabolic rate, and it has sometimes been assumed that this is the source of the advantage (reviewed by Case, 1978). A lowered metabolic rate, for example, might allow naked mole-rats to subsist on fewer calories and therefore suggests a continuing problem in caloric intake that is greater or of a different nature than that encountered by the usual homeothermic mammal.

A potentially profitable way to start thinking about the evolution of ectothermy in a previously homeothermic animal is to consider that it has evidently become more efficient, for whatever reason, for that animal to rely upon an external source of warmth. This situation would seem to prevail whenever such external sources are so reliable and effective that the expense of homeothermic machinery is superfluous. To understand when such conditions might exist, one must consider not only the external source of heat itself, but also the nature of threats, such as inability to obtain food when it is crucial and inability to escape from predators that are able to maintain a high rate of metabolism and predatory ability when external heat sources are minimal or absent. Naked mole-rats, preyed upon largely by snakes in their burrows (Sherman et al., 1991), have apparently been largely relieved of homeothermic predators, and their tropical burrow systems are relatively stable in both temperature and humidity.

Hairlessness and Altriciality.—Altricial juveniles that have given over virtually all protection from predators and supplying of food to their parents (as with naked, ectothermic forms) are in a position to retreat from homeothermy and use their parents as the (primary) external heat source. Such a juvenile might gain from refraining from use of nutrition provided by the parent to maintain a high metabolic rate when the parent is absent, and instead conserve ingested calories for later growth by maintaining a high metabolic rate only when external heat (parent, sun) is available. In this fashion parents become suppliers of calories through not only food itself, but also through providing heat for metabolism. This set of attributes correlates with both nakedness in the juvenile and nakedness in at least the part of the anatomy of the parent that most emphatically contacts the juvenile during brooding (brood patches of birds, naked bellies and mammary glands of mammals). It is probably significant that most naked juvenile birds and mammals occur in litters, and single offspring are rarely naked. Nakedness allows extremely rapid absorption of heat from extrinsic sources such as the sun, bodies of other individuals, and warmed soil on sunny days and into the night. Coincidentally, it also causes or allows rapid heat loss to other individuals, which are always close relatives in the case of parental birds or mammals or colony members in naked mole-rats.

Discussion.—These comparative arguments provide a background for thinking about hairlessness, altriciality, and tendencies to be ectothermic in juvenile mammals and birds, including the human baby. They leave unanswered, however, as I suggested would be the case, precisely why older juvenile and adult humans are relatively hairless. They also fail to answer in a satisfying way the questions why the human baby is (1) apparently the only singly produced mammalian or bird offspring that is naked, and (2) the only nonmarsupial mammalian offspring that is both highly altricial and carried by the parents (newborns of apes are more altricial than those of other primates, but much less so than the human baby: Dienske, 1986; some bat neonates are carried by the mother: Walker, 1975), as opposed to being hidden and left, as with altricial rodent newborns and songbird hatchlings.

Literature Cited

Alexander, R.D. and K.M. Noonan. 1979. Concealment of ovulation, parental care, and human social evolution. *In* N.A. Chagnon and W.G. Irons (eds). *Evolutionary Biology and Human Social Behavior: An Anthropological Perspective* : 436–453. North Scituate, MA: Duxbury Press.

Alexander, R.D., K.M. Noonan, and B.J. Crespi. 1991. The evolution of eusociality. In P. Sherman, J. Jarvis, and R.D. Alexander (eds). *The Biology of the Naked Mole-Rat* : 3–44. Princeton, NJ: Princeton University Press.

Alexander, R.D. 1991b. Unanswered questions about naked mole-rats. In P. Sherman, J. Jarvis, and R.D. Alexander (eds). *The Biology of the Naked Mole-Rat* : 446–465. Princeton, NJ: Princeton University Press.

Campbell, B.G. 1966. *Human Evolution. An Introduction to Man's Adaptation.* Chicago: Aldine Publishing Co.

Case, T.J. 1978. Endothermy and parental care in the terrestrial vertebrate. *American Naturalist* 112:861–874.

Darwin, C. 1871. *The Descent of Man and Selection in Relation to Sex.* Two vols. New York: D. Appleton and Co.

Dienske, H. 1986. A comparative approach to the question of why humans develop so slowly. In J.G. Else and P.C. Lee (eds). *Primate Ontogeny, Cognition and Social Behavior.* 147–154. London: Cambridge University Press.

Gill, F.B. 1990. *Wise Choices, Ape Feelings.* Cambridge, MA: Harvard University Press.

Glass, B. 1966. Evolution of hairlessness in man. *Science* 152: 294.

Greene, D.L. and L. Sibley. 1986. Neandertal pubic morphology and gestation length revisited. *Current Anthropology* 27: 517–518.

Guthrie, R.D. 1970. Evolution of human threat display organs. *Evolutionary Biology* 4: 257–301.

Hamilton, W.J. III. 1973. *Life's Color Code.* New York: McGraw-Hill.

Humphrey, N.K. 1976. The social function of intellect. *In* P.P.G. Bateson and R.A. Hinde (eds). *Growing Points in Ethology* : 303–318. New York: Cambridge University Press.

Hutchinson, G.E. 1965. *The Ecological Theater and the Evolutionary Play*. New Haven, CT: Yale University Press.

Jarvis, J.V.M. 1981. Eusociality in a mammal: cooperative breeding in naked mole-rat colonies. *Science* 212: 571–573.

Keith, A. 1912. *Man, a History of the Human Body*. New York: H. Holt and Co.

Kushlan, J.A. 1980. The evolution of hairlessness in man. *American Naturalist* 116: 72–729.

Kuttner, R. 1960. A hypothesis on the evolution of intelligence. *Physiological Reports* 6: 283–289.

Leaky, R.E. and R. Lewin. 1977. *Origins*. New York: Dutton.

Lyne, A.G. and B.F. Short. 1965. *Biology of the Skin and Hair Growth*. New York: American Elsevier Publishing Co.

Montagu, A. 1961. Neonatal and infant maturity in man. *Journal of the American Medical Association* 178: 56–57.

Morris, D. 1967. *The Naked Ape*. London: Cape Publishing Co.

Napier, J.R. 1970. *The Roots of Mankind*. Washington, DC: Smithsonian Institution Press.

Nice, M.M. 1962. Development of behavior in precocial birds. *Transactions of the Linnean Society of New York* 8:1–211.

Ricklefs, R.E. 1974. Energetics of reproduction in birds. *In* R.A. Paynter, Jr., (ed). *Avian Energetics*. Cambridge, MA: Publication of the Nuttall Ornithological Club No. 15: 152–297.

Ricklefs, R.E. 1975. The evolution of cooperative breeding in birds. *Ibis* 117: 531–534.

Ricklefs, R.E. 1979a. Adaptation, constraint and compromise in avian postnatal development. *Biological Reviews, Cambridge Philosophical Society* 54: 269–290.

Ricklefs, R.E. 1979b. Patterns of growth in birds. V. A comparative study of development in the starling, common tern and Japanese quail. *Auk* 96: 10–30.

Ricklefs, R.E. 1983. Avian postnatal development. *In* D.S. Farrar, J.R. King, and K.C. Parkes (eds). *Avian Biology*. Volume 7: 1–83. New York: Academic Press.

Rosenberg, K. 1986. The functional significance of Neandertal pubic morphology. Ann Arbor, MI: University Microfilms.

Sacher, G.A. 1975. Maturation and longevity in relation to cranial capacity in hominid evolution. *In* R.H. Tuttle (ed.). *Primate Functional Morphology and Evolution*: 418–441. The Hague: Mouton Publishers.

Sacher, G.A. and E.F. Staffeldt. 1974. Relation of gestation time to brain weight for placental mammals: implications for the theory of vertebrate growth. *American Naturalist* 108: 593–615.

Schultz, A.H. 1956. Postembryonic ages changes. *In* H. Hofer, A.H. Schultz, and D. Starck (eds). *Primatologia. Handbook of Primatology. I. Systematik, Phylogenie, Ontogenie*: 887–964. Basel: Basler Druck-und Verlagsanstalt.

Schwartz, G.C. and L.A. Rosenblum. 1981. Allometry of primate hair density and the evolution of human hairlessness. *American Journal of Physical Anthropology* 55: 9–12.

Sherman, P.W., J.V.M. Jarvis, R.D. Alexander (eds). 1991. *Biology of the Naked Mole-Rat*. Princeton, NJ: Princeton University Press.

Smith, B.H. 1989. Dental development as a measure of life history in primates. *Evolution* 43: 683–688.

Trevathan, W. 1987. *Human Birth: An Evolutionary Perspective*. New York: Aldine de Gruyter.

Trinkhaus, E. 1987. On Neandertal pubic morphology and gestation length: reply to Greene and Sibley. *Current Anthropology* 28: 91.

Walker, E.P. 1975. *Mammals of the World* (3rd Ed.). Baltimore: The Johns Hopkins University Press.

West, M.J. and R.D. Alexander. 1963. Sub-social behavior in a burrowing cricket *Anurogryllus muticus* (De Geer). *Ohio Journal of Science* 63:19–24.

Williams, G.C. 1957. Pleiotropy, natural selection and the evolution of senescence. *Evolution* 11: 398–411.

Williams, G.C. 1966. *Adaptation and Natural Selection*. Princeton, NJ: Princeton University Press.

Zeveloff, S.I. and M.S. Boyce. 1980. Parental investment and mating systems in mammals. *Evolution* 24:973–982.

Zeveloff, S.I. and M.S. Boyce. 1982. Why human neonates are so altricial. *American Naturalist* 120:537–542.

Indirect Reciprocity

Reverse Reciprocity

One day at the urinal
He chanced to drop a dollar bill.

It wafted down, he sees it yet,
Lying where the floor is wet,

One end doubtful, the other dry,
At first he thought he'd pass it by.

But then some lowly cunning won,
He folded it and passed it on.

And now a doubt forever lingers
For every dollar bill he fingers.

                    Alexander, *2011, p. 226*

The Basis of Morality, Richard Alexander on Indirect Reciprocity
Karl Sigmund

In Richard Alexander's *The Biology of Moral Systems* (Alexander 1987; hereafter BMS), the concept of indirect reciprocity plays a starring role. The author states (p. 77) that "moral systems are systems of indirect reciprocity," and writes (p. 95) that "systems of indirect reciprocity become automatically what I am here calling moral systems." One chapter of the book is entitled "Moral systems as systems of reciprocity" (p. 93). In this context, Alexander views "moral systems" as guides of actions, or standards of conduct, and carefully separates them from the concept of "morality," which seems much harder to pin down.

When defining indirect reciprocity, Alexander contrasts it with the simpler concept of direct reciprocity. The latter occurs when the return from the social investment in another individual is expected from the actual recipient of the beneficence.... "In indirect reciprocity, the return is expected from someone other than the recipient of the beneficence. The return can come from essentially any individual or collection of individuals in the group" (p. 85). The concept of indirect reciprocity is also defined in *Darwinism and Human Affairs* (1979), where Alexander writes:

> Reciprocity can be divided into two types. Direct reciprocity occurs when rewards come from the actual recipient of beneficence. Indirect reciprocity, on the other hand, is represented by rewards from society at large, or from others than the actual recipient of beneficence. We engage in both kinds more or less continuously. (p. 49)

At the time when Alexander wrote these lines, in the late seventies, evolutionary biology was just beginning to acknowledge the importance of direct reciprocity. In particular, Robert Axelrod and William D. Hamilton were using computers to conduct their famous round-robin tournaments of iterated Prisoner's Dilemma games, and to analyze the merits of Tit for Tat, the epitome of reciprocation (Axelrod and Hamilton 1981, see also Axelrod 1984 and Hamilton 1996). The simplest version of a Prisoner's Dilemma game is obtained if two players can independently decide whether or not to confer a benefit b to their co-player, at cost c to themselves, with $0 < c < b$. The dominating strategy here is to defect, as this maximizes a player's payoff, no matter what the other player does. But if the game is repeated sufficiently often between the same two players, unconditional defection is not a good strategy against a Tit for Tat player, that is, a player who confers a benefit to the co-player in the first round and from then on does whatever the co-player did in the previous round. In particular, Axelrod and Hamilton found that in various computer simulations of repeated Prisoner's Dilemma games, selection led to the emergence of Tit for Tat, and hence to the evolution of cooperation. The work of Axelrod and Hamilton thus confirmed Robert Triver's seminal work (Trivers 1971), which had established reciprocity as the second pillar, next to kin selection, to support altruism in evolutionary biology. In the second edition of Richard Dawkins's *The Selfish Gene*, a chapter was added to celebrate the triumph of Tit for Tat: "Nice guys finish first" (Dawkins 1989).

Indirect reciprocity is considerably more subtle than direct reciprocity. The latter is based on the principle "I'll scratch your back if you scratch mine," whereas the former is based on "I'll scratch your back if you scratch someone else's" (Binmore 1994). The merits of this maxim seem more difficult to grasp. They certainly require some sophistication. In Alexander's words, "indirect reciprocity involves reputation and status, and results in everyone in a social group continually being assessed and reassessed by interactants, past and potential, on the basis of their interactions with others" (BMS p. 85). In another statement, "indirect reciprocity develops because interactions are repeated, or flow among a society's members, and because information about subsequent interactions can be gleaned from observing the reciprocal interactions of others" (BMS p. 77). It is this assessment of the actions of others (even if they are not directed at oneself) which is the basis of moral judgements. It was Richard Alexander who first emphasized the close connection between moral systems and reciprocity, in stressing that systems of reciprocity need not be restricted to dyads of repeatedly interacting individuals.

Every idea has its forerunners, and Alexander points out repeatedly that others before him have dealt with generalizations of reciprocity, which he views as "the binding cement of human social life" (BMS p. 111). Alexander was particularly influenced by Trivers (1971, cf. Trivers 1986 and Trivers 2006). Darwin had also anticipated the idea that assessments by others play a fundamental role in human cooperation. In the *Descent of Man*, Darwin (1871) wrote that (in contrast to other social animals such as bees or ants), "man's motive to give aid no longer consists solely of a blind instinctive impulse, but is largely influenced by the praise and blame of his fellow men." We are all acutely concerned with how we are judged by those around us.

Alexander thus squarely embraced the "misanthropic" tradition of many thinkers before him, who suspected that costly and seemingly altruistic acts often pay, in the long run, and therefore are not altruistic, even if actors themselves may think so. Their reward, or return, can come from various sources, either individuals, or collections of individuals, or even 'society at large'. Let us look at these alternatives in turn.

FROM DIRECT TO INDIRECT RECIPROCITY

Let us suppose first that the return is provided by individuals. In direct reciprocity, A helps B and B helps A. This yields a net benefit for both. For indirect reciprocity, Alexander mentions two mechanisms by way of example (BMS p. 81). One is of the form A helps B, B helps C, C helps A. Such cycles of helping were analyzed by Boyd and Richerson (1989), who found, however, that they were fragile and unlikely to occur, essentially because the "return" is so roundabout that the cycle can easily be broken. The other mechanism suggested by Alexander was the following: A helps B; C, observing, later helps A; A helps C. This means that A was rightly judged to be a reliable partner by C, who uses this information to engage with A in direct reciprocity. In this sense, indirect reciprocity acts as a kind of foreplay for direct reciprocity. The corresponding strategy for the repeated Prisoner's Dilemma was termed Observer Tit for Tat (Pollock and Dugatkin 1992): it only deviates from Tit for Tat in the first round, by refusing to help if the co-player, in the last interaction with some third party, has refused to help.

But actually, the system proposed by Alexander works even if direct reciprocity is explicitly excluded. This was shown by a series of models which assumed that no players would ever meet the same co-player again. The principle is: "A helps B; C, observing, later helps A," which is just as before, except that the appendix "A helps C," which presumes a second meeting between A and C, is omitted. The continuation in the modified version is implicit: "D, observing, later helps C," and so on.

It is worthwhile to explore the simplest models of this type (Nowak and Sigmund 1998a, b; Lotem et al. 1999; Nowak and Sigmund, 2005; Pacheco et al., 2006; Sigmund 2010). Thus, suppose that in a large population two players A and B meet randomly, and each can either provide some help to the other or refuse to help. (In an equivalent version, one player is randomly assigned the role of potential donor of the help, and the other the role of recipient.) If the same two players never meet again, then Tit for Tat strategy makes no sense. But a closely related variant of discriminating cooperation does. Players using this "reciprocating" strategy refuse to help those players who have previously refused to help someone else. In this way, beneficence is channeled toward those players who themselves engage in beneficence. If C observes that A helps B, then C will help A, even if this does not lead to repeated interactions between A and C. In a sense, this is a vicarious return: C returns the help in B's stead.

This model can be made more explicit in various ways. The conditional strategy clearly requires that players have some information about the past behavior of their co-players. Such information can be incomplete, and still lead to cooperation. It is enough, for instance, to require that (a) with a certain probability q, the potential donor knows whether the potential recipient has refused to help on some previous occasion, and (b) in the absence of information, the donor is willing to use the "benefit of doubt". If the probability q is larger than the cost-to-benefit ratio c/b, cooperation can be sustained in the population: exploiters who never provide help will rarely receive help, and do less well than the discriminating co-operators.

Even in this simple toy-model, it is of paramount importance that players can acquire sufficient information about other group members. Clearly, such "social scrutinizing" (to use Alexander's term) is facilitated if individuals have a good memory, and spend much of their time together. But it is likely that direct observation is not enough. In all human groupings, individuals exchange information, and communicate what they observe through gossip. Here, the unique language abilities of our species come into play. ("For direct reciprocity, you need a face; for indirect reciprocity, you need a name."[Haigh, personal communication], ca. 2005). Conversely, the need to exchange information about others may have been a strong, possibly even the major, selective force behind the emergence of the human language instinct (Dunbar 1996).

Both the ability to learn a language and the ability to learn a moral code seem restricted to the human species. Alexander does "not exclude the possibility that indirect reciprocity...will eventually be documented in some primates, social canines, felines, cetaceans and some others" (BMS p. 85). Only few instances of indirect reciprocation have been documented in nonhuman species so far (Bshary and Grutter 2006, Rutte and Taborsky 2007). On the other hand, a large number of economic experiments have shown that humans are highly prone to engage in indirect reciprocity, and that (a) players known to help others usually increase their chances in getting helped by third parties and (b) conversely the propensity to help frequently more than doubles if players know that their decision will be communicated. Help, in this context, is not an altruistic act, but an investment into the social capital of reputation (Wedekind and Milinski 2000, Wedekind and Braithwaite 2001, Seinen and Schram 2001, Milinski et al. 2001). This confirms Alexander's view that morality based on indirect reciprocity may be seen as self-serving, because it causes a sufficient number of persons to regard the actor as a good object of social investment (BMS p. 109).

The moral assessment of other group members can be captured, in its simplest form, by labeling them "good" or "bad," depending on whether they helped or not. Clearly, such a binary assessment leads to a picture in black and white which is much cruder than what we are used to in real life, but it suffices for a proof of principle. Moreover, it raises an intriguing issue. In the most rudimentary form of indirect reciprocity, the discriminating strategy, a cousin of Tit for Tat, extends help to those who are "good" and refuses help to those who are "bad." However, a discriminating player who refuses to help a "bad" player becomes "bad" in the eyes of all observers, and therefore less likely to be helped in turn. In order to maximize the help one receives from discriminators, it would be better to never refuse to help, and thus to stop discriminating. But in a group without discriminators, exploiters go unpunished, and will spread. Cooperation cannot be sustained in the long run.

A remedy coming immediately to mind is to assume that a refusal to help can be justified, if it is directed toward a recipient with a "bad" image (Nowak and Sigmund 1998a, Leimar and Hammerstein 2001, Panchanathan and Boyd 2003). Such a justified refusal ought therefore not to be labeled as "bad." This requires, however, that observers are aware of the reputation of the potential recipient. This, in turn, may require knowing the reputation of the recipient's previous recipient, and so on. It is questionable whether under normal conditions individuals have enough information about their group members, or are sufficiently proficient at coping with this information (Milinski et al. 2001). Moreover, we still have not described the assessment system completely. While it obviously should be good to refuse help to a "bad" player, it seems less clear whether giving help to a "bad" player should be considered as good or bad, for example.

These considerations lead us to consider assessment systems which are not only based on whether help is given or not, but also on the reputations of recipient and donor. There are no less than $2^{2 \times 2 \times 2} = 256$ of them, even under the absurdly oversimplified assumption that a player can only be "good" or "bad," without intermediate grades (Ohtsuki and Iwasa 2004, Brandt and Sigmund 2004). It turns out that only 8 of them are stable, in the sense that a population which adopts them will cooperate and cannot be invaded by unconditional strategies of always giving or always refusing help (Ohtsuki and Iwasa 2006). The competition of these rudimentary "moral systems" under conditions which include the possibility of occasional errors in action or judgement turns out to be remarkably difficult to analyze (Ohtsuki and Iwasa 2007, Uchida and Sigmund 2010, Sigmund 2010, Uchida 2010). Indeed, the status of a given group member will in general be different for observers using different assessment rules. If additionally, the assessment of a player is based on several actions of that player, or if there are more than two labels for a player's reputation (for instance, "good," "bad," and "indifferent"), the complexity of the moral system explodes. Formalizing ethics appears to be harder than formalizing logic. Practical philosophy defies mathematizing.

It is doubtful whether Alexander would view the investigation of formalized systems of moral assessment rules as useful or relevant for evolutionary biology, except as a means for quantifying verbal arguments. The delimitations required for these models risk throwing out the baby with the bathwater. Indeed, models of indirect reciprocity which isolate it from direct reciprocity on the one hand, and from group interactions on the other, are artificial devices, useful for thought experiments but far removed from reality. Alexander is more interested in the fluid, and ever-growing boundaries of systems of reciprocity, and the effect of this development on the human psyche. He suggests that "indirect reciprocity led to the evolution of ever keener abilities to observe and interpret situations with moral overtones" (BMS p. 100). "I regard indirect reciprocity as a consequence of direct reciprocity occurring in the presence of interested audiences—groups of individuals who continually evaluate the members of their society as possible future interactants from whom they would like to gain more than they lose" (BMS p. 93). In particular, "we use motivation and honesty in one circumstance to predict actions in others. . . . Humans tend to decide that a person is either moral or not, as opposed to being moral in one context and immoral in another" (BMS p. 94). And indeed, it is remarkable that ancient philosophers saw virtue as the attribute of a person, whereas contemporary philosophers speak of virtue as the attribute of an act, or a decision.

In most of Alexander's descriptions of indirect reciprocity, the role of reputation, and in particular the image of the potential recipients, is of central importance. However, there seem to exist mechanisms of indirect reciprocity which are not based on reputation. For instance, we could consider a situation where first A helps B, and then B helps a third party C. There are many examples of this. If someone holds the door open for us, we are likely to hold the door open for the next. This type of indirect reciprocity, where the recipient returns the help, but not to the actual donor, has also been observed in economic experiments and seems less easy to explain theoretically.

Moreover, negative interactions may also be reciprocated. Alexander usually speaks of "rewards" being returned, but retaliation clearly is also a widespread form of reciprocation, and he mentions it, for instance in a table describing various forms of indirect reciprocity (BMS p. 86), or by stating that "systems of indirect reciprocity involve promises of punishment as well as reward" (BMS p. 96). In the last two decades, experimental economists have uncovered a widespread propensity to punish those who are perceived as cheaters. So-called peer-punishment is very effective at promoting cooperation (Fehr and Gächter 2002). Interestingly, many individuals are ready to incur personal costs to punish norm breakers, even if they themselves were not affected by the misdeameanour, but merely observed it. This is certainly also a form of indirect reciprocity.

INDIRECT RECIPROCITY AND GROUP COMPETITION

So far, we have considered reciprocity based on returns by individuals. Alexander stressed on several occasions that such a return could also be provided by collections of individuals, as when he writes "indirect reciprocity, whereby society as a whole or a large part of it provides the reward for the beneficence" (BMS p. 105). Societies provide not only rewards for beneficence, but also punishment for free-riding. In fact, most punishment is not meted out by irate individuals in the form of peer-punishment, but rather by institutions (Ostrom 1990; Yamagishi 1986; Sigmund et al. 2010). Institutions can be viewed as tools enabling communities to provide positive or negative incentives. There is a striking similarity between institutionalized punishment of free-riding and the repression of competition encountered in many examples of cooperative groupings encountered in biological evolution (Frank, 1995).

In a particularly interesting aside on indirect reciprocity, Alexander mentions that the reward which an individual obtains for acts of beneficence can simply consist of increased success for the group (BMS p. 94). This observation relates to an aspect which is crucial for the role played by moral systems, in Alexander's view.

Indeed, following Keith (1947), who stressed the competition between groups as a main factor shaping human evolution, Alexander sees "morality as a within-group cooperativeness in the context of between-group competition" (BMS p. 153). According to his celebrated "balance of power" argument, the often lethal competition between families, bands, tribes, and nation was the chief selective force shaping human evolution. "In no other species do social groups have as their main jeopardy other social groups of the same species" (BMS p. 77).

Alexander accordingly writes (BMS p. 194) that "indirect reciprocity is more complex than is usually realized partly because of long-term benefits from being viewed as an altruist, and partly because one must take into account benefits to the individual that accrue from the success of his group in competition with other groups." This latter aspect is not, in general, associated with "indirect reciprocity" nowadays. The former aspect has monopolized the meaning. The "long-term benefits for being viewed as an altruist" help to establish generalized exchange systems working to mutual advantage, even in the absence of strife between groups.

The role of group selection has been hotly debated by evolutionary biologists during the last half-century. Mathematical models often use alternative expressions, such as kin selection, or multi-level selection, sometimes with the intention of avoiding semantic quarrels, and usually with the result of exacerbating them. But there seems no reason to avoid the name "group selection" when one speaks of groups fighting and annihilating each other. Individuals have to balance, in such contests, their well-being within the group with the well-being of their group.

This viewpoint was shared by Darwin, who wrote: "There can be no doubt that a tribe including many members who…were always ready to give aid to each other and to sacrifice themselves for the common good, would be victorious over most other tribes; and this would be natural selection" (p. 166). He did not say: "and this is group selection," but he obviously was aware of the tension between individual and group selection when he wrote "He who was ready to sacrifice his life…would often leave no offspring to inherit his noble nature. Therefore it seems scarcely possible (bearing in mind that we are not here speaking of one tribe being victorious over another) that the number of men gifted with such virtues could be increased through natural selection" (page 163). The term in parentheses clearly indicates that Darwin saw no way of explaining the evolution of such self-sacrificing traits other than by violent intergroup conflict. In another passage, Darwin stressed that "extinction follows chiefly from the competition of tribe with tribe, and race with race" (page 186). To many ears, today, this sounds politically incorrect. But Darwin was very conscious of this dark side to human nature. Some of the most remarkable examples of human cooperation occur in war, and other forms of lethal conflict between groups, and it is well-known that day-to-day solidarity dramatically increases in societies threatened from the outside.

That group selection can favor the emergence of cooperative traits has been shown in countless models. It is all the more remarkable that the models and experiments on indirect reciprocity which have been mentioned so far assume a single, well-mixed population. They show that indirect reciprocity can work even if the population is not structured into competing groups. Individuals who deviate from the cooperative norm will have, on average, their long-term payoff reduced, and thus are unlikely to be copied by others. This confirms Alexander's view that it is fallacious to assume that morality inevitably involves some self-sacrifice (BMS p. 161). However, it confirms it in a setup which is different from Alexander's.

A final remark: Economists have also been investigating extensions of the concept of reciprocity, in parallel to evolutionary biologists. Their point of departure was usually the so-called folk theorem on repeated games, which states that if the probability of another round between the same two players is sufficiently high, cooperation can be sustained by so-called trigger strategies, of which Tit for Tat is but one example. A rational player would forgo the exploitation of a co-player in one round if this jeopardizes collaboration in all future rounds. It is clear that if players interact only once, a cheater cannot be held to account by its victim, and therefore personal enforcement must be replaced by community enforcement. Game theorists have shown that even if information is transmitted only imperfectly, cooperation can be sustained, based on trigger strategies adopted by the whole community. No rational player has an interest in deviating unilaterally (Rosenthal 1979, Sugden 1986, Kandori 1992, Ellison 1994, Okuno-Fujiwara and Postlewaite 1995, Bolton et al. 2004a). The interest of economists in indirect reciprocity has been singularly heightened in recent years by the fact that one-shot interactions between anonymous partners become increasingly frequent in today's society. Web-based auctions and other forms of e-commerce occur between strangers who never meet face-to-face. They are built on rudimentary reputation mechanisms similar to those which were developed in the simplest formal models capturing Richard Alexander's ideas on indirect reciprocity (Bolton et al. 2004b, Keser 2002).

Richard Alexander recognized indirect reciprocity based on reputation and status as major factor in the emergence of moral systems in human societies. It is amazing that the same simple, robust mechanisms that shaped early hominid societies now play a central role in shaping the internet civilisation.

References

Alexander, R.D. 1979. *Darwinism and Human Affairs*. Seattle: University of Washington Press.

Alexander, R.D. 1987. *The Biology of Moral Systems*. New York: Aldine de Gruyter.

Alexander, R.D. 2011. *The Mockingbird's River Song: Poems, Essays, Songs and Stories, 1946–2011*. Manchester, MI: Woodlane Farm Books.

Axelrod, R. 1984. *The Evolution of Cooperation*: New York: Basic Books, New York (reprinted 1989 in Penguin, Harmondsworth).

Axelrod, R, and Hamilton, WD. 1981. The evolution of cooperation. *Science* 211:1390–1396.

Binmore, K. 1994. *Playing Fair: Game Theory and the Social Contract*. Cambridge: MIT Press.

Bolton, G., Katok, E., Ockenfels, A. 2004a. Cooperation among strangers with limited information about reputation. *J. Pub. Econ.* 89:1457–1468.

Bolton, G., Katok, E., and Ockenfels, A. 2004b. How effective are online reputation mechanisms? An experimental investigation. *Manag. Sci*, 50:1587–1602.

Boyd, R., and Richerson, P.J. 1989. The evolution of indirect reciprocity. *Social Networks* 11:213–236.

Brandt, H., and Sigmund, K. 2004. The logic of reprobation: assessment and action rules for indirect reciprocity. *J. Theoret. Biol.* 231:475–486.

Bshary, R., and Grutter, A.S. 2006. Image scoring causes cooperation in a cleaning mutualism. *Nature* 441:975–978.

Darwin, C.R. 1871. *The Descent of Man and Selection in Relation to Sex*. London: John Murray.

Dawkins, R. 1989. *The Selfish Gene*, 2nd edition. Oxford: Oxford University Press.

Dunbar, R. 1996. *Grooming, Gossip and the Evolution of Language*. Cambridge: Harvard University Press.

Ellison, G. 1994. Cooperation in the Prisoner's Dilemma with anonymous random matching. *Rev. Econ. Stud.* 61:567–588.

Fehr, E., and Gächter, S. 2002. Altruistic punishment in humans. *Nature* 425:785–791.

Frank, S.A. 1995. Mutual policing and the repression of competition in the evolution of cooperative groups. *Nature* 377:520–522.

Hamilton, W.D. 1996. *Narrow Roads of Gene Land*, Vol I. New York: Freeman.

Kandori, M. 1992. Social norms and community enforcement, *Rev. Econ. Stud.* 59:63–80.

Keith, A. 1947. *Evolution and Ethics*. New York: Putnam and Sons.

Keser, C. 2002. Trust and Reputation Building in e-Commerce. *IBM Sys. J.* 42:498–506.

Leimar, O., and Hammerstein, P. 2001. Evolution of cooperation through indirect reciprocation. *Proc. R. Soc. Lond. B*, 268:745–753.

Lotem, A., Fishman, M.A., and Stone, L. 1999. Evolution of cooperation between individuals. *Nature* 400: 226–227.

Milinski, M., Semmann, D., Bakker, T.C.M., and Krambeck, H.J. 2001. Cooperation through indirect reciprocity: image scoring or standing strategy? *Proc. R. Soc. Lond. B* 268:2495–2501.

Nowak, M.A., and Sigmund, K. 1998a. Evolution of indirect reciprocity by image scoring. *Nature* 282:462–466.

Nowak, M.A., and Sigmund, K. 1998b. The dynamics of indirect reciprocity. *J. Theoret. Biol.* 194:561–574.

Nowak, M.A., and Sigmund, K. 2005. Evolution of indirect reciprocity. *Nature* 437:1291–1298.

Ohtsuki, H., and Iwasa, Y. 2004. How should we define goodness? Reputation dynamics in indirect reciprocity. *J. Theoret. Biol.* 231:107–120.

Ohtsuki, H., and Iwasa, Y. 2006. The leading eight: social norms that can maintain cooperation by indirect reciprocity. *J. Theoret. Biol.* 239:435–444.

Ohtsuki, H., and Iwasa, Y. 2007. Global analyses of evolutionary dynamics and exhaustive search for social norms that maintain cooperation by reputation. *J. Theoret. Biol.* 244:518–531.

Okuno-Fujiwara, M., and Postlewaite, A. 1995. Social norms in matching games. *Games Econ. Behav.* 9:79–109.

Ostrom, E. 1990. *Governing the Commons.* Cambridge: Cambridge University Press.

Pacheco, J., Santos, F., and Chalub, F. 2006. Stern-judging: a simple, successful norm which promotes cooperation under indirect reciprocity. *PLOS Compu. Biol.* 2:e178.

Panchanathan, K., and Boyd, R. 2003. A tale of two defectors: the importance of standing for evolution of indirect reciprocity. *J. Theoret. Biol.* 224:115–126.

Pollock, G.B., and Dugatkin, L.A. 1992. Reciprocity and the evolution of reputation. *J. Theoret. Biol.* 159:25–37.

Rosenthal, R.W. 1979. Sequences of games with varying opponents. *Econometrica* 47:1353–1366.

Rutte, C., and Taborsky, M. 2007. Generalized reciprocity in rats. *PLoS Biology* 5:1421–1425.

Seinen, I., and Schram, A. 2001. Social status and group norms: indirect reciprocity in a helping experiment. *Euro. Econ. Rev.* 50: 581–602.

Sigmund, K. 2010. *The Calculus of Selfishness.* Princeton, NJ: PrincetonUniversity Press.

Sigmund, K., De Silva, H. Traulsen, A., and Hauert, C. 2010. Social learning promotes institutions for governing the commons. *Nature* 466:861–863.

Sugden, R. 1986. *The Economics of Rights, Cooperation and Welfare.* Oxford: Basil Blackwell.

Trivers, R. 1971. The evolution of reciprocal altruism. *Q. Rev. Biol.* 46:35–57.

Trivers, R. 1986. *Social Evolution.* Menlo Park, CA: Benjamin Cummings.

Trivers, R. 2006. Reciprocal altruism: 30 years later. In: *Cooperation in Primates and Humans: Mechanisms and Evolution*, P.M. Kappeller and C.P. van Schaik (eds.). Berlin, Springer, pp. 67–83.

Uchida, S. 2010. Effect of private information on indirect reciprocity. *Phys. Rev. E* 82:doi 10.1103/PhysRevE.82.036111.

Uchida, S., and Sigmund, K. 2010. The competition of assessment rules for indirect reciprocity. *J. Theoret. Biol.* 263: 13–19.

Wedekind, C., and Milinski, M. 2000. Cooperation through image scoring in humans. *Science* 288:850–852.

Wedekind, C. and Braithwaite, V.A. 2002. The long-term benefits of human generosity in indirect reciprocity. *Curr. Biol.* 12:1012–1015.

Yamagishi, T. 1986. The provision of a sanctioning system a a public good. *J. Pers. Soc. Psychol.* 51:110–116.

CLASSIFYING HUMAN EFFORT

The "Atoms" of Sociality

Excerpt from Alexander, R.D. 1986. *The Biology of Moral Systems*. New York, Aldine Press.

If ethical, moral, and legal systems are ultimately understandable in evolutionary terms, then we ought to be able to explain them eventually, at even the most complex and synthetic levels, by beginning with the "atoms" of sociality proposed out of the life history and effort theories of modern evolutionary biology (see table 9.1). We can note, first, that lifetimes are divisible into somatic and reproductive effort, and, second, that reproductive effort can be subdivided, at least in humans, into mating effort (on behalf of gametes), parental effort (on behalf of offspring), and extraparental nepotistic effort (on behalf of collateral or non-descendant relatives and descendants other than offspring). Next it is useful to emphasize that humans are unusual among organisms in that all of their life effort that is *social* in nature is permeated with *reciprocity*. The possibility of mutually beneficial reciprocal interactions can cause both somatic and reproductive effort to be *socially mediated* (nepotism, for example, can involve reciprocity in which the return benefit goes to a relative of the originally beneficent individual). In turn, reciprocity itself can be direct (A helps B, B helps A) or indirect (A helps B, B helps C, C helps A. Or A helps B; C, observing, later helps A; A helps C). Reciprocity is probably never complete, or balanced. (For further explanation, see below and fig. 9.1; tables 9.1–9.5.) The effects of reciprocity and other forms of social mediation (competition, pseudoreciprocity) multiply the number of different "atoms" of sociality that must be elucidated if a biological approach to human social behavior is to be integrated with the approaches of philosophers and social and political scientists.

   1. Direct Somatic Effort. In this category of phenotypically selfish behavior are included those aspects of somatic effort (i.e., directed toward growth, development, and maintenance of one's own phenotype or soma) that explicitly do not involve benefits routed through other individuals. Direct somatic effort is mediated by neither nepotism nor reciprocity. Examples are eating, drinking, seeking shelter, and avoiding danger, when these actions are carried out without the assistance or positive intervention of others. This category of behavior, which probably corresponds most closely to the moral philosophers' *egoism* (particularly when the latter specifically involves seeking to promote one's own welfare *rather* than

TABLE 9.1 } Kinds of Effort and Their Outcomes[1]

"Atoms" of Sociality	Kinds of Effort	Phenotypically	Genotypically
	I. Somatic Effort	Selfish	Selfish
	A. Direct		
1.	1. Immediate payback		
2.	2. Delayed payback		
	B. Indirect		
	3. Via indirect reciprocity		
3.	a. Immediate payback		
4.	b. Delayed payback		
	4. Via indirect reciprocity		
5.	a. Immediate payback		
6.	b. Delayed payback		
	II. Reproductive effort	Altruistic	Selfish
	A. Mating effort		
7.	1. Directly nepotistic (no social mediation)		
	2. Indirectly nepotistic (social mediation)		
	a. Via direct reciprocity		
8.	(1) Immediate payback		
9.	(2) Delayed payback		
	b. Via indirect reciprocity		
10.	(1) Immediate payback		
11.	(2) Delayed payback		
	B. Parental effort		
12.	1. Directly nepotistic (no social mediation)		
	2. Indirectly nepotistic (social mediation)		
	a. Via direct reciprocity		
13.	(1) Immediate payback		
14.	(2) Delayed payback		
	b. Via indirect reciprocity		
15.	(1) Immediate payback		
16.	1. Directly nepotistic (no social mediation)		
	(2) Delayed payback		
17.	C. Extraparental effort		
	2. Indirectly nepotistic (social mediation)		
	a. Via direct reciprocity		
18.	(1) Immediate payback		
19.	(2) Delayed payback		
	b. Via indirect reciprocity		

Continued

TABLE 9.1 } *Continued*

"Atoms" of Sociality	Kinds of Effort	Phenotypically	Genotypically
20.	(1) Immediate payback		
21.	(2) Delayed payback		

¹ Direct somatic effort refers to self-help that involves no other persons. Indirect somatic effort involves reciprocity, which may be direct or indirect. Returns from direct or indirect reciprocity may be immediate or delayed. Reciprocity can be indirect for two different reasons, or in two different ways. First, returns (payment) for a social investment (positive or negative) can come from someone other than the recipient of the investment, and second, returns can go either to the original investor or to a relative or friend of the original investor.

someone else's), is not as common as might be thought in today's specialized social environment. We can actually do very little for ourselves without some kind of assistance, however indirect, by others.

2. *Indirect (Socially Mediated) Somatic Effort.* This kind of phenotypically selfish behavior is routed through other individuals. Thus it may involve social investments (initial costs, beneficent acts) in *direct and indirect reciprocity.* These investments are expected eventually to be repaid to one's self (as opposed to one's relatives, which would identify them as indirect nepotism). Examples are purchases of goods to be used by one's self (which may represent reciprocity in the form of mutually beneficial exchanges of resources between buyer and seller) and all social and other benefits given to others as investments in reciprocity when the returns are realized through assistance to one's self. There are two kinds of mediation of somatic effort: (a) direct assistance to Ego (i) by a relative (repayment to the relative's phenotype is likely not necessarily to be expected or required) and (ii) by a nonrelative (eventual repayment is likely to be expected, to either the benevolent individual or its relatives) and (b) assistance to Ego as a return on an investment made by Ego as a part of somatic effort (i.e., direct reciprocity, such as, Ego helps a nonrelative, expecting return assistance from the helped person at some later

TABLE 9.2 } Indirect Reciprocity

Rewards (why altruism spreads)
1. A helps B
2. B helps A
3. C, observing, helps B, expecting that
4. B will also help (or overhelp) C
(ETC.)
Or
1. A helps B
2. B does not help A
3. C, observing, does not help B expecting that, if he does
4. B will not return the help
(ETC.)

Table 9.3 } Indirect Reciprocity

Punishment (why rules spread)
1. A hurts B
2. C, observing, punishes A, expecting that if he does not,
3. A will also hurt C
Or
4. Someone else, also observing, will hurt C, expecting no cost (ETC.)

date). We can predict that infants and very young children, especially, are evolved to exhibit effort that elicits the (a) *(i)* kind of assistance above. What Trivers (1971) called "reciprocal altruism" would include both the nepotistic and somatic (egoistic) effects of investments in both direct and indirect reciprocity (thus, while *beneficent,* or initially costly, it would be only temporarily phenotypically altruistic and not *genetically* altruistic or costly at all).

Although also egoistic in its consequences, socially mediated somatic effort that involves initial costly investments (beneficence) may often be misinterpreted as altruistic (that is, as *genetically* costly). Because of divisions of labor and social interdependency in the modern world, this kind of egoism is also more commonplace than direct somatic effort.

3. Direct Nepotism. Included here are all investments in relatives for which the return may be expected in genetic terms, through the reproduction of the assisted relatives.

Mating effort is the most problematic form of effort placed here: it may be viewed as selfish (hence, be confused with somatic effort), particularly in males, who usually, as in mammals, strive to place their sperm inside the body of the female, hence are in a better position to abandon the offspring to its mother's care than vice versa. A gamete, after all, is not an individual, and it possesses no genetic materials other than those of its producer. But gametes are also not merely parts of the phenotypes that produce them, and (except in species with haploid males) are not genetically identical to their producers. Mating effort, moreover, is reproductive effort that involves risks and expenditure of calories. Effort exerted on behalf of gametes lowers the reproductive value of the individual as surely as that exerted on behalf of offspring or other relatives. Perhaps the example is confusing

TABLE 9.4 } Indirect Reciprocity

Deception (why cheating spreads)
1. A_1 makes it look as though he helps B
2. C_1 helps A_1, expecting that A_1 will also help him
3. C_2 observes more keenly and detects A_1's cheating and does not help him (avoids or punishes him)
4. A_2, better at cheating, fools C_2
5. C_3 detects A_2's cheating (ETC.)
$C_1 \rightarrow C_2 \rightarrow C_3$; $A_1 \rightarrow A_2 \rightarrow A_3$ Either learning or evolution (or both)

Motivations

Outcomes of Social Acts	Doesn't know or think about what he is doing		Believes he is selfish and expects to win because of selfishness		Does these things deliberately				
	Considered to be insane or incompetent (i.e, *can-not* know - includes non-humans)	Considered to be lazy or thoughtless - does these things without thinking (could know but doesn't)	Sees his way of life as satisfying; acts this way because he enjoys it	Sees his life as a burden (as compared to other lives possible)	Believes he is altruistic and expects to win *because* of altruism — Sees his way of life as either satisfying or as a burden	Believes he is altruistic but expects to win *despite* altruism			
						Expects reward on Earth		Expects reward in Heaven	
						Satisfying	Burden	Satisfying	Burden
Helps only self	1 Neutral (e.g. baby)	2 Immoral	3 Immoral	4 Unlikely	5 Immoral	6 Immoral	7 Unlikely	8 Unlikely	9 Unlikely
Helps only self and relatives	10 Neutral	11 Immoral	12 Immoral	13 Unlikely	14 Prob[b]	15 Prob[b]	16 Prob[b]	17 Prob[b]	18 Prob[b]
Helps self, relatives, and friends who are likely to reciprocate with interest	19 Neutral[d]	20 Immoral?	21 Immoral	22 Unlikely	23 Prob[c]	24 Prob[c]	25 Prob[c]	26 Prob[c]	27 Prob[c]
Helps self, relatives, reciprocating friends, and others in the presence of potential reciprocators	28 Neutral[d]	29 Moral?[d]	30 Immoral	31 Unlikely	32 Prob[c]	33 Prob[c]	34 Prob[c]	35 Prob[c]	36 Prob[c]
Helps all of the above, and also helps strangers when it is not too costly, even when not in the presence of reciprocators	37 Neutral[d]	38 Moral	39 Unlikely?	40 Unlikely	41 Moral	42 Moral?	43 Moral?	44 Moral	45 Moral
Helps anyone who needs it even if the immediate cost is great	46 Neutral[d]	47 Moral	48 Unlikely	49 Unlikely	50 Moral	51 Moral?	52 Moral?	53 Moral	54 Moral
Helps others indiscriminately while maintaining self at approximately the lowest level consistent with doing this effectively	55 Neutral[d]	56 Moral	57 Unlikely	58 Unlikely	59 Moral	60 Moral?	61 Moral?	62 Moral (Saint)	63 Moral (Saint)

[a] I have speculated as to how each category of act is likely to be judged. Problematic cases illustrate the difficulty of deciding questions of morality when self-interest is broadened to include reproductive (genetic) interests, and when motivation comes to include realizations of the nature of such interests. Squares 25–32 and 41–48 would probably be marked "moral" by those unaware of biological considerations because they seem to involve self-sacrifice. An evolutionary biologist might regard all squares as representing possible behaviors, and as all possibly representing self-interested behaviors, but he might also regard squares 55–63 as less likely than would nonbiologists. Biologists would also be more likely to search for ways in which squares 55–63 could represent behaviors that serve the actor's interests.

[b] Behaviors that will probably be seen by most as immoral because of outcomes and despite motivations. Prob, problematic.

[c] Behaviors that will probably be seen by most as moral because of outcome and motivation combined.

[d] Desirable behavior even if morally neutral.

primarily because mating effort involves interactions of two individuals, and there is a tendency to compare the *relative* "selfishness" of the two. Also, in mating effort, as compared to other reproductive effort, the *genetically selfish* aspects of acts may be relatively more apparent than their *phenotypically altruistic* aspects. We are more likely to regard a male mammal's effort to place his gametes in a warm, safe place where they can fertilize an egg as an act of reproductive selfishness that benefits him in relation to the female involved than we are to see it as an act of phenotypic altruism benefiting his gametes.

4. *Indirect (Socially Mediated) Nepotism.* This category of reproductive effort includes investments in reciprocity in which returns from one's beneficence may reasonably be expected to be realized by relatives rather than one's self (e.g., heroism or good will created by one's benevolent acts may cause benefits to accrue to one's family).

Although the last two categories of behavior are not easily understood as a part of the moral philosophers' category of "egoism," neither are they either indiscriminately or genetically altruistic or utilitarian; if carried out appropriately in evolutionary terms, all of the above four kinds of behavior are *genetically selfish,* even those which are *phenotypically self-sacrificing* or *(temporarily) altruistic* (or *beneficent)* (Alexander, 1974, 1979a).

5. *Reciprocity (Direct and Indirect).* It appears to me that all reciprocity so far documented in nonhuman organisms is appropriately termed *direct reciprocity,* in which the return from a social investment in another (i.e., an act of "temporary" altruism) is expected from the actual recipient of the beneficence, although not necessarily in the same currency (Trivers,1971; Axelrod, 1984). In *indirect reciprocity* (Alexander 1977a, b, 1979a, 1982, 1985b), the return is expected from someone other than the recipient of the beneficence (tables 9.2–9.4). This return may come from essentially any individual or collection of individuals in the group. Indirect reciprocity involves reputation and status, and results in everyone in a social group continually being assessed and reassessed by the interactants, past and potential, on the basis of their interactions with others. I do not exclude the possibility that indirect reciprocity, in this sense, will eventually be documented in some primates—especially chimpanzees (e.g., de Waal, 1982, 1986) social canines, felines, cetaceans, and some others.

What I am calling "indirect reciprocity" Trivers (1971) referred to as "generalized reciprocity." I avoided the latter term because of the way Sahlins (1965) used it—cf. Alexander (1975, 1979a, 1985). Sahlins typified generalized reciprocity as involving one-way flows of benefits in which the expectation of return is vague or nonexistent. He included nepotism, citing the case of a mother nursing her child, and with respect to nonrelatives seemed to be referring to what we would now call genetic or reproductive altruism. Perhaps both terms will survive: Indirect reciprocity for cases in which the return explicitly comes from someone other than the recipient of the original beneficence, and generalized reciprocity for social systems in which indirect reciprocity has become complex and general.

Reciprocity in nonhuman organisms (Trivers, 1971; Axelrod and Hamilton, 1981) may have arisen between bonded males and females in species with both maternal and paternal care. In such cases it may have developed out of mating effort (e.g., a male gives a gift to a female and is allowed to copulate), subsequently as indirect nepotism through effects on offspring produced jointly by the pair (e.g., a female copulates with a male and he gives her a gift which she uses to rear his offspring). Alternatively (and perhaps more likely), reciprocity may have arisen as a modification of direct nepotism in which flows of benefits began to pass in both directions successively rather than just one direction from one relative to another. Benefits can be returned to a social investor as a part of the egoistic behavior of the individual returning it—i.e., at no cost to that individual. In such instances, possibilities for cheating are essentially nonexistent and the overall complexity of the mental activities accompanying evolution of the act will be reduced in comparison to reciprocity per se. Connor (1986) calls such interactions "pseudoreciprocity." In terms of my examples, it would be pseudoreciprocity if the female receiving a gift from a courting male copulated with him strictly because it was to her immediate advantage. If the act was at least temporarily costly—e.g., if she was initiating a longer-term exchange of beneficence—then the step into direct reciprocity would have been taken. I agree with Connor that most discussions of "reciprocity" in nonhuman species probably involve pseudoreciprocity instead.

It will be seen that subdividing somatic effort into direct and indirect—and reproductive effort into mating effort, parental effort, and extraparental nepotistic effort—creates five major "atoms" of sociality. The overlays of reciprocity and pseudoreciprocity, with the possibility of immediate or delayed returns, brings the number of such "atoms" to 21 (table 9.1). Further complicating the picture are additional overlays involving consciousness and deliberateness in different kinds of acts (table 9.5).

Identifying atoms or units of sociality for the purpose of understanding the flow of human social interactions seems to be a matter of locating units or transactions that can substitute for one another in the functioning of nepotism or reciprocity. I mean to suggest units for which cost-benefit analyses can be accomplished more or less independently, units that can be regarded as alternatives or options for the purpose of compensating beneficence or cheating. I believe that without very large numbers of such units, and large numbers of alternative players, reciprocity could not become complex. Human sociality seems composed of countless such units, and I think we are constantly separating the flow of our social interactions into units (acts, interactions, transactions) that we can use for our own purposes (e.g., investments intended to test the readiness of another to repay social debts with interest).

In light of the biological separation of lifetimes into somatic and reproductive effort, it is curious that moral philosophers' views of moral behavior usually require either 100% selfishness or 100% altruism but scarcely ever combinations of the two; with a few exceptions (e.g., Whiteley, 1976; MacIntyre,

1981b), philosophers seem to find it impossible to combine the two. The reason seems to be the view that consistency is required in moral behavior (i.e., to be moral one must advocate for himself only those rights and privileges he will advocate equally strongly for all others), and a dual human nature (i.e., involving both egoistic and altruistic tendencies or acts) has inconsistency built into it. Whiteley (1976) and MacIntyre (1981b) believe it is possible to be consistently (morally) egoistic, but I find it difficult to imagine that a true egoist would be likely to advocate the right of others to resources that egoism would require him to seek for himself. One has to presuppose that resource seeking does not involve conflict and ignore the argument that success is relative. The only time that utilitarianism (promoting the greatest good to the greatest number) is predicted by evolutionary theory is when the interests of the group (the "greatest number") and the individual coincide, and in such cases utilitarianism is not really altruistic in either the biologists' or the philosophers' sense of the term. It seems more likely that restraints on individuals and subgroups serving their own interests occur *solely* because of the likelihood of prohibitive costs being imposed by some part of the rest of society; this is precisely the definition of moral systems I am developing here.

Moral philosophers have not treated the beneficence of humans as a part, somehow, of their selfishness; yet, as Trivers (1971) suggested, the biologist's view of lifetimes leads directly to this argument. In other words, the normally expressed beneficence, or altruism, of parenthood and nepotism and the temporary altruism (or social investment) of reciprocity are expected to *result* in greater (genetic) returns than their alternatives.

If biologists are correct, all that philosophers refer to as altruistic or utilitarian behavior by individuals will *actually* represent either the temporary altruism (phenotypic beneficence or social investment) of indirect somatic effort or direct and indirect nepotism. The exceptions are what might be called evolutionary mistakes or accidents that *result* in unreciprocated or "genetic" altruism, deleterious to both the phenotype and the genotype of the altruist; such mistakes can occur in all of the above categories (see also Alexander, 1979a, table 1, for a discussion of genetic and phenotypic altruism). Part of our analysis (p. 100ff.) will involve the effects of certain kinds of indirect somatic effort and nepotistic altruism on the numbers and significance of such accidents by others. The question involved is whether or not we are *evolved* to promote such mistakes in others, and to resist them in ourselves, and the effects of any such tendencies on the nature of our moral and legal systems.

PHILOSOPHY AND CONFLICTS OF INTEREST

Among social and political scientists, moral philosophers, ethicists, and others who study and think about moral questions, probably everyone would agree that conflicts of interest, and the human attitudes, tendencies, and actions that derive from histories of conflicts of interest, are alone responsible for ethical, moral, and legal questions. On the other hand, not everyone who writes in this arena headlines his discussions with the question of conflicts of interests. I think of such recent and influential volumes as John Rawls' (1971) *A Theory of Justice,* Richard Brandt's (1979) *A Theory of the Good and the Right,* William Frankena's (1973, 1980) *Ethics and Thinking About Morality,* and Lawrence Kohlberg's (1981) *The Philosophy of Moral Development.* One cannot find general discussions of the nature of interests or the quantification of their conflicts; or do phrases like "conflict of interest," "differences of opinion," or even "disagreements" and "interests" appear in either the tables of contents or the indexes. Similarly, if one keeps the question of conflicts of interest, at both individual and various group levels, in mind while reading the essays of Richards (1986b) and the responses to them, I believe some of the seemingly most difficult questions are simplified.

The general theory of interest, called for by Pound (quoted pp. 33–34; see also Pound, 1959) has not been developed, and is discussed by few other authors. Although everyone who writes in this arena may recognize that disagreements and conflicts about interests are what underlie moral systems, for some reason few have been compelled to generalize about them or dwell on their bases.

It is my impression that many moral philosophers do not approach the problem of morality and ethics as if it arose as an effort to resolve conflicts of interests. Their involvement in conflicts of interest seems to come about obliquely through discussions of individuals' views with respect to moral behavior, or their proximate feelings about morality—almost as if questions about conflicts of interest arise only because we operate under moral systems, rather than vice versa.

An excellent example of a philosophical discussion that does not directly confront the question of whether the whole flow of social interactions depends on conflicts and confluences of interest is that of Callahan (1985). Writing on "What Do Children Owe Elderly Parents?" Callahan alludes to interests driving the interaction only twice. On p. 32 he notes "As a piece of practical advice, however, it [honoring fathers and mothers] once made considerable sense. In most traditional and agricultural societies, parents had considerable power over the lives of their offspring. Children who did not honor their parents risked not only immediate privation, but also the loss of the one inheritance [land] that would enable them to raise and support their own families." On p. 36, speaking of "The poor," he comments "...adults with elderly parents ought not to be put in the position of trying to balance the moral claims of their own children against those of their parents, or jeopardizing their own old age in order to sustain their parents in their old age. Though such conflicts may at times be unescapable, society ought to be structured in a way that minimizes them." It is as if Callahan did not consciously consider whether or not conflicts of interest were the central issue of the topic he was discussing.

Most of Callahan's article attempts to resolve the problem of care for the elderly in contemporary society by wrestling from every possible direction with proximate mechanisms, such as the feelings of children toward their parents, of parents toward their children, and of "society" toward both. This exercise ends inconclusively, as if presaged by his statement (p. 35). "I am searching here with some difficulty for a way to characterize the ethical nature of the parent-child relationship, a relationship that appears almost but not quite self-evident in its reciprocal moral claims and yet oddly elusive also."

The trouble is, there is no one "parent-child relationship" and the whole problem of care for the aged has arisen, not so much because lives have been prolonged, but because familial bonds have been fractured, explicitly in ways that have made confluences of interest between aged parents and their children much less likely. Self-interested reasons for children to care for aged parents have all but disappeared, and I suggest that this is why their care is being thrust upon "society" and why it has become so expensive. This expense to society is, in turn, why "society" has become concerned to discuss the question of children's moral and ethical obligations to their parents. I see no likelihood that such questions can ever be elucidated by confining the arguments to proximate mechanisms. Callahan does conclude that "A minimal duty of any government should be to do nothing to hinder, and if possible do something to protect, the natural moral and filial ties that give families their power to nurture and Sustain." He does not delve deeply into the basis for this (or any other) "should," however. Somehow he also realizes that "To exploit that bond by coercively taxing families is, I believe, to threaten them with great harm...It...presupposes a narrower form of moral obligation...than can naturally be defended...[and] promises to rupture those more delicate moral bonds ... that sustain parents and children in their lives together." I think that what Callahan is describing here are the probable repercussions of forcing children to care for their parents, repercussions that will exacerbate the overall problem and cause even greater expense to "society." I am not suggesting that intensive analyses of the feelings (and other proximate mechanisms) that underlie moral dilemmas are not useful; I think it is obvious they are essential. I am, rather, saying that cases like the relationships between parents and children in different societal milieus show that the underlying conflicts of interest drive the proximate mechanisms, at least as significantly as the reverse. To try to understand one without the other seems futile.

In contrast to Callahan's discussion, Strong (1984) develops an excellent and well-documented discussion of conflicts of interest among neonates, parents, and physicians, which I believe shows well how thinking directly in terms of conflicts and confluences of interest of the involved parties can lead to potentially satisfying ways of making decisions. At one point (p. 15) he describes three useful reasons for "...The general principle that the patient comes first":

> One is the utilitarian consideration that people will be more likely to seek health care if they trust doctors, and people are more likely to trust doctors if it is generally perceived that doctors put the interests of patients first. Another consideration is that a trusting relationship has therapeutic

advantages in that it reduces anxiety and enhances compliance. In addition, because physicians in general are known to espouse the principle of putting patients first, patients assume that doctors will act accordingly. As Albert Jansen and Andrew Jameton see it, patients have a right to make that assumption, and doctors are morally obligated not to disappoint them.

When patients are incompetent, society itself prudently (and rightfully) requires physicians to put patients first, since any of us might some day become incompetent patients. In addition, there may be an implied agreement with the incompetent patient's family to do what is best for the patient.

None of these considerations, however, seems to support the premise that the infant comes first even if that means great sacrifice by the family. The therapeutic advantage and the implicit agreement with the patient do not apply in the case of newborn patients. Neither does the prudential reason, since none of us will be neonates again. Besides, the utilitarian goal of encouraging parents to seek medical care for their children would seem to be served just as well, if not better, by the principle of putting the infant first, unless doing so creates a great burden for the family. Furthermore, concerning any implied agreement with an incompetent patient's family, it is doubtful that parents generally agree, either implicitly or explicitly, that the physician is to do what is best for the infant regardless of the burden to parents. I submit, then, that there is no basis for the view that the duty to the patient is absolute in the kind of case we are considering.

There is one curious omission in Strong's discussion: nowhere does he acknowledge that physicians might have interests of their own that could conflict with those of both parent and neonate, or most especially with those of parents in the matter of "aggressive treatment" to save neonates: "Thus, when doctors make unilateral judgments about newborns with the avowed or implicit purpose of protecting the infant's interests, they are not behaving paternalistically toward anyone." But separate interests on the part of the physician are an obvious probability, and I suggest that the forcing apart of physicians' and their patients' interests by the impersonality and nonrepetitiveness of physician-patient interactions in modern urban society is a principal reason for the rises in dissatisfaction, litigation, and the expenses of medical care. For example, excessive use of diagnostic procedures (that are often risky to the patient) are carried out to protect the interests of physicians and hospitals because, I believe, physicians and hospitals have lost the ability to convince patients that they have indeed assumed the medical interests of the patient as if they were their own. Family doctors in rural settings with stable populations of interrelated, interacting people suffered far less from such difficulties, partly because their reputations were constantly on the line. Patients had ways other than litigation to reciprocate inferior treatment, and more reasons to believe they were obtaining the best care the physician could give.

The closest Strong comes to acknowledging that physicians may have separate interests of their own is the following:

> There are vested interests, in that an entire medical subspecialty has developed to care for impaired newborns. Just try to suggest to neonatologists and NICU nurses that their patients are not persons! (p. 14).

Contrary to most authors, a few who have dealt with moral issues have focused quite *directly* on conflicts of interest. Thus, Perry (1954, pp. 87, 165) noted that:

> It is an open secret...that morality takes conflict of interest as its point of departure and harmony of interests as its ideal goal (p. 87).
>
> The ultimate data of moral science are not men's approbations and disapprobations, but conflicts of interest, and the organizations of interests by which they are made non-conflicting and cooperative (p. 135).

The philosopher, Hans Kelsen (1957), is another example, and in his book *What Is Justice?* he stated what is meant by conflicts of interests (pp. 2–4):

> Where there is no conflict of interests, there is no need for justice. A conflict of interest exists when one interest can be satisfied only at the expense of the other; or what amounts to the same, when there is a conflict between two values, and when it is not possible to realize both at the same time; when the one can be realized only if the other is neglected; when it is necessary to prefer the realization of the one to that of the other; to decide which one is more important, or in other terms, to decide which is the higher value, and finally; which is the highest value.

Despite this clear statement and the rather grand essay written 16 years earlier by Roscoe Pound (see pp. 33–34), neither Kelsen nor any other author I have discovered sets out to identify interests and quantify their conflicts and confluences (although Strong, 1984, comes close); most do not even acknowledge this as a matter of importance. Piecemeal efforts are as old as history, but there seems to have been no effort toward a general theory.

Why should it be true that the very disciplines preoccupied with conflicts of interest should deal with them as if they were something other than the central issue? Perhaps the authors involved would respond to this question by saying: "Rubbish! Of course we know that conflicts of interests are the heart of the problem. The reason we don't dwell on them is *because* we know their role and importance so well."

Perhaps. But precisely the same thing is sometimes said about altruism. At a recent meeting at the University of Michigan, a prominent moral philosopher said, in evident puzzlement, "We have been discussing altruism forever. What is all the recent excitement from biology about?" The excitement exists because theories of inclusive-fitness-maximizing (Hamilton, 1964) and reciprocity (Trivers, 1971) enable us to formulate testable hypotheses about aspects of altruism that previously could not be investigated scientifically. A new understanding of what human interests are all about might similarly justify a reexamination of the central question of conflicts of interest in connection with morality and ethics and the general conduct of people. I think we have such a new understanding and, as with our new understanding of altruism, it comes from biology. Grant (1985), for example, remarks that "When Rawls speaks of human beings as rational, he means that they are able to calculate their self-interest.... What men primarily calculate about are those good things which lead to comfortable self-preservation." It is a major argument of this book that this prevalent view simply will not allow us to analyze human sociality to the core. (A more extensive discussion of moral philosophy occurs later on pp. 145ff.)

MORAL SYSTEMS AS SYSTEMS OF INDIRECT RECIPROCITY

The problem, in developing a theory of moral systems that is consistent with evolutionary theory from biology, is in accounting for the altruism of moral behavior in genetically selfish terms. I believe this can be done by interpreting moral systems as systems of indirect reciprocity.

I regard indirect reciprocity as a consequence of direct reciprocity occurring in the presence of interested audiences-groups of individuals who continually evaluate the members of their society as possible future interactants from whom they would like to gain more than they lose (this outcome, of course, can be mutual).

Returns from indirect reciprocity may take at least three major forms:

(1) the beneficent individual may later be engaged in profitable reciprocal interactions by individuals who have observed his behavior in directly reciprocal interactions and judged him to be a potentially rewarding interactant (his "reputation" or "status" is enhanced, to his ultimate benefit); (2) the beneficent individual may be rewarded with direct compensation from all or part of the group (such as with money or a *medal* or Social elevation as a hero) which, in turn, increases his likelihood of (and that of his relatives) receiving additional perquisites; or (3) the beneficent individual may be rewarded by simply having the success of the group within which he behaved beneficently contribute to the success of his own descendants and collateral relatives.

Obviously, various forms of punishment, including ostracism or social shunning, can also be applied to individuals repeatedly observed not to reciprocate adequately or follow whatever codes of conduct may exist.

Typically, then, in interactions solely involving direct reciprocity, individuals may be expected to seek a net gain, although (1) this does not necessarily come at the expense of the other interactant (i.e., both may profit) and (2) the net gain may *only* be realized after a long iterated series of interactions, anyone or fraction of which may actually yield a net loss for the individual in question. Even in directly reciprocal interactions, however, net losses to self (and even explicitly greater losses than those of the partner) may be the actual aim of one or even both individuals, if they are being scrutinized by others who are likely to engage either individual subsequently in reciprocity of greater significance than that occurring in the scrutinized acts. In effect, what goes on in such cases could be termed "social hustling," in which a "player" more or less deliberately (though conscious purpose is not a requirement) loses in order to "set up" the observer for a later overcompensating gain. I am referring to all effects of such social scrutinizing as indirect reciprocity.

If current views of evolutionary processes are correct, reciprocity flourishes when the donated benefits are relatively inexpensive compared to the returns (Hamilton, 1964; Trivers, 1971; Alexander, 1974; West-Eberhard, 1975). This kind of gain is possible under two circumstances: the first is when threats or promises extrinsic to the interactants cause joint similar efforts to be worth more than the sum of their separate contributions, leading to more or less symmetrical cooperation. The second is when the contributions of partners in reciprocity are different, leading to division of labor. The second situation can arise out of different abilities or training in different contributors, or from differences in their accumulated resources. Systems of indirect reciprocity as expressed in humans require memory, consistency across time, the application of precedents, and persistent and widely communicated concepts of right and wrong (Trivers, 1971, 1985; Alexander, 1977–1982; Axelrod and Hamilton, 1981). They become, automatically, what I am here calling moral systems. I believe that the feeling that human behavior is out of reach of biological analyses arises, not so much because we are confused by the existence of "culture" or socially heritable learning (the reason usually given), but because culture includes systems of indirect reciprocity (moral, ethical, and legal systems) by which the costs and benefits of acts deemed by others to be socially positive or negative can be manipulated. The "problem of culture," then, may not be so much one of ontogenetic disjunction (as most authors seem to imply—meaning that at some point learning or culture takes over and the human organism begins to operate independently of its evolutionary heritage), but rather a failure to appreciate the pervasiveness and the consequences of indirect reciprocity. Systems of indirect reciprocity, and therefore moral systems, are social systems structured around the importance of status. The concept of status implies that an individual's privileges, or its access to resources, are controlled in part by how others collectively think of him (hence, treat him) as a result of past interactions (including observations of interactions with others). Status can be determined by physical prowess, as in those nonhuman (animal) dominance hierarchies in which coalitions are absent, or (as in humans) by mental or social prowess. Mental and social prowess, in this sense, includes (as in moral systems) effectiveness and reliability in reciprocity and cooperation.

Once social interactions become instrumental in establishing status, then testing and practice, as in the forms typically called "play," may become prominent. Play represents practice for later status-affecting interactions, but it also affects current and projected status. In humans, social-intellectual play in the form of humor has become uniquely elaborate. This kind of play, moreover, unlike most forms of play, is prominent across all of adult life. As with physical play in humans (Alexander, 1974), humor has also come to include unique group-against-group forms (i.e., "ostracizing" humor, such as ethnic and racial jokes) (Alexander, 1986b).

WHERE DO RULES COME FROM

As soon as planning, anticipating, "expecting" organisms are interacting without complete overlap (confluence) of interests, then each of two interactants may be expected to include in (add to) its repertoire of social actions special efforts to thwart (intercept, alter) the expectations of the others—explicitly in ways designed to be costly to those others and beneficial to himself. Interruption of another's expectations will be costly to that other when the expectation involves some investment (cost), or when pursuing it has been done in a way that is less costly because it did not allow for the possibility of the kind of interruption that the first party caused. I suggest, then, that the expense of investing in expectations, and the possibility of doing so with less expense if certain kinds of interruptions are forestalled, are the essential reasons for the invention and maintenance of *rules*. Rules are aspects of indirect reciprocity beneficial to those who propose and perpetuate them, not only because they force others to behave in ways explicitly beneficial to the proposers and perpetuators but because they also make the future more predictable so that plans can be carried out. This proposition is subtly, but significantly, different from that implied by Rawls (1971, p. 6):

> In the absence of a certain measure of agreement on what is just and unjust, it is clearly more difficult for individuals to coordinate their plans efficiently in order to ensure that mutually beneficial arrangements are maintained. Distrust and resentment corrode the ties of civility, and suspicion and hostility tempt men to act in ways they would otherwise avoid. So while the distinctive role of conceptions of justice is to specify basic rights and duties and to determine the appropriate distributive shares, the way in which a conception does this is bound to affect the problems of efficiency, coordination, and stability.

The significance of indirect reciprocity has to do not only with *rules*, but with intent (table 9.5), and the general levels of altruism prevailing in the society. Because systems of indirect reciprocity involve promises of punishment as well as reward they lead to avoidance of selfishness as well as positive acts of altruism. This is why, I believe, humans tend to decide that a person is either moral or not, as opposed to being moral in one time or context and immoral in another, and why intent is said to be "nine-tenths of the law." We use motivation and honesty in one circumstance to predict actions in others.

Seeing morality as self-serving because of indirect reciprocity enables us to visualize it in two stages or as involving two kinds of outcomes for acts widely regarded as immoral:

1. If I do that ("immoral thing") to someone I am apt to suffer costs greater than the benefits, imposed on me by the other members my group as a result of indirect reciprocity (within-group indirect reciprocity).
2. If I do that ("immoral thing") to someone I will foster, or at least not restrain, the development of a pattern of within-group behavior that will

eventually impose a cost on me greater than the benefits. These costs will be imposed on me either by a change of rules that I precipitate or promote against my long-term interests or by the ability of members of groups other than my own to injure or destroy my group because of its lack of unity (between-group indirect reciprocity). Awareness of this later possibility by my group members will cause them to be even more watchful of my behavior within the group, thus increasing the costs of "immoral" actions monitored by within-group reciprocity.

The consequences of indirect reciprocity, then, include the concomitant spread of altruism (as social investment genetically valuable to the altruist), rules, and efforts to cheat (tables 9.2–9.4). I would not contend that we always carry out cost-benefit analyses on these issues deliberately or consciously. I do, however, contend that such analyses occur sometimes consciously, sometimes not, and that we are evolved to be exceedingly accurate and quick at making them (table 9.5).

DISCRIMINATE AND INDISCRIMINATE BENEFICENCE

The beneficence involved in human nepotism and direct reciprocity is discriminative: different relatives, and relatives of different needs, are distinguished. Friends are treated individually. As yet, no organism outside clones has been shown to display population-wide indiscriminate beneficence, and I will argue that, with the exception of clones and certain kinds of eusociality (especially, in termites, ants, wasps, and bees), or in human systems of indirect reciprocity does a modicum of essentially indiscriminate beneficence or social investment exist in large groups

Indirect reciprocity must have arisen out of the search for interactants and situations by which to maximize returns from asymmetrical, hence highly profitable direct social reciprocity. One consequence of large complex societies in which reciprocity is the principal social cement and indirect reciprocity is prevalent is that opportunities for such mutually profitable asymmetrical reciprocal interactions are vastly multiplied. This situation, in turn, fosters the appearance of tendencies to engage *indiscriminate social investment* (or indiscriminate beneficence)—which I define as willingness to risk relatively small expenses in certain kinds social donations to whomever may be needy—partly because of the prevalence of interested audiences and keenness of their observation, and the use of beneficent acts by others to identify individuals appropriate for later reciprocal interactions. In complex social systems with much reciprocity, being judged as attractive for reciprocal interaction may become an essential ingredient for success. Similarly, to be judged harshly because of failure to deliver small social benefits indiscriminately in appropriate situations may lead to formidable disadvantage because of either direct penalties or lost opportunities in subsequent reciprocal interactions.

I suggest that indirect reciprocity led to the evolution of ever keener abilities to observe and interpret situations with moral overtones. In such a milieu, I would argue, a modicum of indiscriminate beneficence would arise, as social investment, because of benefits to individuals who are viewed as altruists. Beneficence can approach being indiscriminate in two ways: (1) *some acts* can be indiscriminately beneficent and (2) *all acts,* or social behavior in general, can tend in the direction of being indiscriminately beneficent. These two different expressions of beneficence will not necessarily have the same consequences for the actor.

Population-wide indiscriminate beneficence might also evolve when small "populations" are regularly composed of relatives related to a similar degree, and if the individuals of other populations are never contacted and therefore not discriminated against. This may be an unlikely situation for mammals or even vertebrates in general. This kind of indiscriminate beneficence would require no special proximate mechanisms—no social learning; but there is yet no undisputed evidence for unlearned recognition of relatives in any species (see reviews by Sherman and Holmes, 1985; Alexander, 1985).

Figure 9.1 describes hypothetical social stages through which the evolving human species might have passed (many times), and seeming to lead (but, as argued here, probably not actually doing so) toward a utilitarian or idealized model of morality in which *all* social investment becomes indiscriminate beneficence.

As already noted, complete and indiscriminate beneficence, as in the utilitarian system of philosophers (i.e., systems promoting the greatest good to the greatest number), would not always be a losing strategy for individuals, even in evolutionary terms. Indiscriminate beneficence would not lose, for example, when the interests of the group and the interests of the individuals comprising it are the same. Such a confluence of interests would occur when all group members were equally and most closely related to the individuals destined to reproduce. It could also occur (temporarily) whenever the group was threatened externally in such fashion that complete cooperation by its members would be necessary to dissipate the threat, and when failure of the group to dissipate the threat would more severely penalize any remaining individuals than would the group's survival after that individual had used all of its effort to support the group. In such cases it is not trivial to consider the different treatments likely to be accorded those individuals who contributed wholeheartedly and unselfishly to the well-being of the group when it was threatened, and those who did not, or who in fact selfishly betrayed their fellows or ignored their needs.

General encouragement of indiscriminate beneficence, and general acceptance of its beneficial effects, results in a society with high social unity. This encouragement and acceptance is expected to occur partly because of the likelihood, much of the time, that nearly everyone benefits from living in a unified society (as opposed to a socially divisive one), but also partly because individuals gain from portraying themselves as indiscriminate altruists, and from thereby inducing indiscriminate beneficence in others (and often from inducing degrees of it that are deleterious to those others). This means that whether or not we know it when we speak favorably to our children about Good Samaritanism, we are telling them about a behavior that has a strong likelihood of being reproductively profitable. In a small social group this can be true for the Good Samaritan even if he or she is never identified, but Good Samaritan acts seem likely to be most profitable to the actor if his responsibility for the act is discovered accidentally, and most importantly through no effort of his own.

I would postulate that self-serving indiscriminate social investment—because it was seen as net-cost (i.e., genetic as well as phenotypic) altruism and was interpreted wrongly as part of a real trend toward universal indiscriminate (net-cost) altruism-provided the impetus for the idealized modern model of morality portrayed in figure 9.1. The value of self-serving beneficence, it seems to me, is what sets the stage for the evolution of the ability and tendency to develop a conscience, which I have interpreted (Alexander, 1979a) as the "still small voice that tells us how far we can go in serving our own interests without incurring intolerable risks."

The implication is that approaches to morality are expressed consistently, and to the degree they are usually realized in society, because there is continual pressure to bring about a condition of more nearly ideal morality. If so, this pressure is likely to be applied by each individual so as to cause his neighbor, if possible, to be a little more moral than himself. Stated differently, it would be to the advantage of each individual in a society that other individuals, especially those not most closely related to him, actually achieve or approach the ideal of completely moral behavior. Ideally moral (indiscriminately beneficent) people would tend to "help" others in the society, however slightly, to achieve the goals that evolutionists believe have driven evolution by natural selection. They would contribute slightly to everyone else's interests by (1) helping their interactants directly, (2) hurting themselves in relation to others, and (3) setting an example that others may follow, thereby contributing to the interests of the group as a whole. Accordingly, one expects that the individuals in a society would gain from exerting at least a little effort toward encouraging other individuals to be more moral (altruistic, beneficent) than they otherwise might have been. Among the many ways of accomplishing this is included the setting up of an idealized morality as a model or goal, and the encouragement of everyone (else) to become like that. One way of promoting this situation is to designate as heroes those who approach the ideal moral condition. One expects that sainthood may be awarded to individuals who spend their lives on explicitly antireproductive behavior. The prevalence among saints of asceticism, self-denial, celibacy, isolation from relatives, devotion to the welfare of strangers, and otherwise indiscriminate tendencies to be altruistic supports this hypothesis. So does the fact that sainthood is generally awarded (long) after the death of the awardee (thus, the awardee cannot personally gain from this heroic designation).

The long-term existence of complex patterns of indirect reciprocity, then, seems to favor the evolution of keen abilities to (1) make one's self seem more beneficent than is the case; and (2) influence others to be beneficent in such fashions as to be deleterious to themselves and beneficial to the moralizer, e.g., to lead others to (a)

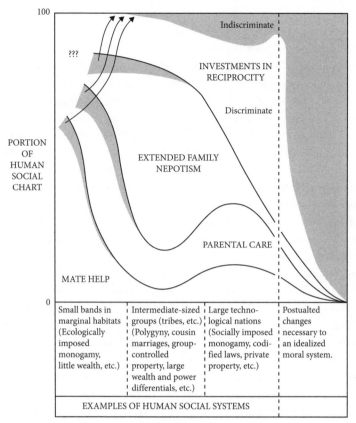

FIGURE 9.1 A speculation about the relative importance of different kinds of social interactions in some different kinds of societies. The principal purpose is to show the probable origins of indiscriminate altruism, its probable significance in different societies, and the changes from existing societies that would be necessary to realize an idealized model of morality in which everyone was indiscriminately altruistic.

invest too much, (b) invest wrongly in the moralizer or his relatives and friends, or (c) invest indiscriminately on a larger scale than would otherwise be the case. According to this view, individuals are expected to parade the idea of much beneficence, and even of indiscriminate altruism as beneficial, so as to encourage people in general to engage in increasing amounts of social investment whether or not it is beneficial to their interests. They may be expected to locate and exploit social interactions mimicking genetic relatedness leading to nepotistic flows of benefits (e.g., to insinuate themselves deceptively into the role of relative or reciprocator so as to receive the benefits therefrom). They may also be expected to depress the fitness of competitors by identifying them, deceptively or not, as reciprocity cheaters (in other words, to moralize and gossip); to internalize rules or evolve the ability to acquire a conscience, interpreted (Alexander, 1979a) as the ability to use our own judgment to serve our own interests; and to self-deceive and display false sincerity as defenses against detection of cheating and attributions of deliberateness in cheating (Trivers, 1971, 1985; Campbell, 1975; Alexander, 1974, 1977b, 1979a, 1982, 1985).

So we are provided with the general hypothesis that tendencies toward moral behavior, and the establishment of moral systems, are vehicles for promoting the goals of society as a whole: that they develop because all of the individuals of society often share the same goals, that, ultimately, these goals involve competition with and defense against other human groups; that, except in times of severe external threats recognized by everyone as requiring extreme cooperativeness, the ideal of universal indiscriminate beneficence is not met within groups; that the ideal morality has never even been approached between societies or nations; and that, because some degree of within-society competition probably occurs nearly all of the time, every individual may be expected to use the impetus toward realizing the goals of everyone to his own advantage by promoting a slightly greater degree of "morality" in his neighbor than in himself (i.e., a net-cost or genetic altruism).

The question may be raised, however, why anyone should be vulnerable to manipulation unduly far in the direction of beneficence, if we have been subjected to such manipulations a very long time? Why, in other words, should moralizing ever be effective? I think there are at least four contributing factors. First, the degrees of beneficence that are actually reproductively appropriate will vary dramatically as societies move between periods of extreme danger and relative security, making it difficult to know how to behave. When will a specified degree of failure to accede to exhortations to be beneficent cost more than it yields, because of (a) failure of the group on which one depends for success, or (b) responses within the group to one's failure to be beneficent? Second, individuals may be expected to take advantage of this dramatic shifting to deceive others about the degree of danger so as to induce unduly beneficent behavior in others. Sometimes aspiring leaders may use such deception to promote their own leadership as an antidote to the supposed threat, and as a promoter of unity. Third, we may expect that the individuals in a society such as we have been describing will evolve to deceive others about the degree of beneficence they themselves are exhibiting: Everyone will wish to appear more beneficent than he is. There are two reasons: (1) this appearance, if credible, is more likely to lead to direct social rewards than its alternatives; (2) it is also more likely to encourage others to be more beneficent. If one's associates are beneficent, then he can afford to be (or is forced to be) more beneficent than if they are not (we may note the additional feedback from these facts that would cause everyone to be concerned that everyone else appear beneficent so that people in general will feel comfortable with a higher degree of beneficence than would otherwise be the case). Fourth, if kin recognition is *learned* (Alexander, 1977a; 1979a; 1985a; Greenberg, 1979; Sherman and Holmes, 1985; Mintzer, 1982), mistakes are likely, and one may insinuate himself into the role of a relative so as to receive inappropriate nepotism, or even pretend to be nepotistic so as to receive the appropriate beneficent responses. Playing upon the tendency of everyone to strive to appear more beneficent than he is, and using the other ploys just described, may lead to much success in social manipulation. The introduction of indirect reciprocity, whereby society as a whole or some large part of it provides the reward for benefi-cence and the punishment for selfishness (fines for running stop signs; tax rebates for donating to charity, etc.), simultaneously served both society and the individu-als comprising it, and provided a vehicle for manipulating individuals socially to levels and kinds of beneficence detrimental to them (or to their reproductive suc-cess). It is somewhat paradoxical that the tendencies and pressures in the direction of idealized moral systems should serve everyone in the group up to a point, but then be transformed by the same forces that molded them, into manipulations of the behavior of individuals that are explicitly against the interests of those being manipulated and in the interests of those ostensibly contributing to everyone's interests by promoting trends toward morality in the system.

One consequence (and a saving grace) of the pressure within societies for everyone to be a little more moral than would pay, and of keen abilities by people in general to determine tendencies and willingness to behave beneficently or not, is that no one can afford to lag too far behind relative to everyone else. As benefi-cence continues to be promoted, *everyone* has to follow along, so that the most selfish individuals would be forced to be less selfish, and perhaps as well variance in readiness to be beneficent would narrow as the "front" of beneficence advances in the direction of the ideal of indiscriminate investment and true justice in the form of equal opportunity and equal treatment under the law. It is possible that the societies in which moral philosophers operate have actually changed so much in this direction during the past two or three centuries as to alter the preoccupa-tions of moral philosophy toward a greater concern for utilitarian ideals and a less jaundiced view of humanity. If so, there is likely no greater irony than the fact that modern technological societies, whatever the degree of egalitarianism or approach to the philosophical ideal of morality within some of them, are teetering us on the brink of disaster as a result of their interactions with one another. The prob-lem has become one of inducing *between* and *among* societies the same processes of moralizing pressure and democratization that have developed so intricately within them.

It is also true that, in the absence of overriding power differentials, dramatic departures from usual levels of beneficence in the direction of serving one's own interests are virtually certain to result in net losses as a result of shifts of reputation or status, or subsequently diminished beneficence from others. On the other hand, dramatic departures from usual levels of beneficence in the direction of *indiscrimi-nateness* may raise status and multiply subsequent benefits so as to produce a net return to the actor. This asymmetry of effects would seem likely to cause acts of cheating and selfishness to depart minimally from norms while promoting dra-matic or extreme acts of heroism, charity, and saintliness. Perhaps, to some extent, the asymmetry of these two effects is involved in the "creep" of certain kinds of closely knit, stable societies toward more highly cooperative, democratic opportu-nity-equalizing structures.

I believe that the various factors discussed here are the essential elements that produce and maintain what we commonly call moral systems, and moral behavior in individuals. Understanding them represents the means for resolving the existing paradoxes with respect to morality, eliminating the aura of mystery that has surrounded the concept, and understanding not only why moral systems have always fallen short of our ideals but why we nevertheless establish and maintain such ideals. If accurate, these arguments may also clarify the routes by which we can most closely approach what are seen as idealized moral systems, and perhaps most confidently avert moral disasters.

> Moral talk is often rather repugnant. Leveling moral accusations, expressing moral indignation, passing moral judgment, allotting the blame, administering moral reproof, justifying oneself, and above all, moralizing—who can enjoy such talk? And who can like or trust those addicted to it? The most outspoken critics of their neighbors' morals are usually men (or women) who wish to ensure that nobody should enjoy the good things in life which they themselves have missed and men who confuse the right and the good with their own advancement (Baier, 1965, p. 3).

> Seeking to protect the autonomy that we have learned to prize, we aspire ourselves *not* to be manipulated by others; seeking to incarnate our own principles and stand-point in the world of practice, we find no way open to us to do so except by directing towards others those very manipulative modes of relationship which each of us aspires to resist in our own case (MacIntyre, 1981b, p. 66).

Literature Cited

Alexander, R.D. 1974. The evolution of social behavior. *Ann. Rev. Ecol. Syst.* 5:352–383.

Alexander, R.D. 1977a. Evolution, human behavior, and determinism. *Proc. of the Biennial Meeting of the Philosophy of Science Association* (1976), 2:3–21.

Alexander, R.D. 1977b. Natural selection and the analysis of human sociality. In *Changing scenes in the natural sciences*, C. E. Goulden (ed.). 1776–1976 Bicentennial Symposium Monograph, Philadelphia Academy of Natural Sciences, pp. 283–337.

Alexander, R.D. 1979a. *Darwinism and Human Affairs*, University of Washington Press, Seattle xiv + 317 pp.

Alexander, R.D. 1982. Biology and the moral paradoxes. *J. Social Biol. Struct.* 5: 389–395.

Alexander, R.D. 1985a. Genes, consciousness, and behavior theory. In *A Century of Psychology as Science*, S. Koch and D. E. Leary (eds.). New York: McGraw-Hill, pp. 783–802.

Alexander, R.D. 1985b. A biological interpretation of moral systems, *Zygon* 20: 3–20.

Alexander, R.D. 1986b. Biology and Law. In *Ethology and Sociobiology* 7:167–173. R. Masters and M. Gruter (eds.).

Axelrod, R. 1984. *The evolution of cooperation*. New York: Basic Books.

Axelrod, R. and W.D. Hamilton. 1981. The evolution of cooperation. *Science* 211:1390–1396.

Baier, K. 1965. *The moral point of view: A rational basis of ethics.* Abridged ed. New York: Random House.

Brandt, R.B. 1979. *A theory of the good and the right.* Oxford: Clarendon Press.

Callahan, D. 1985. What do children owe elderly parents? *Hastings Center Rep.* 15:32–37.

Campbell, D.T. 1975. Conflicts between biological and social evolution and between psychology and moral tradition. *Amer. Psychol.* 30:1103–1126.

Connor, R. 1986. Psuedo-reciprocity: Investing in mutualism. *Animal Behaviour* 34:1562–1566.

de Waal, F. 1982. *Chimpanzee politics: power and sex among apes.* New York: Harper and Row.

de Waal, F. 1986. Who pays for social instability? Scapegoating in primates, and a fatal fight in the Arnhem chimpanzee colony. *Ethol. Sociobiol.* 7.

Frankena, W.K. 1973. *Ethics,* 2nd. ed. Englewood Cliffs, NJ: Prentice Hall.

Frankena, W.K. 1980. *Thinking about morality.* Ann Arbor, MI: Univ. Michigan Press.

Grant, G.P. 1985. *English-speaking justice.* Notre Dame, IN: Univ. Notre Dame Press.

Greenberg, L. 1979. Genetic component of bee odor in kin recognition. *Science* 206:1095–1097.

Hamilton, W.D. 1964. The genetical evolution of social behavior, I, II. *J. Theoret. Biol.* 7:1–52.

Kelsen, H. 1957. *What is justice? Justice, law, and politics in the mirror of science. Collected essays.* Berkeley, CA: Univ. Calif. Press.

Kohlberg, L.L. 1981. *Essays on moral development. I. The Philosophy of moral development.* San Francisco, CA: Harper and Row.

MacIntyre, A. 1981b. *After virtue: a study in moral theory.* Notre Dame, IN: Univ. of Notre Dame Press.

Mintzer, A. 1982. Nestmate recognition and incompatibility between colonies of the acacia ant, *Pseudomyrmex ferruginea. Behav. Ecol. Sociobiol.* 10:165–168.

Perry, R.B. 1954. *Realms of value: A critique of human civilization.* Cambridge, MA: Harvard Univ. Press.

Pound, R. 1959. *Jurisprudence.* St. Paul, MN: West Publ. Co.

Rawls, J. 1971. *A theory of justice.* Cambridge, MA: Harvard Univ. Press.

Richards, R. 1986b. A defense of evolutionary ethics. *Biol. Phil.* 1:265–293.

Sahlins, M.D. 1965. On the sociology of primitive exchange. In M. Banton (Ed.), *The relevance of models for social anthropology,* pp. 139–236. London: Travistock.

Sherman, P.W. and W.G. Holmes. 1985. Kin recognition: Issues and evidence. In B. Holldobler and S. Lindauer (Eds.), *Experimental behavioral ecology, Fortschritte der Zoologie,* Vol. 31, pp. 437–460. NY and Stuttgart: G. Fischer Verlag.

Strong, C. 1984. The neonatologist's duty to patient and parents. *Hastings Center Rep.* 14:10–16.

Trivers, R.L. 1971. The evolution of reciprocal altruism. *Quart. Rev. Biol.* 46:35–57.

Trivers, R.L. 1985. *Social evolution.* Menlo Park, CA: Benjamin/Cummings.

West-Eberhard, M.J. 1975. The evolution of social behavior by kin selection. *Quart. Rev. Biol.* 50:1–33.

Whiteley, C.H. 1976. Morality and egoism. *Mind* 85:90–96.

Evolution of Human Intelligence

Reality and the Human Enterprise

Science discovers and documents reality,
at its best enthusiastically,
imaginatively, meticulously,
and necessarily dispassionately.

Music, Art, and Literature elaborate on reality,
at their best enthusiastically,
imaginatively, elegantly,
and necessarily passionately.

Politics, Law, and Morality stabilize reality,
at their best, wisely and justly.

Humor mocks reality,
at its best, constructively.

Realities can be physical,
string, particle, or wave to a universe
or biological I (animal and plant)
or biological II (human, and a special sociality).

Human social realities can be completely accurate,
because complete agreement is the only criterion.
And they can be contrary to physical realities,
deliberately, even, yet serve the believers
better than other realities, but only sometimes,
and that sometimes is when cooperation,
joining to achieve a common goal,
is required to be all that is important.

Animal and plant realities can be affected by humans,
subordinated or not to social realities.

Human realities regarding self have so far
been subordinated to human social realities;
we have not been willing yet to know
everything about ourselves,

not the things we are evolved to suppress,
or those that seem to make us unhappy;
not the self-deception we use to further our own
desires and goals, and that generates social reputations
but only temporarily and shakily accurate versions.

Art, Music, Literature,
Politics, Law, Morality,
and Humor, too, are restricted
to the human social kind of reality.

Mocking individual realities can
save the humorist too,
but only when it works
for him in the eyes of others.

Failing that,
the effort buries him.

But a humor that mocks
social and political realities
may also save the world.

Alexander, 2011, p. 260

INTRODUCTION

Reflections on the Evolution of the Human Psyche
R.I.M. Dunbar

Commentary on: R. D. Alexander (1989). Evolution of the human psyche.
In: P. Mellars & C. Stringer (eds.) *The Human Revolution*, pp. 455–513.
Edinburgh University Press.

The edited volume *The Human Revolution* was published in 1989 with the aim of summarizing the background to the appearance of modern humans. Dick Alexander contributed a lengthy piece to this volume on the evolution of the modern human mind—in effect, what it is to be human and how we came to be that way. His article (it is 60 pages long) divides naturally into three basic components. The first part sets out the background and agenda, and argues a trenchant case for the claim that the origins of the human mind lie in the social domain. The second part is as an interlude and revisits the thorny old debate surrounding the application of evolutionary theory to human behavior and psychology—a reflection, perhaps, of the fact that, barely a decade after the "Sociobiological Revolution," the vitriolic debate that had followed the publication of Ed Wilson's *Sociobiology* and Dawkins's *Selfish Gene* was still alive and kicking. (Sadly, although things are much improved in this respect three decades on, a deep-rooted antipathy to Darwinian ideas still colors attitudes in some parts of the humanities and social sciences.) The third part is the heart of the paper and addresses the question of how and why the human mind came to be so different. Dick placed a strong emphasis on this in the paper: The "cultural" activities of animals not-withstanding, humans *are* in a different league to everyone else in that they do things (like building cathedrals and writing plays) that no other species does.

The essence of his argument is summed up in a three remarkably prescient sentences at the outset of this third section:

> I think the key argument....is that consciousness represents a system of
> (1) building scenarios or constructing possible (imagined) alternatives; (2)
> testing and adjusting them according to different projected circumstances;
> and (3) eventually using them according to whatever circumstances actually
> arise. [In a previous paper] I referred to such abilities as the capacity to over-
> ride immediate rewards and punishments in the interests of securing greater
> rewards visualized in the future....In this view, consciousness, cognition,
> and related attributes—which probably represent the core of the problem in

understanding the human psyche—have their value in social matters, and the operation of consciousness can be compared to the planning that takes place in a game in which the moves of the other players cannot be known with certainty ahead of time. (p. 477)

Play loomed large in this argument, but for Dick it had a special meaning over and above the more conventional interpretation of play as physical practice: central to his argument was what he termed "social/intellectual play," the capacity to ruminate on and mull over alternative scenarios in the light of others' possible responses to one's actions. Play, in his view, allowed individuals to learn how to predict others' strategic responses as adults, and so gave rise to the human capacity to evaluate the outcomes of alternative courses of action and thus choose between them more effectively—a capacity sometimes subsequently referred to as "mental time travel" (Barrett et al., 2003; Suddendorf et al., 2009).

In a singularly influential paper published a decade earlier (Alexander, 1974), Dick had argued that the evolution of group-living in primates (and other species) arose in response to one of two core problems: defence against predators or the opportunity to monopolise resources. Both these suggestions were taken up by others in later decades (e.g., van Schaik, 1983, Wrangham, 1980) and they continue to exercise debate even today. However, in his *Human Revolution* contribution, Dick eschews resource defence as a likely selection factor for group living and comes down hard in favour of predation risk. Resource defence, he argues, could only have arisen *after* group living had come about, and so must be a consequence, not a cause, of group living. In this respect, his arguments have largely been borne out, at least in the case of primates and ungulates (see, e.g., Shultz and Dunbar, 2006; Adamczak & Dunbar, 2008).

Predation as the driver for group living is the bedrock on which Dick built his argument. The key steps can be summarized as: predation selects for living in groups; living in groups inexorably gives rise to competition between groups (thus, opening up the opportunity for resource defence); such competition selects for the capacity to cooperate in mounting a defence (or attack) against rivals over resources; cooperation selects for consciousness (i.e., the cognitive abilities needed to manage effective group-level cooperation), which in turn selects for increased brain size etc. The problem to be solved was, as he put it:

[W]hat kinds of mechanisms enabled group-cooperative humans to conduct intergroup aggression cooperatively. How did humans manage the coordination necessary to carry out raids efficiently, especially against enemies belonging to their own species and possessing the same general abilities and tendencies? (p. 498)

In the article, Dick makes a great deal of kin selection and tactical deception in this respect—something that was perhaps inevitable given that it was written in the immediate aftermath of the launch of the Machiavellian Intelligence

Theory of primate brain evolution (Byrne & Whiten, 1988) and an increasing interest in the evolution of deception (e.g., Krebs & Dawkins, 1984). Kin selection, of course, remains a central plank in all evolutionary analyses, but the Machiavellian Intelligence Hypothesis has mutated into the softer form of the Social Brain Hypothesis (Dunbar, 1992a, 1998; Barton & Dunbar, 1997) in which affiliative bonding (and/or social learning) is emphasized at the expense of the deviousness implied by the term "Machiavellian" (Dunbar, 2011). Nonetheless, deception remains, of course, an important issue in evolution, as evidenced by the fact that it has spawned a number of important counter-strategies in terms of human cognitive evolution (notably the cheat detection mechanisms of Cosmides [1989; Cosmides & Tooby, 1992]). Re-reading the original text, however, reminds me how much it anticipates the emergence of theory of mind (or mentalising) as a central theme both in child development and in primate comparative psychology:

> [One of] the most effective ways to deal with human competitors [is]....the ability to see ourselves as others see us, so as to cause them to see us as we would like them to rather than as they would like to. The human psyche is evidently evolved to excel at such practices" [i.e., "the ability to see ourselves as others see us"]. (p. 491)

If this isn't an explicit reference to theory of mind, it's hard to see what else it could be.

This emphasis on the importance of cooperation in animal and human social evolution has, over the past decade or so, given rise to a minor industry in both experimental and modelling studies of the conditions under which cooperation can arise, with the numbers of papers generated by this theme running into many hundreds. For myself, however, I found his appreciation of the fundamental importance of the need for community cohesion in human evolution particularly prescient. Contemporary versions of the Social Brain Hypothesis place a singular premium on the problem of freeriders, the effect these have on destabilizing the kinds of implicit social contracts on which cooperation necessarily depends and the mechanisms needed to prevent freeriders from overwhelming cooperators (Enquist & Leimar, 1993; Nettle & Dunbar, 1977; Dunbar, 1999, 2009a). Prominent among the mechanisms being discussed in the contemporary literature are ethnic markers and the role of religion, both of which are explicitly mentioned by Dick:

> Acceptance of unifying myths or information or goals depends on the individual's acceptance of the value of group unity, including the position or status of himself that will result, or other effects on himself and his intimates (children, spouse, relatives, reciprocants). Even myths widely regarded as counterfactual may be accepted, repeated, and elaborated if their effect is seen as [socially] unifying. (p. 493)

My own view has come to be that the central problem we have faced throughout our evolutionary history has been how to neutralise the destructive effects of freeriders as we have sought to push social community size above the limits set by social grooming (the standard mechanism that primates use to bond their social groups) (Dunbar, 1992b, 2008). There are two related issues here. One is the fact that time constraints impose what amounts to a glass ceiling at around 50 (the mean group size typical of the most social of the monkeys and apes, including baboons and chimpanzees) on the number of individuals with whom one can form coherent relationships. The second is that as community size increases, so the pressure to defect on the social contract that underpins primate sociality increases proportionately. If some mechanism is not found to counteract this effect, the community will collapse back to the minimum size that primates can maintain by grooming. A solution to the first problem has to be able to solve the second.

What is needed is some process that effectively bridges across that barrier to reach more individuals, and laughter, music and religion (or at least the rituals of religion) appear to have been the mechanisms of choice—mainly because all three are extremely good at triggering the release of endorphins (the same mechanism that allows grooming to be an effective mechanism for social bonding: Dunbar, 2009a, 2010). This almost certainly has a long history, building successively through these three processes from around two million years ago (the first appearance of the genus *Homo* and the point at which the grooming time constraint first kicks in) until the appearance of fully fledged religion and language with the advent of anatomically modern humans around 250,000 years ago (Dunbar, 2003, 2009b). With the advent of language, shared myths (or worldviews) come to play a central role during the course of later human evolution because language allows us to coordinate religious rituals and provide them with a meaningful *raison d'être*.

Central to this process is emotion, not least because endorphins have a positive effect on affect and are responsible for creating that sense of wellbeing and warmth that goes with successful social interactions (Zubieta et al., 2003). Dick makes a great deal of emotion and, unusually perhaps for an evolutionary biologist, of the physiological processes that underpin emotions. However, his discussion is largely couched in terms of the manipulation of others' emotions in order to deceive them—or avoid being deceived by them. As he saw it:

> Once individuals became capable of recognizing emotional states in other individuals, then it seems virtually certain that selection would alter both this ability and the emotional states themselves, or the external evidence of them, in ways that would be called communicative.... This is the point (in evolution) at which it would become important for us to know about our own "feelings" or emotions—because we could then manipulate them to affect use by others of evidence about them.

It has taken a long time, but emotions have finally begun to loom large in discussions of human cognitive evolution (see, for example, Gamble et al., 2011; Gowlett et al., in press).

The key to all this lies in a growing intensity of inter-group conflict, giving rise, in effect, to a perpetual state of war between neighboring groups. War was undoubtedly out of fashion in the 1980s, notably among archaeologists and paleoanthropologists who tended to take a rather benign view of human psychology and behavior (Keeley, 1996; Wrangham & Peterson, 1997). However, war has begun to surface as a likely driver of human demographic and psychological evolution and is currently undergoing a major resurgence of interest among evolutionary biologists and psychologists (Bowles, 2009; Mathew & Boyd, 2011; Gneezy & Fessler, 2012).

Persistent states of conflict of this kind imply competition for resources, and that in turn implies high population densities. Indeed, it is difficult to imagine why neighboring communities should behave aggressively toward each other in the absence of competition for resources, since the costs of war are inevitably rather high. One question that troubled Dick was whether human paleo-populations had ever been dense enough to instigate significant competition between groups. The word on the street among paleoanthropologists tended to be "No". Resources were in plentiful supply, population densities low and many (if not most) populations were continually expanding into new territories. Dick felt obliged to conclude that:

> densities *per se* [are] not critical [to my argument], or else that estimates of densities [are] wrong. It is probably more important to know what kinds of social groups people lived in, and why, than to know densities *per se*.

However, maybe there was an alternative explanation. For ecologists, the term *resources* inevitably means ecological resources. This need not always be so, of course, since women remain a potentially limiting resource for men. Males might, therefore, compete to defend land in order to monopolize access to women, or compete to defend groups of females, rather than defend land for the sake of the food resources it contains. If competition was for women rather than ecological resources, this might have solved the problem that troubled Dick. Competition for women could in principle generate just the kind of arms' race that his hypothesis requires. Although it is not always clear what motivates warfare in small scale societies, women are often a trophy of such activity (see, for example, Chagnon 1968) and might be seen as the ultimate objective even where the *causus belli* is something more mundane. However, this is, I think a minor issue. The fact, as Dick himself was at pains to point out, is that traditional societies revolve around an in-group/out-group effect in which those from other communities are invariably viewed with suspicion and as game for exploitation.

Interestingly, in an aside, he was "led to wonder—entirely without empirical evidence—whether or not orangutans and gorillas once lived in social groups more like those of chimpanzees and humans than is presently the case—in other words, in larger multimale groups, "perhaps multi-male groups in which the males were cooperative in hunting, or even in intergroup aggression" (p. 502). The speculation turns out to have some validity. In the models that we have developed of great ape socioecology—models that use the time costs of foraging to model the constraints on social group size—it has become clear that gorillas and orangutans represent alternative solutions to the chimpanzees' problem of coping with the ecological costs of sociality (Lehmann et al., 2008; in press). The orangutan's solitariness is simply the limiting condition in the chimpanzee's attempt to dissipate the costs of large social communities through a fission-fusion form of social organization, while the gorilla seems to represent an attempt to solve the same problem by increasing body size and shifting to a more folivorous diet. Once, both almost certainly had the same social system as the chimpanzee.

One central question remains outstanding: why have only humans, of all the many species of primates and other mammals, needed to go so far in evolving their unique psyche? Associated with this are the subsidiary evolutionary questions as to what was the trigger for all this and when exactly did these traits first appear? These tantalizing questions remain as opaque now as they were in 1989. Whatever happened on the long road from our common ancestor with the great apes some eight million years ago to the final appearance of modern humans around a quarter of a million years ago, the history of brain evolution in our lineage suggests that it probably happened quite late—during the last 800,000 years marked by the appearance of archaic humans. In looking for potential drivers of evolutionary change around this time, the most obvious feature of this period is the onset of climatic instability with rapid fluctuations between cold and warm periods that eventually give rise to increasingly deep ice ages. The cool, dry conditions that prevailed even in the tropics must have been very challenging for these early humans, requiring novel strategies to cope with them.

Like all ecologists, paleoanthropologists tend to reach instinctively for the foraging innovations solution. After all, the Paleolithic record provides us with more handmade stone tools than anything else, and, at the very least, tools imply the dismembering of carcasses. So conventional wisdom assumes that early humans survived because they evolved the intellectual skills to respond with novel ways of extracting nutrients from an increasingly challenging environment. But the question is whether these solutions to the challenges of survival related solely to foraging strategies (and, hence, essentially involved individual trial-and-error learning) or to something more social as implied by the Social Brain Hypothesis (such as mutual exchange networks that allowed communities to gain access to others' foraging territories as a buffer against local environmental catastrophes). It is difficult to see anything in the first option that would demand the doubling of brain size that occurred during the last half-million years of human evolution. Certainly, as Wynn (1988) pointed out at around the time Dick was writing his article, the toolkits manufactured by successive populations fail to suggest that the answer lies in technological inventiveness (see also Gowlett et al., 2012). While brain volume increased exponentially through time, tool quality improved on something closer to a power curve: tools hardly changed in appearance or quality for the first nine-tenths of the period, and then underwent a dramatic development from about 50,000 years ago in a dazzling display of technical expertise—the very phenomenon that attracted attention as the "Human Revolution" of the title of the volume in which Dick's paper was published. Meanwhile, if we are to believe the quantitative implications of the Social Brain Hypothesis, community sizes would have been increasing in the same exponential fashion as brain volume. Bigger groups mean more stress, both in ecological terms and, more importantly perhaps, in social, reproductive, and demographic terms.

Why would one need such large social groups if it was not to provide *social* solutions to the ecological problems that these early humans faced? The answer must surely have been a form of ecological buffering that exploited extended social networks whose geographical extension was sufficient to provide access to areas that still had resources even in the worst of times. Extended networks of this kind would surely have provided exactly the kinds of competitive regimes that Dick envisaged in his model, especially under environmental conditions in which populations frequently faced resourcing challenges. If so, then it implies that Dick's warfare hypothesis is of relatively recent origin.

In sum, then, the 1989 article has stood up remarkably well to the passage of time. One can always criticize the details, but these are issues that Dick had to make the best of with the rather more limited knowledge available to us 20 years ago. Hindsight is always a wonderful thing. The main theme of his argument, in contrast, appears to be in good shape and as robust as it was when it was fresh ink on the page. Yet, despite this and all the research that has been done since 1989, one question remains obdurately unanswered: Just *why* are humans *so* different from other species of animals? While we are beginning to unpack the mechanistic explanations that provide an answer to how we are different (the neuropsychology of mentalising competences, for example), there is still no obvious answer to why we needed these competences, what environmental conditions selected for them, and when exactly these evolutionary events were set in train. These tantalizing questions remain to be solved by those who follow in Dick's footsteps.

References

Adamczak, V., and Dunbar, R.I.M. (2008). Variation in the mating system of oribis and their ecological determinants. *African J. Ecol.* 45: 197–206.

Alexander, R.D. (1974). The evolution of social behaviour. *Annu. Rev. Ecol. System.* 5: 325–383.

Alexander, R.D. (2011). *The Mockingbird's River Song: Poems, Essays, Songs and Stories, 1946-2011.* Manchester, MI: Woodlane Farm Books.

Barrett, L., Henzi, S.P., and Dunbar, R.I.M. (2003). Primate cognition: from 'what now?' to 'what if?' *Trends Cog. Sci.* 7: 494–497.

Barton, R.A., and Dunbar, R.I.M. (1997). Evolution of the social brain. In: A. Whiten and R. Byrne (eds.), *Machiavellian Intelligence II.* Cambridge: Cambridge University Press, pp. 240–263.

Bowles, S. (2009). Did warfare among ancestral hunter-gatherers affect the evolution of human social behaviors? *Science* 324: 1293–1298.

Byrne, R., and Whiten, A. (eds.) (1988). *Machiavellian Intelligence: Social Expertise and the Evolution of Intellect in Monkeys, Apes, and Humans.* Oxford: Oxford University Press.

Chagnon, N.A. (1968). *Yanomamö: The Fierce People.* New York: Holt Rinehart and Winston.

Cosmides, L. (1989). The logic of social exchange: has natural selection shaped how humans reason? Studies with the Wason selection task. *Cognition* 31: 187–276.

Cosmides, L., and Tooby, J.H. (1992). Cognitive adaptations for social exchange. In: J.H. Barkow, L. Cosmides, and J.H. Tooby (eds.), *The Adapted Mind.* Oxford: Oxford University Press, pp. 163–228.

Dunbar, R.I.M. (1992a). Neocortex size as a constraint on group size in primates. *J. Hum. Evol.* 22: 469–493.

Dunbar, R.I.M. (1992b). Coevolution of neocortex size, group size and language in humans. *Behav. Brain Sci.* 16: 681–735.

Dunbar, R.I.M. (1998). The social brain hypothesis. *Evol. Anthro.* 6: 178–190.

Dunbar, R.I.M. (1999). Culture, honesty and the freerider problem. In: R.I.M. Dunbar, C. Knight and C. Power (eds.), *The Evolution of Culture.* Edinburgh: Edinburgh University Press, pp. 194–213.

Dunbar, R.I.M. (2009a). Mind the bonding gap: constraints on the evolution of hominin societies. In: S. Shennan (ed.), *Pattern and Process in Cultural Evolution.* Berkeley: University of California Press, pp. 223–234.

Dunbar, R.I.M. (2009b). Why only humans have language. In: R. Botha and C. Knight (eds.), *The Prehistory of Language.* Oxford: Oxford University Press, pp. 12–35.

Dunbar, R.I.M. (2010). The social role of touch in humans and primates: behavioural function and neurobiological mechanisms. *Neurosci. Biobehav. Rev.* 34: 260–268.

Dunbar, R.I.M. (2011). Evolutionary basis of the social brain. In: J. Decety and J. Cacioppo (eds.), *Oxford Handbook of Social Neuroscience.* Oxford: Oxford University Press, pp. 28–38.

Enquist, M., and Leimar, O. (1993). The evolution of cooperation in mobile organisms. *Anim. Behav.* 45: 747–757.

Gamble, C., Gowlett, J.A.J., and Dunbar, R.I.M. (2011). The social brain and the shape of the Palaeolithic. *Cambr. Archae. J.* 21: 115–135.

Gneezy, A., and Fessler, D.M.T. (2012). Conflict, sticks and carrots: war increases prosocial punishments and rewards. *Proc. R. Soc. Lond. B:* 279: 219–223.

Gowlett, J.A.J., Gamble, C. and Dunbar, R.I.M. (2012). Human evolution and the archaeology of the social brain. *Curr. Anthrop.* 53: 693–722.

Keeley, L.H. (1996). *War before Civilization: The Myth of the Peaceful Savage.* Oxford: Oxford University Press.

Krebs, J.R., and Dawkins, R. (1984). Animal signals, mind-reading and manipulation. In: J.R. Krebs and N.B. Davies (eds.), *Behavioural Ecology,* 2nd ed. Oxford: Blackwell, pp. 380–402.

Lehmann, J., Korstjens, A.H. and Dunbar, R.I.M. (2008). Time and distribution: a model of ape biogeography. *Etho. Ecol. and Evol.* 20: 337–359.

Lehmann, J., Korstjens, A.H. and Dunbar, R.I.M. (in press). Apes in a changing world–the effects of global warming on the behaviour and distribution of African apes. *J. Biogeog.*

Mathew, S., and Boyd, R. (2011). Punishment sustains large-scale cooperation in prestate warfare. *Proc. Nat. Acad. Sci. USA*, 108: 11375–11380.

Nettle, D., and Dunbar, R.I.M. (1977). Social markers and the evolution of reciprocal exchange. *Curr. Anthro.* 38: 93–99.

van Schaik, C.P. (1983). Why are diurnal primates living in groups? *Behaviour* 87: 120–144.

Shultz, S., and Dunbar, R.I.M. (2006). Chimpanzee and felid diet composition is influenced by prey brain size. *Biol. Lett.* 2: 505–508.

Suddendorf, T., Addis, D.R., and Corballis, M.C. (2009). Mental time travel and the shaping of the human mind. *Phil. Trans. R. Soc. Lond.* 364B: 1317–1324.

Wrangham, R.W. (1980). An ecological model of female-bonded primate groups. *Behaviour* 75: 262–300.

Wrangham, R.W., and Peterson, D. (1997). *Demonic Males: Apes and the Origins of Human Violence*. New York: Houghton Mifflin.

Wynn, T. (1988). Tools and the evolution of human intelligence. In: R. Byrne and A. Whiten (eds.), *Machiavellian Intelligence: Social Expertise and the Evolution of Intellect in Monkeys, Apes, and Humans*. Oxford: Oxford University Press, pp. 271–284.

Zubieta, J.-K., Ketter, T.A., Bueller, J.A., Xu, Y., Kilbourn, M.R., Young, E.A., and Koeppe, R.A. (2003). Regulation of human affective responses by anterior cingulate and limbic μ-opioid neurotransmission. *Arch. Gen. Psychiat.* 60: 1145–1153.

EVOLUTION OF THE HUMAN PSYCHE

Richard D. Alexander

Alexander, R.D. Evolution of the Human Psyche 1989. In P. Mellars and C. Stringer (eds). *The Human Revolution. Behavioral and Biological Perspectives on the Origins of Modern Humans*. Princeton, NJ: Princeton University Press, pp. 455–513.

The gap (between us and our nearest living relatives, the apes...) is largest, and most difficult to comprehend, in terms of mind... As human beings are distinguished so much by their minds,...those minds must be a legitimate object of evolutionary studies (Gowlett 1984: 167 and 188).

Introduction

The purpose of this essay is to develop and test hypotheses about the process and pattern by which the human psyche evolved, and to seek to understand why humans, and humans alone, differ strikingly in mentality from their closest relatives—and evidently from all other organisms. Understanding the human psyche is a key to understanding human sociality (1) as it relates to the behaviour of individuals in different circumstances and after different kinds of learning experiences or developmental events and (2) as it yields variations in cultural patterns in different environments, and following different histories, including extreme and complex phenomena such as the rise of nations.

By the human 'psyche' I mean the entire collection of activities and tendencies that make up human mentality. I include concepts such as (1) *consciousness* and all of its correlatives or components, such as subconsciousness, self-awareness, conscience, foresight, intent, will, planning, purpose, scenario-building, memory, thought, reflection, imagination, ability to deceive and self-deceive, and representational ability; (2) *cognition* (i.e. learning, logic, reasoning, intelligence, problem-solving ability); (3) *linguistic ability;* (4) *the emotions* (grief, depression, elation, excitement, enthusiasm, anger, fear, indignation, embarrassment, despair, guilt, uncertainty, etc.); and (5) *personality traits* (stubbornness, pliancy, subservience, timidity, persistence, arrogance, audacity, etc.).

One can analyse human mentality by (a) morphological and physiological studies of the brain and its functions; (b) psychological and psychoanalytical investigation of behaviour and its underlying motivations and other correlates; (c) inquiries into artificial intelligence, including modelling with machines or mathematics; (d) archaeological and anthropological analysis of fossils and artifacts (focusing primarily on the most direct possible evidence and the *pattern* of evolutionary change); or (e) comparative study of humans and other animals, especially close relatives, combined with adaptive modelling (utilizing primarily predictiveness from knowledge of the *process* of evolutionary change). The last method is the one principally employed here.

First, a unique selective situation is postulated to account for humans departing as far as they have, psychically and in other regards, from their closest living relatives, and it is compared to alternative hypotheses. The human psyche is then characterized in terms of the probable reproductive significance of its different aspects, thereby generating additional hypotheses about its selective background. Finally, an effort is made to test the hypotheses generated by the first and second parts of the discussion.

The Postulated Selective Situation

BACKGROUND OF THE HYPOTHESIS

There is probably general agreement that explaining human evolution is to a large extent a question of understanding how human mental attributes evolved. The problem is not only why brains evolved to be bigger and intellects to be more complicated, but also why they became so dramatically different from those of our closest living relatives. Humphrey (1976) suggested that the selective situation was primarily a social one, with evolving humans providing their own selective challenge; as with others who have made suggestions in this direction, however (see references below), he did not explain what forces caused humans to continue to live under the social conditions responsible for the expenses of intense competition and the resultant manipulations, deception, and favouring of social cleverness that he and others postulate. Thus, he did not account for the fact that humans alone have followed an evolutionary pathway leading to what he called a 'runaway intellect'.

Humans are not just another unique species, rather they are unique in many and profound ways—that is, in many attributes, and also in ways that unexpectedly set them apart from all primates, all mammals, or even all life (e.g. Alexander and Noonan 1979; Tooby and DeVore 1987; Wilson 1975; Wrangham 1987). For example, rapid evolution usually means more speciation, but humans, whose brains are regarded as having evolved according to an 'autocatalytic' model—increasingly rapidly—during at least the past two million years (Godfrey and Jacobs 1981; Stringer 1984), have no living close relatives. By this I mean that there are no closely similar or sister species, no congeneric species, no interfertile species, no species, even, with the same number of chromosomes. Why? Why do we have to go back two, five (or more) million years to find the most recent phylogenetic juncture with our nearest *living* relatives (Ciochon 1987)?

Human social groups are also unique (currently) in being huge and socially complex, while also having all individuals both genetically unique (excepting monozygotic twins) *and expecting to reproduce;* and only humans (apparently) play competitively, group-against-group (currently on a large scale). Although we became a virtually world-wide, highly polytypic species with numerous geographic variants, until recently those different variants have not easily mixed or lived in sympatry; and there seems to be no universally accepted evidence that multiple species of hominids ever lived together. With increases in the world population of humans, moreover, we did not come to live in a single, huge, dense, amorphous, universally beneficent population; rather, we have always lived with tense national boundaries, patriotism, xenophobia, and almost continual and destructive inter-group competition and conflict. Today we have a terrible international arms race as a central horror in our lives. We have at least been primed by evolution so as to allow these things to happen. How were we so primed?

COMPONENTS OF THE HYPOTHESIS

A. The most unpredictable and demanding aspects of the environment of evolving humans have always been its *social* aspects, not the physical climate or food shortages, as is often implied. The human psyche was designed primarily to solve *social* problems within its own species, not physical and mathematical puzzles, as educational tests and some concerns of philosophers might cause us to believe. Darwin (1859, 1871), Keith (1949), Bigelow (1969), Wilson (1973), Hamilton (1975), Alexander and Noonan (1979), and Alexander (1967–1988) have all suggested some parts or versions of this model, but Humphrey (1976) probably described it most clearly (independent hints toward it are also numerous—e.g. Trivers 1971; Fox 1980; Kurland and Becker 1985; Box and Fragaszy 1986; Burling 1986). This hypothesis implies that even the solving of mathematical, physical, and non-human biotic problems had its central significance (in the broadest sense, its reproductive rewards) in social contexts. (For example, Lenneberg (1971) argues that 'mathematical ability may...be regarded as a special case of the more general ability that also generates language...' and Burling (1986) that 'the...evolution of the...capacity to learn and use highly complex language is unlikely to be explained primarily by any subsistence or technological advantages that language offers. Rather, language probably served social purposes'.) In other words, this hypothesis rejects the notion that complex intellects evolved because they saved early humans from starvation, predation, climate, weather, or some combination of such challenges. All other organisms have solved these kinds of problems, in a variety of ways, without complex human-like intellects. If humans solve such ordinary problems in extraordinary ways, I am suggesting, it is because they are using an intellect evolved in a different context (For comparative purposes, Isaac (1979) lists six hypotheses bearing on human evolution: Dart's

(1949, 1954) 'hunting' hypothesis; Jolly's (1970) 'seed-eating' hypothesis; Tanner and Zilman's (1976) 'gathering' hypothesis; Isaac's (1978) 'food-sharing' hypothesis; Parker and Gibson's (1979) 'developmental' hypothesis, based primarily on food skills; and Lovejoy's (1981) 'shortened birth interval' hypothesis). Dart's is the closest to that espoused here; the others all depend on non-human biotic or physical threats as primary forces. As I am restricting it, my hypothesis also requires that human proficiency in tool construction and use is also a secondary or incidental effect of the evolution of an intellect designed to be effective in social contexts. Wynn (1979) and Gowlett (1984) discuss the relationship between the manufacture of tools, and especially the transport of materials involved in their construction, to the evolution of planning and foresight. In this connection it is relevant that chimpanzees show evidence of planning, and perhaps scenario-building, in such tool-using behaviour as the selection, preparation and carrying of termiting sticks (Goodall 1986; Ghiglieri 1988). It is obviously important to my hypothesis that they also seem to show considerable foresight in social activities (Goodall 1986; Ghiglieri 1988).

B. Human mental abilities evolved as a result of *runaway social competition,* an unending within-species process dependent upon interminable (and intense) conflicts of interest, compared (below) to Fisher's (1958) concept of runaway sexual selection (see Alexander 1987).

C. *Balance* (or *imbalance of power races*) between social groups, either within or between (very similar) species, facilitated runaway social competition by favouring complex social living, and abilities to behave cooperatively and competitively within (and between) social groups. Such races 'trapped' humans into social interdependence, led to within-group amity and between-group enmities, and in part created the selective situation that gave rise to our creative intellects. Humans may not be the only species to engage in social reciprocity and cooperation-to-compete (with conspecifics), but they are probably the only one in which this combination of activities is a central aspect of social life.

D. These processes became paramount partly because the *ecological dominance* of evolving humans diminished the effects of 'extrinsic' forces of natural selection such that within-species competition became the principal 'hostile force of nature' guiding the long-term evolution of behavioural capacities, traits, and tendencies, perhaps more than in any other species. The evidence for this having happened is the current ecological dominance of humans; the only problem is when and how it came about. One might ask if (1) the ecological dominance of humans allowed the evolution of complex intelligence or (2) complex intelligence enabled humans to become ecologically dominant. I would argue, rather, that the two went hand in hand, reinforcing one another at every stage, and I suggest (below) that, aside from the human presence, chimpanzees may already have attained the required dominance.

The reference to 'extrinsic forces' above is to Darwin's Hostile Forces of Nature – parasites, predators, diseases, food shortages, climate, and weather—as an exhaustive list of the features of natural selection that determine the reproductive success and failure of different genotypes. Darwin (1859, 1871) distinguished between natural selection and sexual selection, so he did not include in the list, as I do, mate 'shortages' (meaning, ultimately, variations in mating success, including quality as well as quantity of mates). Darwin's emphasis on sexual selection to account for the evolution of many human traits is in accord with the idea presented here, if the context is expanded to include other kinds of social competition, and I believe that the current idea supports his general suggestion.

> E. The combination of (1) *balance-of-power races* between human social groups, (2) *runaway social competition* and the emphasis on creative and manipulative intellects, allowed or facilitated by (3) *human ecological dominance,* can also be used to help explain the changes in social structure that occurred as human social groups expanded toward their present sizes and took the forms (bands, tribes, and nation-states –'egalitarian', despotic, totalitarian, or democratic societies) represented across human history.

> F. The central evolved function of the human psyche, then, is to yield an ability to anticipate or predict the future—explicitly the social future—and to manipulate it in the (evolutionary, reproductive) interests of self's genetic success. In the hypothesis developed here, all other effects or properties of the psyche are secondary to this strategic function. This general situation came about because evolving humans (a) came to live in highly cooperative social groups and (b) became ecologically dominant, these two conditions together (i) reducing the significance of hostile forces of nature other than conspecifics and (ii) leading to cooperation to compete against conspecifics who were doing precisely the same thing. In this fashion the combination of an unending runaway social competition and an unending balance-of-power race was set in motion, which continues within and among human populations today. This general situation allowed and caused the radical departure of humans from their closest relatives, in psychical and other attributes.

A central feature of the human psyche is the construction of alternative scenarios as plans, proposals, or contingencies in a manner or form perhaps appropriately termed *social-intellectual* practice for social interactions and competitions (practice which lacks a prominent physical component). This hypothesis of *scenario-building* sheds light simultaneously on a collection of human enterprises that have seemed virtually impossible to connect to evolution—such as humour, art, music, myth, religion, drama, literature, and theatre—because they are involved in *surrogate scenario-building,* a form of division of labour (or specialization of occupation) that may be unique to humans (partly because language is required for communication of mental scenarios between individuals). The centrality of scenario-building in human sociality (which will be related to the concept of *play*) is connected with the appearance of *rules* (hence, moral and legal systems) through (in part) the value of limiting the extent to which the elaborate and expensive scenarios (plans) of others can be thwarted by selfish acts (Alexander 1987, see also below). Finally, part of the game of human social competition involves concealing how it is played, and some of such concealment involves concealing it from one's self (*self-deception*). This in turn compounds the problem of understanding ourselves because of the difficulty of bringing into the conscious items that have been kept out of it by natural selection, most particularly items involving social motivations.

The hypothesis assumes that some version of these social functions initially drove the evolution of consciousness and other aspects of the human psyche, and that other uses of the psyche, such as in predicting or dealing with aspects of the physical universe, are (or were initially) incidental effects (in the evolutionary sense). This scenario does not preclude adaptive functions of the psyche in dealing with nonsocial phenomena throughout human evolution, only that such functions could not have caused the evolution of consciousness, cognition, linguistic traits, and the emotions as a set of human attributes. The emphasis on manipulation and deception is because the hypothesis holds that the human psyche would not have evolved in a world dominated by truth-telling, so that its complexity is tied to its use in deception. Once efforts at deception are widespread, successful, and complicated, truth-telling also becomes difficult to identify or prove. The argument is that truth is approached only when necessary—that is, cost-effective.

GENERAL COMMENTS ON NATURAL SELECTION

Ultimately, there must be compatibility between our view of the functions of the human psyche and our understanding of the selective background that gave rise to it. I am going to develop the argument from the beginning, because there can be no agreement, or adequate evaluation of arguments, unless common ground has been established from the outset.

If we accept the view of modern biology that natural selection is the principal guiding force of evolution, this means, first, that to understand traits we must concentrate on their reproductive significance and discard most of the old notions about adaptive function, such as survival of the individual (at all cost—i.e. even when survival is opposed to reproductive success), benefit to the population or species (again, when there is conflict with benefit to the individual's reproduction), progress, or any kind of goal-oriented or orthogenetic trend. We are not free to assume that genetic drift or other random events can account for elaborate attributes, just because they seem to give an unprejudiced, amoral, or value-free aspect to evolution or because they can account for minor differences between populations (Alexander 1979, 1987).

RULES FOR APPLYING SELECTIVE THINKING

Continuing from this initial assumption, I assume five general rules in applying natural selection to the attributes of organisms (for the first two, see Williams 1985):

> *First,* we must consider the question of *adaptation,* not according to some notion of optimality or ends to be achieved, but rather according to the now widely accepted usage, from Williams (1966), of simply *better versus worse in the immediate situation.* This view implies that long-term trends occur because particular selective forces remain in place for long times, so that step-by-step small changes sometimes give a false retrospective appearance of goal-oriented or orthogenetic trends. As Williams (1966) emphasized, we must also distinguish between incidental effects of traits and their evolved functions or evolutionary 'design'.

> *Second,* natural selection must always work from 'last year's model'—a fact often referred to by modern biologists under concepts like phylogenetic and ontogenetic inertia, or structural laws of development and evolution. This particular rule implies that phenomena like allometry or neoteny are in general maintained as a result of selection and not in spite of it; that when such phenomena cause some kinds and degrees of evolutionary 'inertia', they must be presumed to have developed the potential for such effects as a result of past selection. To invoke physiological, developmental, or phylogenetic constraints to explain evolved phenomena is thus an argument of last resort.

> *Third,* random events such as mutations and drift introduce noise into the adaptive process but do not guide long-term directional change.

> *Fourth,* selection is more potent at lower levels in the hierarchy of organization of life (Williams 1966, 1985; Hamilton 1964, 1975; Lewontin 1970; Dawkins 1976–1986; Alexander and Borgia 1978; Alexander 1979, 1987), so that, as Williams (1966) first argued convincingly, 'most of the characteristics of organisms, including social behaviour, must be the result of differential fitness at the level of individual genotypes' (Lewontin 1966).

> *Fifth,* to understand traits, it is effective, and parsimonious, to seek or hypothesize singular selective causes (or contexts or changes) in evolution, as opposed to accepting multiple ones too readily. This is so because (1) it is difficult to falsify individual causes when multiple contributing factors are accepted uncritically; (2) single causes can be sufficient, even when multiple contributory factors are known; and (3) once a particular event, such as group-living, has occurred, then secondary effects will

appear that (especially without attention to the possibility of single sufficient causes) can be confused with the primary cause (in other words, single *different causes* may occur *in sequence* without violating these arguments).

Hypothesizing singular causes, I believe, is a way of making one's ideas maximally subject to falsification, if they are incorrect, and therefore of advancing knowledge most effectively. It is a way of going most forcefully after the actual driving forces in evolutionary change, and of unravelling most quickly and completely the actual patterns of change. This 'rule' for applying selection is the most controversial one, and the controversy arises primarily because some see it as a way of over-simplifying causation in human social affairs. This criticism, in turn, is prevalent among those who believe that knowledge (or supposed knowledge) of history yields ideology for the future. This problem arises in large part out of ignorance about the relationship between genes and phenotype, heredity and development, rigidity and plasticity in expression of behaviour. It cannot be erased, however, by emasculating the procedures of science. Rather, it must be removed from importance by explaining why the argument that history justifies ideology cannot be sustained.

I have given no rules regarding the mechanisms by which the phenotype is acquired, such as learning or maturation. There are two reasons: First, the operation of natural selection can be understood without knowing about, implying, or eliminating any particular ontogenetic or physiological mechanisms. Second, predictions about expected proximate mechanisms in particular evolutionary situations are extremely difficult and complicated. Although ignorance about design mechanisms makes it more difficult to be Confident about interpreting function, so long as no particular restrictions are assumed (e.g., that a behaviour is 'innate', learned in some particular fashion, etc.), analysis of such mechanisms can usually be postponed without necessarily causing error.

GROUP SELECTION

Group selection (*versus* selection at lower levels in the hierarchy of organization of life: Alexander and Borgia 1978) is an issue that is central, yet seems to remain complex and confusing. Two different situations may be implied by the term 'group selection'. The first appears to be relatively unimportant, for the reasons given. The second is the one to which I referred (Alexander 1974) when I suggested that humans are an excellent model for group selection (see also Alexander and Borgia 1978).

  1. *Group Selection when there are Conflicts of Interest between Individual and Group Levels.* In this kind of selection, the spread or maintenance of alleles is determined at group levels, regardless of conflicts of interest within groups, because selection at the group level is simply more powerful than

that at lower levels. In other words, the maintenance and spread of alleles is determined primarily by the differential extinction or reproduction of groups, regardless of what is happening at individual or other levels. This is the kind of selection that Wade (1976, 1978), D. S. Wilson (1975, 1980), and others seek to validate theoretically (as feasible or likely in natural populations) and demonstrate in laboratory experiments. Their work, however, actually indicates that group effects are weak in the face of the strength of selection at lower levels (Williams 1985; Dawkins 1986; Alexander 1987). Thus, to create potent group selection they have been forced to postulate populations with attributes much like those of individuals. They invoke groups that are founded by one or a few individuals (thus as near as possible to being single broods of offspring), and last about one generation (thus have the same generation time as individuals). In the laboratory they create populations (sometimes highly artificial) with minimal within-population genetic variance and maximal between-population genetic variance. The effort is to maximize genetic variance and minimize generation time at population levels, because the intensity of selection depends on these attributes (Fisher 1930; Lewontin 1970), which are virtually always more favourable to selection at lower levels. There are multiple indications, in the behaviour and life histories of organisms, that this kind of group selection does not often prevail (Williams 1966; Alexander 1979, 1987).

Group selection that is weaker than individual selection may often affect the *rate* of selective change as a result of the differential reproduction of individuals, but group selection probably is only rarely potent enough to affect the *direction* of selection within species in natural situations when it runs counter to selection at lower levels. Selection that results in one of two competing species becoming extinct is this kind of group selection, but such interspecies selection occurs under conditions when the differences between groups cannot be compromised as a result of interbreeding. Interbreeding and gene flow between adjacent groups reduce the potency of group selection within species because they reduce genetic differences between adjacent populations (Hamilton 1964, 1975; Alexander and Borgia 1978).

Group selection of the sort just posited ultimately will result in individuals that sacrifice their genetic reproduction for the good of the group because differential extinction and/or reproduction of alleles at the group level exceeds in importance that at the individual level. This is an effect postulated by Wynne-Edwards (1962, 1986) to explain what many ecologists saw a few decades ago as 'intrinsic' population regulation that adaptively avoids over-use of the environment. It has been invoked in ways implying that there are no special difficulties in the postulated sacrifices. Many people believe (erroneously) that this kind of 'genetic altruism' (Alexander 1974, 1979, 1987) would necessarily typify a species, human or otherwise, that had evolved through a process significantly involving group selection. Only a predominance of this kind of selection, it would seem, could prime the individuals of a species to participate readily in the kind of 'greatest good to the greatest number' utilitarianism proposed by many social philosophers. The required kind of sacrifice of one's reproduction, however, is not expected to have evolved, and no natural cases of this kind of selection have been documented or are widely suspected.

2. *Selection with Confluences of Interest at Individual and Group Levels.* In this kind of situation the individual who gives his life for the group (all or any part of the population) gains genetically from the act because his individual interests are identical with those of the group. Identity of evolutionary interests will be temporary in sexually recombining organisms because individuals in such species will for the most part be genetically unique. Identity of interests, on the other hand, is permanent in clones (barring mutation). Altruism between genetically non-identical individuals can thus evolve when neighbours (interactants) are alike genetically—compared to more distant conspecifics (as in Hamilton's (1975) 'viscous population'—cf. Nunney (1985))—to a degree that over-compensates the increased competitiveness for resources likely to result from proximity.

Even when individual and group interests are identical, presence and prevalence of alleles are likely to be more affected by lower levels of selection because of the greater potency of selection there (Williams 1966; Lewontin 1970; Dawkins 1976; Alexander 1979). Such selection would, however, be enhanced by selection in the same direction at group levels. In sexually reproducing organisms, such as humans, confluences of interest within groups are likely to occur when different groups are in more or less direct competition. As a result, the kind of selection alluded to here would be expected to produce individuals that would cooperate intensively and complexly *within* groups but show strong and even extreme aggressiveness *between* groups. Such tendencies characterize modern humans and chimpanzees, and there are no convincing reasons for believing that they did not characterize humans throughout their evolution. The kinds of inter-group interactions that occur among modern humans and chimpanzees probably occurred throughout the evolution of the great apes and hominids. In humans, especially, the impression of group selection can be given because of costs imposed socially for failures to be altruistic.

One reason why humans have been described as a good model for within-species evolution by group selection is that opposing groups often behave toward one another as if they were different species. Social rules, morality, law, and a great deal of culture represent the imposition of costs and benefits on the actions of individuals and subgroups designed to force a convergence of their interests with those of the social group as a whole, thus tending to create the second kind of situation described above.

It seems to me that, after more than 50 years of discussion and analysis, the appropriate conclusion from all the arguments on the topic of self-sacrifice, heroism, and altruism toward others is still that suggested by Fisher (1930: 265): 'The mere fact that the prosperity of the group is at stake makes the sacrifice of individual lives occasionally advantageous, though this, I believe, is a minor consideration compared with the enormous advantage conferred by the prestige of the hero upon all his kinsmen'. Even the extreme cases of Japanese kamikaze pilots in World War II, in which all of the men in sizeable units volunteered willingly, and even insistently and competitively (Morris 1975)—if they are to be queried in evolutionary adaptive terms—must be analysed in the light of ceremonies, honours, and other described effects having to do with relatives of the volunteers, and connected to the costs of cowardice and rewards for heroism, during the long-term cultural history of Japan.

A NOTE ON COMPETITION

Humans live in groups, and individual interests are expressed in cooperation and competition at all levels of social organization. I interpret human social organization of virtually all kinds to be cooperation, either for the explicit purpose of direct competition with other humans also living in groups or as a part of the indirect competition of non-intentional or non-interactive differential reproduction. With respect to the general process of evolution, the concept of competition must be taken in this broad sense, in which it stands alone, with no real or existing opposite. Thus, cooperation, and all parallel activities, cannot be regarded as *alternatives* to competition; as such they could not evolve. Cooperation must exist because it has aided reproductive competition, however indirectly. One individual or group can be said either to compete or cooperate with another; but the cooperation, if it depends on evolved tendencies, also represents either direct or indirect competition with still other such units. In evolution, only genetic altruism (Alexander 1974, 1979, 1987) could be regarded as opposing competition, and for this reason such altruism will not evolve and will not be maintained if it appears incidentally.

DISTINCTIVE ASPECTS OF SELECTION ON HUMANS

The nature and all-inclusiveness of organic evolution requires us to assume that the human psyche evolved as a vehicle serving the genetic or reproductive interests of its individual possessors. These interests are expressed in individual humans via success in (1) survival and social integration across juvenile (and adult) life; (2) mate-seeking and -holding as an adult; (3) offspring-production and -tending; (4) beneficence (and the seeking of beneficence) with respect to collateral kin; and (5) various forms of (direct and indirect) social reciprocity to both kin and non-kin (or close and distant kin), which means beneficence dispensed in situations in which returns with interest are expected (Trivers 1971, 1985; Axelrod and Hamilton 1981; Alexander 1979, 1987).

WHY LIVE IN GROUPS?

To understand any group-living species, such as humans, we must first ask why organisms live in groups and then ask, in turn, why humans live in groups. I have previously discussed these two questions in detail (Alexander 1974, 1979: 58–65; 1987: 79–81). Partly because the above view of natural selection is fairly new in biology, the answers are not the same as those that have been prevalent in anthropology and the other social sciences.

Sociality by definition can exist only in organisms that live in groups. Efforts to understand sociality for a long time rested upon the intuitive view that groups exist for the good of the species, and individuals for the good of the group. If selection is more potent at individual than at group levels, however (see above), we should expect organisms to behave as if their own reproductive success is what matters (and they do—see examples in Alexander 1979, 1987).

Except in clones, the life interests of individuals within groups are rarely identical. If behaviour evolves because it helps its individual possessors, then group-living inevitably entails expenses to individuals, such as increased competition for all resources, including mates, and increased likelihood of disease and parasite transmission. Why, then, should animals live in groups? Why be social, beyond the minimum required to mate and raise a family (indeed, why even keep one's offspring near for a time)? If the answer is that individuals living in groups reproduce more than individuals not living in groups, what are the reasons?

Theoretically, the causes of group-living include enhanced access to some resource, or enhanced ability to exploit a resource, which more than offsets, for individuals, the automatic detriments. I have previously argued (Alexander 1974) that reasons for group-living are few in number:

> (1) lowering of susceptibility to predation either because of aggressive group defence or because of what Hamilton (1971) called *selfish herd* effects (e.g. a more effective predator detection system or the opportunity to place another individual between one's self and a predator); (2) cooperative securing of fast, elusive, aggressive, or hard-to-locate prey; and (3) localization of resources (food, safe sleeping sites, etc.) that simply forces otherwise solely competitive individuals to remain in close proximity (see also Alexander 1975, 1979).

It is obviously useful to distinguish, when possible, between the primary causes of group-living and its secondary results. Postulated causes of group-living other than those listed above are probably secondary. For example, it seems unlikely that cooperative defence of clumped resources could ever be a primary cause of group living, since clumping of resources would at first yield the third kind of group-living just listed (Alexander 1974). My own opinion is that groups, such as in primates, that cooperate now, whether to defend food, females, or territory, probably evolved originally because of predator influence (even if what was at first involved was only one or a small group of females protecting their young) and secondarily (after having evolved cooperative tendencies or abilities in other contexts) began to defend food resources in those groups. Similarly, I would suppose that defence of large prey items in cooperatively hunting canines and felines is also secondary, evolving after group hunting. Again, group hunting most likely took the initial form of one or two parents hunting with their own offspring. It is obviously easier to understand cooperative interactions within groups of relatives, especially parents and offspring, than among nonrelatives.

Wrangham (1980), Cheney and Wrangham (1987), and others have downplayed the role of predation in causing group-living in primates, mainly because of the paucity of observations of predation. Nevertheless, predation is probably responsible for herding in ungulates and colonial life in a great many vertebrates that live in groups similar to the groups of related females discussed in various primates by Wrangham (1980), and that have little possibility of being explained as defenders of food bonanzas or any other resources. Moreover, Cheney and Wrangham (1987) list estimated predation rates on 30 primate species as averaging about 6.5% per annum.

Humans have affected predators negatively by their own actions more than prey species, and as well may deter predators completely by their presence as observers, so the above figures can scarcely be regarded as insignificant. Indeed, the only other significant source of primate mortality discussed in Smuts et al. (1987), is infanticide by conspecifics. Predation is also rarely observed in many species where the observers do not doubt that it has been responsible for chemicals, powerful senses, mimicry, cryptic coloration, or the patterning of social life or group living (e.g. Alcock 1984). If one wishes to answer the question of why groups form and persist, moreover, observed rates of predation on normally structured social groups are far less significant than observations of the fates of individuals outside their social groups, or in groups that, for example, lack large males or are abnormally small.

CAUSES OF HUMAN GROUPING

Early groups of humans are widely believed to have been group hunters. This idea is not incompatible with evidence that gathered foods provided most of the diet of early humans (e.g. Lee 1968; Tooby and DeVore 1987); but gathering seems less likely to have led to complex cooperative tendencies (for perhaps the best case for the alternative possibility, see Kurland and Beckerman 1985). If arguments given earlier are correct, the predecessors of the earliest hominid group hunters almost certainly lived in groups because, as with probably all modern group-living nonhuman primates, they were the hunted rather than the hunters. By all indications, humans are the only primate that became to some significant extent a group-hunter—the only group-living primate who, at least for a time, has escaped having its social organization essentially determined by large predators (chimpanzees are nearest to being an exception in both regards, but they obtain far less meat from hunting, and they are obviously subject to human and other kinds of predation: see King 1976; Goodall 1986; Cheney and Wrangham 1987). For this reason it is not surprising that we empathize to some degree with canine and feline social groups.

The human brand of sociality appears to have been approached by various other primates because they are our closest genetic relatives, but by canines and felines (lions) because they most nearly (among nonprimates) do, socially, what we did for some long time.

But the organization and maintenance of recent and large human social groups cannot be explained by a group-hunting (or gathering) hypothesis (Alexander 1974, 1979). The reason is that the upper size of a group in which each individual gained because of the group's ability to locate and secure large game or other food bonanzas would be rather small. Indeed, according to such a hypothesis, as weapons, skills, and cooperative strategies improved, group sizes should have gone down, owing to the expenses (to individuals) of group-living, which tend to be exacerbated as group sizes increase (acceptance of this argument depends on the assumption of powerful selection below the group level). Cooperative group hunters among nonhumans tend to live in small groups (canines, felines, cetaceans, some fish, and pelicans); and most large groups (e.g. herds of ungulates) are probably what Hamilton (1971) called 'selfish herds', whose evolutionary *raison d'être* is security from predation. (The relevant security is to individuals, but not necessarily to the species as a whole; as Hamilton pointed out, the population can actually suffer higher overall predation, but because existence of groups causes predation on lone individuals to become even more severe, group-living in selfish herds can nevertheless continue to evolve.) Even groups evidently evolved to cooperate against predators are typically small (chimpanzees, baboons, musk ox). But maximum human group sizes, beginning at times and places not easily ascertainable, went up—right up, eventually, to nations of hundreds of millions.

Modern human groups are unique, suggesting that explaining them will call for selective situations that are in some sense unique. Thus, other complexly cooperative groups either tend to be small, as with cooperatively hunting groups of canines or felines, or else they are structurally unlike human groups. In human groups, for example, coalitions exist at many levels of stratification, and in many functional contexts. Clones are often very large, as are modern human groups, but clones are composed of individuals with continuously identical evolutionary interests, while human groups are not. Eusocial insect colonies, such as those of ants and termites, are often both huge (up to 15–20 million: Wilson 1971) and complexly social, as with humans. But they represent variations on a nuclear family theme. In them, a single female (usually) and one or a few males produce all the offspring, and all but these few reproductive individuals tend to be full or half siblings to one another. Even in a eusocial colony of millions, every individual is closely related to every other one. Completely unlike this, large human groups are composed of close and distant relatives, and nonrelatives; and every individual expects to reproduce unless special circumstances intervene. Moreover, in the huge modern eusocial insect colonies, such as with ants and termites, the interests of the different colony members, whether queen or workers, may be virtually identical. The reason is that all will realise reproductive success only through the rather small group of reproductively mature individuals that will emigrate and found new colonies. Not only are the workers and the queen likely to be similarly related to these reproductives (especially in forms with diplo-diploid sex determination, as with termites), but they are likely to have no other opportunity to reproduce. A good comparison in familiar terms would be represented by a species in which the male and female tend to be obligately monogamous, bonded for life. If opportunities for differential assistance to nondescendant relatives, and for philandering, are rare or non-existent, then, even though the male and female may be completely unrelated, their reproductive interests are identical. Each will reproduce only via the offspring they produce together; and in most cases the two parents will be equally related to the offspring. In such cases the male and female are expected to behave as though their evolutionary interests are identical, as with members of clones, and queen and workers in some eusocial colonies (Alexander 1987).

In this light, conflicts of interest take on special significance in the huge modern social groups of humans: such conflicts are more or less continual, and they can become exceedingly complicated. They wax and wane in response to competitive and cooperative interactions of groups with one another, as well as in response to the interactions of individuals within groups. As we shall see, chimpanzees live in multi-male groups that in some ways parallel those of humans, and this is significant for efforts to understand human evolution and the nature of the human psyche.

In trying to explain how modern humans developed such huge and unified political groups, and became involved in the current international arms race, it is possible to argue that the early benefits of group-living—whatever they might have been—were so powerful that they produced humans with such strong tendencies to be socially cooperative that the huge groups of recent history developed as more or less incidental effects, despite wide-spread deleterious effects on the reproduction of the individuals that comprised them. Such a view may at first seem correct, in the sense that living in small, highly cooperative groups may have produced humans who readily adopt competitive or adversarial attitudes toward members of other groups, and continually compare the relative strengths of groups and strive to produce or maintain strength (through extension and intensification of cooperation) in their own group (Hamilton 1975). But this view allows continual readjustments that return positive effects on the reproduction of individuals living in groups. In any other sense, the argument implies a degree of rigidity, or a kind of genetic control, of behaviour that I would like to regard as an argument of last resort. Moreover, if this latter kind of rigidity were the correct view, then we should not be able to identify widespread advantages from living in large groups currently, and alternative hypotheses to explain modern human groups should be difficult to apply. Neither is the case.

IMBALANCES OF POWER AND RUNAWAY SOCIAL COMPETITION

The general hypothesis that I support to account for the maintenance and elaboration of group-living and complex sociality in humans, described earlier, derives from a theme attributable to Darwin (1871) and Keith (1949), and developed by a succession of more recent authors (Bigelow 1969; Carneiro 1970; Wilson 1973; Pitt 1978; Strate 1982; Betzig 1986; Alexander 1967–1988; Alexander and Noonan 1979; Alexander and Tinkle 1968). It includes group-against-group, within-species competition as a central driving force, leading to balance-of-power races with a positive feedback upon cooperative abilities and social complexity. It implies that the only plausible way to account for the striking departure of humans from their predecessors and all other species with respect to mental and social attributes is to assume that humans uniquely became their own principal hostile force of nature.

This proposition is immediately satisfying, for it (perhaps alone) can explain any size or complexity of group (as parts of balance-of-power races). It accords with all of recorded human history. It is consistent with the fact that humans alone play competitively *group-against-group* (and, indeed, they do this on a large and complex scale)—if play is seen, broadly, as practice. And it accords with the ecological dominance of the human species and the disappearance of all its close relatives.

No other sexual organisms compete in groups as extensively, fluidly, and inexorably as do humans. In no other species, so far as we know, do social groups have as their main jeopardy other social groups in the same species—hence, the unending selective race toward greater social complexity, intelligence, and cleverness in dealing with one another at every social level. No other species deals in war so as to make it a centre-piece of social cooperation and competition. I am not aware of hypotheses other than that given above which can deal with all of these issues.

Most of the evolution of human social life, I am hypothesizing, and the evolution of the human psyche, has occurred in the context of within- and between-group competition, within-group competition shaped by between-group competition, and the centrality of social competition resulting from the ecological dominance of the human species. Once cooperation among individuals (and subgroups) became the *central means of within-species competition,* the race toward intellectual complexity was on. Without the pressure of between-group competition, within-group competition would have been relatively mild, or at least dramatically different, because groups would have remained small and would have required less unity and different kinds of cooperativeness. There would have been no selective pressure that could produce the modern human intellect.

The situation I am postulating is not simply that described by Darwin's observation that, because of their similarity to one another, the members of a species are their own worst competitors for food, shelter, mates, and so forth. Rather, this view calls for a species that has so dominated its environment that all other hostile forces have been manipulated and modified into relative trivialities, *compared to the effects of competitive and cooperative conspecific neighbours.* If there are forces that remain potent (for humans, parasites are the most obvious example), then my argument would suggest that they could not be neutralized effectively by human effort in ways that led to major long-term trends in behaviour that could account for the evolution of the human intellect (the reasons might be erratic or infrequent appearance, rapid evolution, invisibility, or other causation that has somehow been outside human knowledge or capabilities of thwarting). In present circumstances the AIDS virus, a 'social' disease, might be seen as a counterexample to my argument. It seems beyond doubt that this disease is at least temporarily modifying human sexual and social behaviour in a significant fashion. Particularly interesting is the extent to which people who previously regarded as immoral actions that increase the likelihood of contracting AIDS use this jeopardy to promote their particular views of morality. To place AIDS-like diseases in an appropriate perspective with regards to human evolution, however, one has to consider what the reaction to them would have been without modern technology and knowledge from it. It seems unlikely that connections would be easily established between sexual interactions and physical deterioration many years later, or that sexual behaviour would have been severely modified prior to modern medical knowledge.

Runaway social competition can be understood by considering three features: (1) interminable conflicts of interest that cause social competition to be unending; (2) runaway aspects that can come into play most powerfully when the competition is within species; and (3) minimizing of brakes or direction changes because of the ecological dominance of the human species.

Interminable conflicts of interest cause unending evolutionary races. Such races occur, for example, between predators and their prey, so long as two species remain in this relationship to one another. They also occur within species, as between males and females, when conflicts of interest exist between the sexes in regard to the social interactions that sexual reproduction requires them to undertake. For example, males in many insect species use their genitalia in ways contrary to the interests of females, such as by holding the female longer than is to her advantage during copulation, so as to decrease the likelihood of another copulation and competition from the sperm of another male. Females may be expected to evolve to extricate themselves sooner, males to evolve to hold them more effectively; and the race is potentially unending. Similarly, in many mammals, males can maximize their reproduction only by mating with multiple females and showing little or no paternal care, while females in the same species can only maximize their reproduction by securing more paternal care than is advantageous for males to give. Such conflicts can lead to rapid evolutionary change, and sometimes unending races, in which each party evolves in response to the particular changes that occur in the other: what is beneficial for a male will depend on what countering changes occur in females, and *vice versa* (although extrinsic environmental changes may also be crucial in both of these examples).

One of the relevant facts from the 'balance (or imbalance) of power' argument described above for humans is that in social-intellectual-physical competition (such as physical competitions in which intelligence and the ability to gain and use support from others are important), conspecifics are likely to be—as no other competitors or hostile forces can be—inevitably no more than a step behind or ahead in any evolving system of strategies and capabilities. (The exception is when geography restricts contact, hence prevents more or less continual transfers of information via either aggression or cooperation and allows either cultural or genetic divergence or both.) Evolutionary unending races are thus set in motion that, because of the presumed paucity or absence of hostile influences extrinsic to the human species, have a severity and centrality as in no other circumstance. In other words, human social competition may be expected to involve a 'runaway' aspect, comparable to Fisher's runaway sexual selection, that is not likely in evolutionary races between, say, predators and prey. Indeed, the postulated process could be more extreme than runaway sexual selection.

Fisher (1930: 58) used the term 'runaway' sexual selection for situations in which females (usually) begin to favour extremeness of traits in males, leading to greater mating success by males that possess extremes of traits that are deleterious in every other respect (Trivers 1972). Within-species social competition is likely to take on 'runaway' aspects for three reasons: (1) the interdependence of the adversarial parties causes the significance of change in one to depend on the traits of the other; (2) the traits involved in the competition are likely to be arbitrary (and deleterious) in all other contexts; and (3) within-species groups of an ecologically dominant species such as humans are relatively immune to effects from other selective agents. When one's adversary continually remains similar or identical to one's self in all but the particular trait that is at the moment changing, when changes in one party depend solely upon changes in the other, and when other hostile forces are insignificant, then there are few or no brakes on change in the traits used in the competition, and little extrinsic guidance (cf. West Eberhard 1979, 1983). I believe that the current human arms race is the prime example of such a process, and as well a logical outcome of the history that such a process suggests for the human species (Alexander 1987).

Runaway social competition would (perhaps alone) account for the fact, stressed earlier, that human evolution has resulted in a single species, with all the intermediate forms having become extinct along the way. (I also speculate that the evolving human line has for a long time been a severe predator and competitor of apes, and is at least partly responsible for the low number of surviving Pongidae.) Indeed, unlike any other hypothesis so far advanced, it appears to *require* this outcome. It could also account not only for an acceleration of the relevant changes in the psyche, in social organization, and in culture in general, at certain stages of human social evolution, but as well for deceleration or even reversal of the direction of evolution of the psyche at other stages (Alexander 1971; Pitt 1978). Stringer (1984) summarizes the evidence that '... the autocatalytic model of endo-cranial volume increase seems most appropriate since there is an increasing rate of change until the late Pleistocene, when endocranial capacity values stabilize or even decline'. All that is required for the presumed stabilization or decline is that

(1) social change eventually creates large societies in which the kinds of abilities and actions that preserve the entire group are possessed and used appropriately by smaller proportions of the society's members (in their own interests); and (2) group success and the social structure somehow lessen the reproductive disadvantages previously suffered within (and between) societies by those who lack the qualities of such leaders or governors. As numbers of leaders diminish in relation to numbers of followers—with increases in the sizes of social and political groups—the probability of producing a sufficient number of individuals with the necessary qualities to lead or govern effectively (whatever these qualities may be) would not necessarily diminish, owing partly to effects of genetic recombination. This condition (absence of advantage for increased complexity of mental activity) may exist in all large societies today (e.g. Vining 1986; compare with Alexander 1988). Whether or not it accounts for what Stringer describes (which could also reflect changes in brain structure consistent with the previous trend towards greater brain size, body size changes, or other forces) is another question.

In human intergroup competition and aggression, there are two prominent facilitators that unbalance the power of competitive groups, leading to more dramatic outcomes of confrontations and a greater likelihood of significant group selection in the form of unilateral extinction or one group taking over another's women and resources (Alexander 1971). These are (1) social and cooperative abilities that allow or cause larger (hence, variable) group sizes, and more concerted and effective group actions; and (2) culture and technology, which can provide one side or the other with superior competitive ability through means as diverse as language, weapons, and patriotic or religious fervour or perseverance. Changes in these regards can repeatedly adjust balances of power and fuel the kind of runaway social competition here postulated.

This argument may be compared to Gowlett's (1984) comment that 'It has become widely accepted that...biological evolution and cultural evolution affect one another in a positive feedback relationship, thus providing both change and its cause.'

'INITIAL KICKS' OR WHAT STARTED THE PROCESS?

As Wilson (1975: 568) notes: 'Although internally consistent, the autocatalysis model contains a curious omission—the triggering device.' Gowlett (1984) also argues that any such view as the above '...accounts for a process that is going on, rather than for the start of it. It does not satisfy the desire for a single perceptible factor, an "initial kick" which starts off human evolution. Consequently hypothesis after hypothesis has emerged in which such a kick has been found, and in some of them its momentum dominates the whole story.' In other words, even if the above scenario were acceptable, it is still essential to know how, when, and why the ancestors of modern humans became involved in an evolutionary, balance-of-power, runaway social competition.

In chimpanzees and a few other primates that live in multi-male groups—as in humans—females rather than males move between groups (Wrangham 1979; Pusey and Packer 1987; Cheney 1987). This change allowed males to bond, in connection with defending the home area as bands of relatives. Cheney (1987) and Manson and Wrangham (n.d.) believe that the initial resource involved in intergroup aggression was females; and they cite chimpanzees as most like us in these regards. King (1976) regards the pivotal resource as territory, but the point is the same: male cooperativeness and territoriality, coupled with female transfer, lead to intergroup aggression. Following King, Manson and Wrangham also lay great stress on variability in sizes of attacking and attacked groups, which they say is likely in human hunter-gatherers and chimpanzees partly because they live in 'communities', in which (as Reynolds 1965, pointed out) 'travelling parties are almost always *subgroups* of the politically autonomous unit'. Wilson (1975) speaks of 'the fluidity of chimpanzee social organization' being 'truly exceptional'. King refers to this fluidity as temporary fragmenting of otherwise stable societies, while Manson and Wrangham use Kummer's (1971) phrase 'fission and fusion'. In other words, through temporary alliances with individuals and subgroups with whom relationships have been maintained in the larger 'community', chimpanzees—as with humans on an immensely larger scale—meet threats, some of which are posed by other cooperating groups of conspecifics (Goodall 1986; Manson and Wrangham, n.d.).

Chimpanzees are evidently also most like humans psychically, using as criteria their performances at linguistic and other tasks in the laboratory, their use of tools, their tendencies to hunt cooperatively and to cooperate against 'enemies', and the evidence that they possess a self-awareness in some sense paralleling our own (which, however, at least orangutans share—Gallup 1970; Suarez and Gallup 1981; see also Premack and Woodruff 1978). Manson and Wrangham (pers. comm.) are sceptical that the cognitive abilities shared by humans and chimpanzees are relevant to the conduct of intergroup aggression. On the other hand, cooperation, coordination of emotions by displays, bluffing, and anticipation (if not planning) all seem to occur in connection with chimpanzee attacks on members of other groups (references in Goodall 1986; Smuts et al. 1987). My view is that the evidence for (1) cooperation to compete against conspecifics and (2) social reciprocity, foresight, and deception in both chimpanzees and humans implies that this combination of attributes has functional significance.

In general, the above arguments are compatible with the scenario developed here and earlier for the driving forces in human evolution (Alexander 1967–1988; Ghiglieri 1987, 1988). They help clarify the similarity between humans and chimpanzees, and that similarity implies that the processes suggested here were initiated quite early, before the primates involved would have been termed 'human' by modern investigators (see also Wrangham 1987).

These arguments tend to support an affirmative answer to Gowlett's (1984) question whether or not 'the earliest known men were hunters' (see also Tooby and DeVore 1987). The question remains whether chimpanzee (and pre-human) males initially cooperated to defend against predators, to hunt, to defend territory (or females or both), or to 'export' aggression to other males (Manson and Wrangham, n.d.). My scenario has these behaviours occurring in the order just given (Alexander 1979: 223), and suggests that, *in a primate physically, socially, and psychically like chimpanzees or ourselves, cooperative hunting or defence of territory or females are all adequate 'initial kicks' for intergroup balance-of-power races if extrinsic hostile forces are sufficiently insignificant.*

All of these comparisons imply to me that if by some chance the human species should be extinguished while chimpanzees were not, there is a fair chance that chimpanzees would embark upon an evolutionary path paralleling in some important regards that taken by human ancestors across the past million years or so. Indeed, they also imply that chimpanzees have been kept in their current status by the predatory and competitive actions of humans (Alexander 1974: 335), and that if they were even more like humans than they are, they would have long ago suffered the same fate that I believe had to befall the closer relatives of modern humans: extinction by their closest relatives, the evolving human line. Cheney and Wrangham (1987) refer to humans as 'particularly dangerous predators', of baboons, so classified with lions, and remark that '...human predation no doubt accounts for the greatest number of primate deaths, even in areas where hunting methods are still primitive. The regular hunting of primates by hunter-gatherers suggests that humans were important predators of nonhuman primates long before the advent of firearms and that human hunting may have exerted an influence on the evolution of antipredator behaviour and even social structure...' (see also Tenaza and Tilson 1985). I emphasize that the scenario I am developing does not require that the relevant competition be restricted to members of the same species; rather, it would be expected that any species similar to a species in which intergroup competition had become regular and intense would also be in jeopardy.

Female dispersal, and male relatives defending territory cooperatively in both chimpanzees and humans, cause even more intrigue to be attached to the relationship between male competition for females—both within and between groups—and the perplexing problem of (1) the rise of despotism (including extreme polygyny for despots) in societies of intermediate sizes and social complexity (tribes, chiefdoms,), yet (2) the institution of socially-imposed monogamy, reverence for the nuclear family, and suppression of extended families and kin networks in still-larger human societies (nation-states) (Alexander 1979, 1987, in press; Betzig 1986; Harpending 1986). Various authors (Levi-Strauss 1949; Irons 1981; Flinn and Low 1986) have pointed out that, in humans alone, males treat females as commodities, to be bargained with and for, in connection with their movement between groups. (Two major differences between chimpanzee and human societies are that human males show paternal care, while chimpanzee males do not; and human females conceal ovulation, thus affording males some confidence of paternity, while chimpanzee females advertize their ovulation and mate promiscuously—Goodall 1986; Alexander and Noonan 1979). Whether or not women were the resource that led to the initiation of intergroup aggression, and even if sexual selection has remained prominent in the activities of war and the admonitions given to young men of fighting age (Alexander 1987; Manson and Wrangham, n.d.), it is most unlikely that women are still a central resource at issue in the international arms races that baffle us all (see also below).

> The nettlesome question, of course, is why are [chimpanzees] territorial? Where is the survival advantage in risking one's life for land? The answer appears to be that winning more habitat enhances a group's mating success. Because ecological resources limit the number of females who can live in any region, the success of males in expanding, or at least holding, their territory determines the upper limit of their reproductive potential. No wonder they are territorial; if they were pacifists, or even individualists, their more coordinated neighbours would carve their territory into parcels and annex them. Thus armies are introduced into the natural arms race. Once this happens, solidarity between a community's males becomes essential (Ghiglieri 1987: 70).

DIFFICULTIES IN ADVANCING EVOLUTIONARY UNDERSTANDING OF OURSELVES

Evolution proposes to explain the explainers themselves. The difficulty of this proposition lies not only in the fact that some of the traits to be explained must be used in their own explanation, but also that one trait of the explainers is that they do not always wish to be explained—at least not too completely to anyone else—and that humans, more than any other species, are evolved to be exceedingly clever at deceiving other humans. Moreover, there may be no task of learning or teaching that is imaginably more difficult than that of bringing into the conscious items that have been kept out, not incidentally but by natural selection, and most particularly by selection that has disfavoured conscious knowledge of *motivation* as a social strategy.

There are probably two other main reasons for the slow progress of evolutionary understanding, especially with respect to ourselves:

> 1. Applying evolution to the understanding of organisms is not easy, even if the process itself is deceptively simple. Such understanding is difficult because organisms are exceedingly complex. It calls for understanding the relationships between gene action and development as well as physiological, morphological, and behavioural outcomes and their variations. It is helped greatly by a repeated or continual necessity of dealing with these problems, a difficulty that many biologists face on a daily basis while most nonbiologists do not.
> 2. The preceding difficulty is exacerbated by two facts: first, most people don't care about evolution, believing that it has little effect on their personal

everyday lives; and, second, some people care too much. Evolutionary arguments seem to many to threaten cherished beliefs about humans and their history. Evolutionary arguments about humans also come from biology—a field distinct from the social sciences and the humanities, and one traditionally preoccupied with nonhuman species. Moreover, evolution has had a notoriously poor record in explaining humans in the past: during the decades when the social sciences were developing, biologists simply did not know how to apply selection to understand behaviour. For the most part they didn't even try, and so the social sciences developed more or less independently of biology and evolutionary theory. Finally, science and the humanities—disciplines preoccupied, respectively, with searches for undeniable facts and meaning or values—clash when biologically oriented scientists begin to analyse human actions in terms of their functions or effects, because such analyses seem to infringe on questions of meaning and value, hence to represent ideologies (Alexander 1988).

Cooperation also seems always to be a matter of congeniality and pleasantness. The concept of competition, and the idea of evolution by natural selection, on the other hand, imply nastiness. As already noted, however, in evolutionary terms cooperation and competition are not opposites, but rather, cooperation is *necessarily* a form of competition, which can be either indirect and remote, or quite direct. We have evidently evolved tendencies to develop proximate feelings that make it convenient to ignore the competitive effects of cooperative activities that give pleasure to us and our associates. It is also widely believed that group selection implies peaceful and non-competitive existence, as compared to selection at lower levels. As the definitions and descriptions of group selection given earlier indicate, this is not necessarily so: group selection and its mimics, as with some of the most pleasurable and intimate forms of cooperation, can also imply the most heinous and destructive kinds of between-group competitive interactions. The vilest of discriminatory jokes can be told in an atmosphere of warmth and conviviality; and, as Bigelow (1969) noted, 'A hydrogen bomb is an example of mankind's enormous capacity for friendly cooperation'. He suggested, with irony, that its successful construction might seem an occasion for us to 'pause and savor the glow of self-congratulation we deserve for belonging to such an intelligent and sociable species'.

Only with such considerations in mind, I think, is there likelihood of a thorough understanding of the human psyche, or other distinctive human features. Even if it seems inappropriate to emphasize the competitive and not-so-honourable sides of human action, motivation, and history, neither is it helpful to dismiss summarily the possibility of distasteful kinds or intensities of reproductive competition from human history as over-simplifications or reductionisms not deserving of consideration as causative forces in determining our present psychological makeup and socio-cultural forms.

The Nature Of The Human Psyche

There have been few efforts to characterize the human psyche in terms useful to those who would understand and reconstruct its functional aspects from a modern evolutionary viewpoint (but see Premack and Woodruff 1978; Griffin 1978; Savage-Rumbaugh et al. 1978, and the accompanying commentaries). I think the key argument (Humphrey 1976, 1978, 1983; Alexander 1979, 1987) is that consciousness represents a system of (1) building scenarios or constructing possible (imagined) alternatives; (2) testing and adjusting them according to different projected circumstances; and (3) eventually using them according to whatever circumstances actually arise. Earlier, I referred to such abilities as the capacity to over-ride immediate rewards and punishments in the interests of securing greater rewards visualized in the future (Alexander 1987). In this view, consciousness, cognition, and related attributes—which probably represent the core of the problem in understanding the human psyche—have their value in social matters, and the operation of consciousness can be compared to the planning that takes place in a game in which the moves of the other players cannot be known with certainty ahead of time. In other words, by this hypothesis, the function of consciousness is to provide a uniquely effective foresight, originally functional (*sensu* Williams 1966) in social matters, but obviously useful, eventually, in all manner of life circumstances. I will argue (below) that the emotions, linguistic ability, and personality traits are primarily communicative devices, hence, also social in their function.

The above view of the psyche is compatible with that of cognitive psychologists, such as Neisser (1976). Cognitive psychologists, however, concentrate more on mechanisms than on function, and so the idea that the use of cognition might have evolved explicitly in the context of social competition seems not to have emerged in their arguments. Nevertheless, Neisser's insistence on use of the concept of 'schemata' as plans, representing what is here called scenario-building, is a close parallel to Humphrey's arguments and my own. It is clear that a merging of ideas is likely to be easy, and profitable.

FUNCTIONS OF THE PSYCHE

Learning would appear to include two forms: (1) accumulating memory banks; and (2) modifying memory banks, when 'memory bank' means a store of information that influences abilities and tendencies to act. In some sense all phenotypes are memory banks, in which some (genetic) information carried over from the previous generation (and, to a decreasing extent, from increasingly distant ancestors) has been 'interpreted' ('read out') by the environment of the phenotype (organism), including its associates (e.g. parents) as a result of what is commonly called epigenesis, or ontogenetic or experiential plasticity.

Humour and Play. One can learn (1) by trial and error or successive approximation of the actual performance that is useful or desired, or of surrogates of it (practice? play?); (2) by observing and then imitating or avoiding; or (3) by being told about (taught) or by thinking about (and imitating or avoiding). The last two methods, at least, imply 'observing in the mind'. Learning by observing in the mind parallels the concept of play as practice. Play can be solitary-physical (as with a cat practising predation by playing with a twig or a bunch of dry grass); social-physical (as in practice-fighting or play-fighting); or social-intellectual (i.e., without a prominent physical component, as with building of social scenarios through thinking, dreaming, planning, humour, art, or theatre). Presumably, there are also intellectual (or mental) components to both solitary-physical and social-physical play (e.g. for the latter, in team sports involving complex strategies, bluff, and deception or trickery).

I agree with Fagen (1981) in regarding the concept of *practice* (including low-cost testing) as representing the best general theory of play, and I so use the concept of play throughout this paper (for a best-case dissenting argument, see Martin and Caro (1985) who note that 'at present, there is no direct evidence that play has any important benefits, with the possible exception of some immediate effects on children's behavior'). Fagen concludes (p. 388) that 'Current understanding of the functions of animal play suggests that individuals play in order to obtain physical training, to train cognitive strategies, and to develop social relationships'. He also reviews an extensive literature attempting to connect play behaviour to human deception, self-deception, dance, music, literature, painting, and sculpture (pp. 467 ff.; see also Wilson 1975). He describes 'hints at essential relationships between play and creative thought' in the words of Einstein and the thoughts of some other scientists, noting that 'these unsatisfactory metaphors are the best currently available links between play and human creation'. Klopfer's (1970) brief comment probably comes closest to the discussion of social-intellectual play developed here. Describing aesthetics as 'the pleasure resulting from biologically appropriate activity' and play as 'the tentative explorations by which the organism "tests" different proprioceptive patterns for their goodness of fit', Klopfer suggested that 'thought and abstraction in man is but a form of play' and 'Abstractions may be the play through which we learn how to think well' (Klopfer 1970: 402–403).

To Fagen's conclusions (above), I would add that play sometimes represents low cost repetitions and out-of-context or pretend 'run-throughs' in the interests of (1) practising for predictable situations that cannot actually be experienced beforehand; (2) preparing for different preconceived alternatives in unpredictable situations; and (3) assessing skills and abilities of one's self and others. As Humphrey (1986) says, '...play is a way of experimenting with possible feelings and possible identities without risking the real biological or social consequences'. It is also obvious that playing individuals can learn about one another and establish (accept) dominance relationships in low cost situations which may persist into high- cost situations; conversely, they may also learn how to reverse such relations in their own interests. Symons (1978b; pers. comm.) argues that 'dominance rankings are very unlikely to be established during play'. But I know from personal experience that, at least in humans, they can be either established or altered during play; and that play may be entered into with such goals explicitly in mind. I have done both, and I suspect that few humans do not share this experience.

Loizos (1967) exemplifies the authors who present objections to the general theory that play is practice (see also Martin and Caro 1985; for arguments very similar to mine, see Fagen 1982; Symons 1978a). One of Loizos' objections distinguishes play from practice: '... it is not necessary to play in order to practise—there is no reason why the animal should not just practise'. But I regard play as a form of practice, and so believe that the mistake is precisely the other way around; a playing animal is 'just practising'. Second, Loizos, and Martin and Caro, note that not just juveniles but also adults play. But adults also practise extensively, and there is no reason to expect that this particular kind of practice should be absent in adults, especially long-lived adults with complex sociality who may be subjected to new social situations almost endlessly. Third, Loizos believes that '... it is simply not necessary to play in order to learn about the environment'. I would say, however, that it is often *useful* to play to learn about the *social* environment. Loizos notes that '... it is inevitable that during play, or during any activity, an animal will be gaining additional knowledge about what or who it is playing with; but if this is the major function of play, one must wonder why the animal does not use a more economical way of getting hold of this information'. I suggest that, with regard to the *social* environment, there often is no more effective and inexpensive way of securing information (again, Humphrey 1983: 76–79, comes closest to saying the same thing).

Martin and Caro (1985) argue that because play 'has only minor time and energy costs', is 'highly variable and labile', and 'is curtailed or absent under many naturally occurring conditions, it seems unlikely that it is essential for normal development'. Leaving aside the conservatism of the phrase *essential for normal development*', however, the low costs of play can be cited as reasons for its use in developing social capabilities and increasing predictability of social outcomes. Moreover, feeding is curtailed in the presence of predators and sexually receptive mates, and planning is curtailed when immediate circumstances demand attention; but this does not mean that either feeding or planning is functionless. Any activity having its significance primarily in social behaviour is expected to be variable. Their estimates that play uses 4–9% of a kitten's calories (from Martin 1984) and 1–10% of total time in most species (from Fagen 1981) do not seem convincing for the purpose for which they use them. Thus, one might ask what per cent of calories and time are spent by various species in, say, the act of copulation.

Martin and Caro also question whether play should be suspected, as is commonly the case, of having its primary benefits later in life. They seem to disparage the notion that juvenile life has evolved as a preparation for success in adulthood; but there is no other *raison d'être* for juvenile stages (Alexander, 1990). Moreover, benefits that occur a long time in the future are those most likely to be difficult to identify and evaluate.

Expanding primarily from the arguments of Humphrey (1976–1986), I would relate the evolution of the psyche, and the representational (scenario-building) capacity of the human mind, to social play, as practice. I suggest that, during their 'runaway', group-against-group, social-intellectual evolution, humans went from social-physical play (typical of all social species) eventually to social-intellectual play (as scenario-building and practice), which probably occurs in at least rudimentary forms in all complexly social mammals, and team competitions (evidently unique, as play, in humans). Social-intellectual play I hypothesize to be practice for later, more consequential social-intellectual (and physical) competitions (that is, direct competition for mates or resources), just as solitary- and social-physical play represents practice for later, more consequential solitary- and social-physical activities or competitions (cf. Smith 1982).

I suggest that social-intellectual play led to an expanding ability and tendency to elaborate and internalize social-intellectual-physical scenarios. Along with the increasing elaborateness of internal scenario-building came an increasing elaborateness of social communication, including language and the evolution of linguistic ability. Every trait and tendency that represents or typifies the human psyche—every mental, emotional, cognitive, communicative, or manipulative capability of humans—I regard as a part of, derived from, or influenced by the elaboration of social-intellectual-physical scenario-building, and of the use of such scenarios—and of the emotions, language, and personality—to anticipate and manipulate cause-effect relations in social cooperation and competition. This would happen ultimately in the context of winning or losing both as an individual within a social group and as a member of a social group, the survival of which depends, in the end, on success in group-against-group competitions within the species.

Just as I believe that the evident radical departure of the human psyche from the mentalities of the closest relatives of humans can only be explained by assuming that humans themselves kept driving the selection in a peculiar way, I also believe that only other humans represent a sufficiently complex and unpredictable force to drive the evolution of the psyche in regard to its special ability to make and test social predictions. In other words, we became progressively better at practising for our social competitions through internal scenario-building because our adversaries and competitors were doing precisely the same thing. And because we all belonged to the same species, so that those with differently useful expressions of the psyche were parts of the same interbreeding population, there has for a very long time been a positive feedback involved in the evolution of increased human mental capacities, with the 'losers' or 'followers' never more than a step or two behind the 'winners' or 'leaders'.

To the extent that social-intellectual play can be carried out by observing in the mind, it can also be (1) accomplished (secondarily) in solitary (e.g. we laugh at jokes when alone); and (2) concerned with not only social-intellectual striving or competition but also social-physical or even solitary-physical striving (again, secondarily: note Neisser's (1976) relating of his 'schema' to locomotion). Moreover, effects of what previously was 'pure' play, as practice, can begin to influence actual contests over resources; a simple example would be carry-overs, into the resource competition of adults, of dominance rankings established during play among juveniles. Such secondary effects ought not to confuse our identification of primary causes.

Humans are probably not the only organisms capable of social-intellectual practice or play that does not have prominent physical concomitants. Perhaps all organisms that give evidence of dreaming utilize scenario-building of some sort in their social activities. It is easy to suspect, as Darwin (1871) did, that dogs, as well as apes and some other primates (e.g. Humphrey 1983: 90), do these things. But it is possible that humans alone engage in. what I see as the next stage of evolution of the intellect in respect to scenario-building, and that is to reward or compensate others for building surrogate scenarios that are even more condensed (less time-consuming), more elaborate (hence, more effective), and more risk-free than one's own efforts. Once scenario-building has become widely useful, status and livelihoods can be secured by intellectual-social as well as other forms of occupational specialization—not merely by taking on intellectually demanding or specialized tasks, but by using unusual abilities and experiences to develop and conduct scenarios for others–hence, actors, artists, musicians, writers, comedians, orators, shamans, chiefs, generals, scientists, priests, preachers, teachers, and even professional players in sports. In this fashion, a number of human activities, which have until now seemed inaccessible from an approach stressing evolution or reproductive success, can be understood as a part of explaining the reproductive significance of the human psyche.

Humour and Play. Expanding from previous arguments (Alexander 1986, 1987), I explicitly identify humour as a form of social-intellectual *play,* unusual because of its emphasis among adults, which influences resource competition directly through *status shifts.* To illustrate my arguments here about the social use of intellect in regard to scenario-building, humour can be seen as operating in several different ways:

1. It can represent social practice for later competitions that will be more direct or more expensive because they will involve the actual resources of reproduction (jobs, money, mates, etc.). Such practice, as noted above, can be accomplished secondarily even in solitary, just as one can practise the moves of chess either while alone (even within one's mind) or while playing with others (i.e., one can laugh at a joke, and gain from the practice afforded, even if alone).

2. It can sometimes represent the actual competition for the resources, in the sense that the people engaging in the humour may be those with whom one will actually compete later for significant resources; the competition may involve reputation or status that can be demonstrated so convincingly beforehand, using humour, as to turn aside expensive interactions that

would otherwise have occurred.

3. It can involve surrogate scenario-building, in which one solely or primarily learns through observing scenarios built by another, such as a clown, comedian, or writer. Such professional humourists are compensated for building scenarios for others, more elaborately, more rapidly, or less expensively than these others can do it for themselves.

4. The vicarious aspect of humour can be carried further, in the sense that observers can alter their status among friends and associates (competitors and cooperators) by the kinds of humour they exert effort to observe (or use), and by how they respond to surrogate scenario-building via particular forms of humour.

All of the above four uses of humour involve only its directly competitive effects within groups. In the context of indirect competition, through within-group affiliation, humour can also operate in testing, promoting, or ensuring compatibility, and willingness to cooperate, and simultaneously in establishing group limits and thereby identifying competitors outside the group (Alexander 1986, 1987).

5. Humour can be directed against one's self, in a version of Zahavi's (1975) Handicap Principle, in which the humourist demonstrates that he can denigrate himself, or reveal embarrassing information that causes humour in others, and still maintain superior status. As with the superior racehorse handicapped with extra weight or the golfer handicapped with extra strokes, both of which may still manage to win the contest, the ultimate effect can be an enormous rise in status, worth far more than the prize for the particular contest being waged. In these examples—and particularly in the case of self-directed humour—even if the handicapped individual loses the immediate contest, it can win (because of the rewards for status in human societies) in the long run because of how well it did in spite of the handicap.

Elsewhere (Alexander 1988, n.d.) I have argued that the physical incompetence of the human baby (its *physical* helplessness or *altriciality*), as well as that of certain other organisms, is a correlate of *precociality* in respect to attributes that will improve its performance as an adult; and for humans this precociality is largely social-intellectual. I speculate that the early and astonishing acquisition of complex language ability in the juvenile human is related to its freedom (from the necessity of protecting itself) to devote itself to acquiring the necessary skills and knowledge of social communication, including practice and the analysis and acquisition of strategies, in the interests of becoming a socially and intellectually more capable adult.

Observing in the mind implies *consciousness* and scenario-building. It also implies being able to view and modify the memory bank with the option of saving changes or not, as if two copies existed of the memory bank during the scenario-building process; the question might be raised whether consciousness is somewhat like a viewing screen (relating it, perhaps, to the concepts of short-term and long-term memory). To observe (involve) one's self in scenarios in the mind is, I think, what is called *self-awareness*. To practise by observing one's self in the mind must in some sense be a description of the source of *foresight, purpose, planning, intent, and deliberateness*. Such practice gives rise to the concept of *free will* as freedom to choose among alternatives visualized in the future. This view of free will contrasts with the more widely discussed alternative implying questions about the presence, absence, or nature of physical causation (Alexander 1979, 1987).

A parallel view, expressed in different terms, is that of Neisser (1976, especially p. 20):

> In my view, the cognitive structures crucial for vision are the anticipatory schemata that prepare the perceiver to accept certain kinds of information rather than others and thus control the activity of looking. Because we can see only what we know how to look for, it is these schemata (together with the information actually available) that determine what will be perceived. Perception is indeed a constructive process, but what is constructed is not a mental image appearing in consciousness where it is admired by an inner man. At each moment the perceiver is constructing anticipations of certain kinds of information, that.enable him to accept it as it becomes available. Often he must actively explore the optic array to make it available, by moving his eyes or his head or his body. These explorations are directed by the anticipatory schemata, which are plans for perceptual action as well as readinesses for particular kinds of optical structures. The outcome of the explorations—the information picked up—modifies the original schema. Thus modified, it directs further exploration and becomes ready for more information.

Once planning, anticipating, 'expecting' organisms are interacting without complete overlap (confluence) of interests, then each individual may be expected to include in its repertoire of social actions special efforts to thwart the expectations of others, explicitly in ways designed to be beneficial to himself and, either incidentally or not, costly to the others (not necessarily consciously in either case). The expense of investing in one's scenarios, or expectations, and of having such scenarios thwarted, are involved in the invention of rules (see also Rawls 1971: 6; Alexander 1987: 96). Rules are aspects of indirect reciprocity (Alexander 1979, 1987) beneficial to those who propose and perpetuate them, not only because they force others to behave in ways explicitly beneficial to the proposers and perpetuators but because they also make the future more predictable so that plans can be carried out. One of their effects, especially as the rule-makers and -enforcers come to represent larger proportions of the group (e.g. through democratic processes), is to converge the interests of individuals and group.

Cognition, or problem-solving ability, can, I think, easily be related to the above arguments about the function of consciousness. Logic rationality, and cognition— as ability to perceive cause-effect relations correctly—can be viewed in the contexts of dealing with either (1) social possibilities (which entails assessing probable responses of living actors); or (2) nonsocial puzzles (some of which involve only the somewhat more predictable logic of physical laws). The process of selecting the most profitable (self-beneficial) among possible social alternatives involves *conscience,* as ability to recognize and evaluate consequences (ultimately, reproductive costs and benefits), especially as a result of the existence of rules. But in the sense or to the extent that conscience is linked to being good or bad (moral or immoral)—and to a failure to be conscious that one's motivation is to serve one's own reproduction—either ignorance or *self-deception* (or both) is an obligate concomitant. Trivers (1971, 1985) and Alexander (1979, 1987) have argued that self-deception, via the subconscious, is a social phenomenon, evolved as a system for deceiving others, most generally through denial of pursuit of self-interests, in turn through denial of any broad or precise knowledge of the nature of self-interests.

I regard *the emotions and their expression,* as well as self-deception and *personality traits,* as, in the main, an extraordinarily complex system evolved in the interests of deceiving or manipulating competitors. Deception is a crucial aspect of competition, because only through deception can the predictable outcomes of contests between competitors of unequal strength or resource-holding-power be altered (Parker 1974). The possibility of deception, moreover, and the difficulty of determining its effectiveness, can almost unimaginably complicate predictiveness about the outcomes of contests.

Because humans are, like most other organisms, sexual reproducers, they have evolved to behave, as individuals and families and collections of related families, as if their life interests (which translate as genetic or reproductive interests) are unique—different from those of other such units. Differences of interest between genetically unique individuals may be small (as between close relatives or between spouses in monogamy), but they do not disappear except under special circumstances, and then only temporarily. Understanding such considerations, and the long history of human interactions, provides the only way, I believe, for comprehending why individuals, families, social groups, and nations compete today—fiercely, continuously, and unendingly—even when no seemingly valid or sufficient reasons are evident, or can be given by the participants. (These arguments are expanded in Alexander 1987.)

To summarize, I have suggested that social-intellectual play, as scenario-building without extensive physical concomitants, is restricted to a small number of intensely or complexly cooperative mammals, such as group hunters, and may often be indicated by evidence of dreaming; in humans it is demonstrated by the communication of representational ability. Surrogate scenario-building, or the rewarding of others to build some of our scenarios for us, is probably restricted to humans, as is evidently also true of rules. Morality, I have suggested, represents the placing of more or less agreed-upon restrictions on actions that interfere too severely with the social-intellectual scenarios and plans of other societal members, and leads to convergence of individual and group interests.

The idea of fantasizing as play, and as problem-solving, is by no means original here. Piaget (1945: 131) saw all imaginative thought as "interiorized play". Symonds

(1949), Singer (1966), and Klinger (1971) all saw fantasy as related to play and to later problem-solving. Novel here are (1) the association of scenario-building with social problems and deception; (2) the primacy of scenario-building as social-intellectual practice, leading to the prominence of surrogate scenario-building in human sociality; and (3) the argument connecting these activities to a history of intergroup competition.

THE ULTIMATE MYSTERY: DISCREPANCIES BETWEEN THE FUNCTIONS OF THE PSYCHE AND OUR KNOWLEDGE OF ITS FUNCTIONS

Part of the difficulty in understanding ourselves arises out of the fact that if the human psyche is evolved to promote inclusive fitness maximizing (i.e., genetic reproduction via both descendant and nondescendant relatives: see Hamilton 1964; below), it clearly is *not* evolved to tell us precisely that this is its function and ours. This discrepancy makes it difficult to understand what the psyche is evolved to do, and difficult to construct a statement about what humans are evolved to do and not to do, that makes any sense to humans themselves, in terms of their conscious knowledge. I want to approach this problem indirectly. I am interested first in constructing the most general and explicit statement possible about how organisms— eventually, and in particular, humans—are expected, from evolutionary theory, to behave. Specifically, I wish to describe, in the most general terms, that sense in which behaviour is evolutionarily determined, or to describe what I expect organisms, because of their evolutionary history, *are not able to avoid doing* (Alexander 1979; 1987). In this fashion I propose to get at the question of what the human psyche *is* evolved to do. The reason for interest in what organisms are not able to avoid doing is roughly as follows: It is obvious that genes contribute to the behaviour of organisms. They determine how particular environments affect the developing phenotype. Equally obviously, it is not accurate to say that any particular behaviour of any particular organism is 'genetically determined'. The reason is that, unless it refers explicitly to the differences between variant behaviours being genetically determined, any such statement leaves out the effects of the environment. Thus, if such a statement were made, outside the context of causes of behavioural *variations,* it is quite probable that some one could eventually identify a change in the environment that would alter the behaviour, thus proving, in some sense, that the statement was wrong and the behaviour was in fact not 'genetically determined'.

Alternatively, one might say that a particular behaviour—say, how to recognize or behave differentially toward kin—is learned, if he knows that particular social experiences are necessary to cause the behaviour. But one could then ask: Was the tendency to accept or use the learning situation in that particular way also learned? Such questions then continue, like the turtles under the turtles in a storied Eastern philosopher's conception of the universe which had it 'after that, turtles all the way down'. But we know very well that, without some very special definitions, it cannot be *'learning* all the way down' because, even if they are only potentials to action, there are genes down there, in the form, sometimes, of alternative alleles giving rise to potentials for different actions.

IN WHAT SENSE IS BEHAVIOUR EVOLUTIONARILY DETERMINED?

Human zygotes give rise to human organisms, honeybee zygotes to honeybees, etc. This means that, regardless of environmental variations, particular sets of genes produce particular phenotypes that do not vary beyond certain limits on any axis. It is fair to say that no one expects a human zygote to give rise to anything but a human phenotype. What is it fair to say, generally, about the limits of variation in behaviour, given a primacy for an evolutionary process guided principally by natural selection and effective primarily at the genic or individual level, or some other low level in the hierarchy of organization of life?

The most general statement is this: *No organism is expected to act in a way contrary to its genetic interests, except through error or miscalculation.*

To understand the significance of this statement, we must first establish precisely what are an organism's genetic interests, so that we can recognize whether it is doing as we expect. For example, we must understand that genetic survival is ultimately all that counts in evolution, and that genetic survival results from reproduction. Success in reproduction typically involves producing more lasting copies of one's genetic materials than are produced by competitors and potential competitors, leading to numerical preponderance, and among other things reducing the likelihood of accidental extinction. Moreover, copies of one's alleles appear in collateral relatives as well as descendant relatives, and not only in offspring but in subsequent generations as well. As a result we expect social organisms to measure abilities of relatives available for assistance to translate assistance into increases in reproductive success. We expect them to judge alternative ways of using life effort, comparing their reproductive costs and benefits accurately. In short, we expect the organism continually to be evolving to behave so as to maximize its *inclusive fitness* (Hamilton 1964). Even more explicitly, we expect organisms to behave so as to maximize, on average, the likelihood of survival of their genes—even, that is, if some copies of their genes are in other genomes, and even if some individuals die in the attempt because they are taking risks that are perfectly appropriate. I mean by this that, even if a behaviour results in some copies of alleles being lost, the behaviour may nevertheless maximize the likelihood that not all copies will be lost (cf. Dawkins' 1982 concept of the "extended phenotype").

Second, we must also recognize all of the possible ways that an organism can miscalculate, and the antecedent events that adjust its likelihood of miscalculation. For example, we must understand the concept of evolutionary novelty, and realise that organisms may make reproductive mistakes because they have been subjected to learning experiences or other events which alter their phenotypes yet were not encountered by their ancestors during all the time that the traits and tendencies expressed today were being moulded by selection. We must realise that competitors and predators will evolve to *cause* miscalculations in their adversaries and their prey. We must understand that unpredictable events may catch organisms in unprepared states. We must recognize that selection against 'errors', especially as side effects of adaptive behaviour, can only be effective if the cost of the mistake is greater than the value of the adaptive extreme that leads incidentally to it. We must understand that because organisms are (evidently) selected to maximize inclusive fitness only via the accomplishment of a wide array of more proximate ends or goals (such as avoiding pain, ingesting sufficient food of the right kinds, favouring one mate over another, or risking survival to save a brood of offspring—see below), there are innumerable ways for inclusive-fitness-maximizing to be sidetracked. Finally, we must recognize that we, as observers, will sometimes be able to identify actions more reproductive than those taken by organisms, but for historical or other reasons not available to the organism.

APPLYING THE 'EVOLUTIONARY DETERMINISM' QUESTION TO HUMANS

When the problem of making a general statement about evolutionary determinism is applied to humans, perplexing complications arise. Humans have what we call an 'awareness' of at least some of their intentions and their purposes in life: they can anticipate and reflect upon their activities and their goals.

If this conscious understanding about personal behaviour were tuned precisely in the interests of inclusive-fitness-maximizing through direct and conscious seeking of explicitly that goal, then our task would probably not be greatly complicated. In such event, to make the initial statement above most meaningful for humans, we might modify it to say: *No human is expected knowingly to act in a way contrary to its own genetic interests, except as a result of error or miscalculation.*

Unfortunately, we know immediately that this is not the correct prediction from evolutionary theory, because we know that whatever it is that humans have brought into their consciousness, it is not a maximizing of understanding, or even a steadily increasing understanding, of the process of inclusive-fitness-maximising. Humans are not only unaware of this process until it is explained to them, they are instantly reluctant to believe that they are engaging in it. Even if they see some aspects of their behaviour as according with probable predictions from evolutionary theory, the most enthusiastic among them are unlikely to believe for a moment either that they always behave so as to maximize their inclusive fitness or that this is what they are evolved to use their consciousness to achieve. Most humans would assert immediately that they frequently act in ways contrary to their own interests, even though it might be possible to show that some of the acts for which they believed this to be true were in fact precisely according to their *genetic* interests, others were more or less predictable results of evolutionary novelty in their environment, and still others were side effects of adaptive behaviour or simple errors as a result of deficient information.

Paradoxically, we cannot say both 'knowingly' and 'genetic interests' in the above statement, because humans in general do not know what their genetic interests are or how to maximize them. But they do *think that* they know what their interests are, just as they may think, without careful reflection, that everything important about their behaviour must be conscious or readily available to consciousness. Because consciousness is the only way of considering behaviour, there is something that seems illogical to a conscious being about behavioural knowledge being inaccessible to conscious consideration. This disparity is the source of our greatest problem in understanding ourselves. It causes us to wonder what our brains were designed to accomplish, and to suppose that there are no challenges in everyday life that are sufficient to explain them. Jaynes (1977: 23) created an apt analogy: "It is like asking a flashlight in a dark room to search around for something that does not have light on it. The flashlight, since there is light in every direction it turns, would have to conclude that there is light everywhere. And so consciousness can seem to pervade all mentality when actually it does not".

Together with the reasons for the physical altriciality or helplessness of the human juvenile, and our response to them (Alexander 1988, n.d.), the above difficulty may have helped cause two prominent evolutionary theorists (Hutchinson 1965; Williams 1966) to advance the notion that the intellect of humans evolved solely or primarily to assist the juvenile in social interactions, with effects on adults mere incidental 'overshoots'.

The picture, then, does not fall into place in the way we would expect it to if human consciousness had evolved as a steady improvement of personal understanding of one's own behaviour in the light of the goal of inclusive fitness maximizing. We can be certain that the reason we do not yet know all about inclusive fitness maximizing, and how the human psyche works, is not simply that the psyche has not had time to evolve far enough in that direction. In fact, it has evidently been evolving in some different direction.

CONSCIOUSNESS AND EVOLUTIONARY DETERMINATION OF BEHAVIOUR

Consciousness is the part of the human psyche that enables us to know what we know – or so it might seem. In actuality, it may be designed to enable us to know *certain* things but not others, and to keep from us some of the things that we nevertheless do 'know' in the more general sense of being able to act on possessed information (see Chomsky's 1980: 69 concept of 'cognizing'). The psyche may be designed specifically to keep us from knowing precisely (in the conscious sense) what we know and what is the evolutionary significance of our existence. That kind of information we may have to learn from the evolutionarily novel approach of science and technology.

If consciousness is indeed evolved, then it must be evolved to enable its bearer to maximize inclusive fitness. If it is not evolved to bring the realisation of its own purpose into the conscious understanding of its bearer (it is not necessary for humans to understand Darwinian theory for some version of such understanding to be present), then it has to be evolved to bring something different into the conscious understanding of its bearer. To identify this something else is surely a first step in understanding the evolution of the human psyche.

One procedure for determining the evolutionary function of the human psyche would be to construct a model of a psyche evolved to deliver into conscious understanding the direct goal of maximizing inclusive fitness and then seek to describe the ways in which the human psyche actually deviates from this model.

Kin Recognition (i.e. measuring r in Hamilton's, (1964) formula: k>1/r, suggesting the situations in which beneficence can profitably—in terms of reproduction—be given to a relative; r refers to relatedness, k to the environmental costs and benefits of the situation).

First, we might expect that a psyche evolved to render the human individual acutely conscious of the goal of inclusive fitness maximizing would develop the ability to measure the relative genetic overlap between itself and its various relatives. A growing body of evidence suggests that the human psyche is indeed evolved to accomplish this end, although not in an explicitly conscious way (i.e., the psyche does not automatically deliver to the bearer the conscious realisation of the purpose of the ability, or even, necessarily, the existence of the ability). In other words, any human in a normal social situation can usually identify which of any two of its relatives is more closely related to it. In part, at least, this accomplishment appears to be carried out by some kind of counting of genealogical links. Such counting generally works perfectly well, since each additional link halves (on average) the likelihood of any genes possessed by one of the two relatives also being possessed by the other as a result of their relatedness through immediate descent.

Similarly, the fact does not seem explicitly revealed to our conscious selves that our closest relatives are either approximately or precisely 50% likely to carry any particular gene in our own genomes as a result of relatedness through immediate descent (meaning, at least when an allele first appears in the population— Alexander 1979: 129), and that each link reduces this percentage by one half. It is likely that we are somehow programmed, developed, or instructed to treat relatives as if these things were true; but we are not consciously aware of any such instructions. It is difficult to think of a way in which we could gain by being conscious of such details (again, it is not necessary to be aware of the facts of meiosis or the particulate nature of inheritance to approach this kind of realisation, or to possess a ready acceptance of the significance of such facts when they do become available to us). This realisation highlights the facts that (1) there must be a great deal of knowledge that will do us no more good if conscious than if not; and (2) conscious time may be restricted and valuable, so that different potentially conscious items may compete for the available circuits. These possible kinds of limitations on consciousness, however, are not the ones that most concern us here. We are primarily interested in whether or not items or connections have been excluded from consciousness specifically in the interests of preventing the conscious picture from being complete and accurate, not simply because they are no less effective outside conscious circuits. Said differently, we are interested in the extent to which self-deception is a social phenomenon—a system of deceiving others through restriction of self-understanding and corresponding adjustments of social signals.

Environmental Costs and Benefits (Measuring k). Continuing our description of the hypothetical (but unreal) human psyche, designed to understand inclusive fitness maximizing and to know about it, we might also expect such a psyche to be evolved to develop into a superb and acutely conscious evaluator of the costs and benefits involved in helping relatives, spouses, and friends. Again, although it would appear that we are capable of such judgments, there is every evidence that, when we do it, the operation is not typically brought into or kept precisely in our consciousness. Sometimes we do indeed seem to make conscious judgements—especially if the contemplated act is quite expensive, the returns are not anticipated soon, or the potential recipient of substantial beneficence is a casual interactant or a distant relative (i.e., there is considerable risk involved). Even then it does not seem likely that all aspects of the judgement are manipulated on conscious circuits, or that the eventual reasons for decisions are fully conscious.

Again, it could be argued that no advantage is to be gained by bringing such details into presumably expensive conscious circuits. At some point, however, we must begin to wonder if there are kinds of information that can be made conscious only at a (reproductive) cost so high that selection works to exclude them even when there may be available conscious time that is not very expensive.

On the other hand, although we have gone through the central items in Hamilton's (1964) formula for inclusive fitness maximizing, we seem not to have identified any items yet for which the evolution of consciousness would be particularly advantageous or required. Nonhuman as well as human organisms maximize inclusive fitness, and neither human nor nonhuman forms appear to have evolved a conscious realisation of the fact. So we are still equally intrigued by the items that supposedly make consciousness an advantage in inclusive fitness maximizing and other items that would be disadvantageous if conscious.

Let us, then, consider the question from a different direction. Rather than continue trying to identify the kinds of items that we might expect to have been placed into the consciousness of humans, let us see if we can characterize those that have indeed been placed there, particularly in light of the manner in which inclusive fitness is maximized and how we think about the operation of natural selection as a result of information from the modern science of evolutionary biology.

WITH WHAT, THEN, IS THE HUMAN PSYCHE PREOCCUPIED?

Do we use our consciousness (psyche) primarily to learn how to do the things that incidentally maximize inclusive fitness, explicitly when other humans are the main competitors and adversaries, and social skills are the kind we need (that is, when other humans are the main hostile forces of nature)? Do we learn by extremely sophisticated and complicated kinds of social-intellectual play (as practice) how to do the calculating necessary to fulfil Hamilton's formula, and then perhaps move many of the actual calculations into the nonconscious, as with the actions of our fingers in playing musical instruments (Lieberman 1984 calls this "automatization")? Do we also practise continually at a kind of sincere hypocrisy (Campbell 1975) (self deception) in which we strive to present some picture other than one of seeking to serve our own interests as individuals? If so, then how do we answer the original question about the evolved limits of behaviour? Perhaps as follows: *No human is expected to behave in ways contrary to his own genetic interests, but all humans are expected to believe that they do so (if asked) because, in human sociality, a continual effort to serve one's own interests in a conscious deliberate way is not the best way to accomplish the purpose.* It is sometimes *reproductively* disadvantageous (but, I stress, *not necessarily disadvantageous or undesirable in any other terms*) not only to know that one is serving one's own interests, but even to know what those interests are. Because social interactions are typically long-term and repetitive, because multiple potential (alternative) partners in reciprocity are typically available, and because motivations are often used to judge suitability of individuals for later interactions, the most effective ways to deal with human competitors are: (1) sincerity achieved through self- as well as other-deception; and (2) the ability to see ourselves as others see us, so as to cause them to see us as we would like them to rather than as they would like to. The human psyche is evidently evolved to excel at such practices. I hypothesize that the human psyche achieves and maintains this excellence through continual social-intellectual play—and other practice—in such forms as humour, partying, oratory, theatre, soap operas, planning, purpose, and other kinds of social exchanges and scenario-building. Examples are outdoing a competitor in a game or in banter, being first in any competition that yields the prestige of being thought best at that particular game, etc. Such play or practice involves payoffs and expenses that are typically trivial compared to those for which successful practice can eventually reward the player more handsomely and directly—such as the reality of getting the job, wife, husband, friend, contract, commission, tenure, grant, award, raise, business, farm, inheritance, etc.—thus, actually winning a disproportionate share of the resources of reproduction and the freedom to use them in your own interests or as you see fit. The social functions of the psyche are thus realised in (1) partial-cost play or practice episodes; and (2) full-cost or 'real-life' episodes. I would venture that recognizing the social function of the intellect, and the centrality of self-interest, is a largely unexploited opportunity for those who would analyse human mentality in terms of operations paralleling those of machines (i.e., via 'artificial intelligence'); at the least, the actual nature and complexity of the activities of the psyche seem most likely to be revealed through analysis of social manipulations motivated by self-interest (in the form, of course, of being ultimately genetic and reproductive).

DECEPTION AND THE BACKGROUNDS OF INFORMATION IN THE SUBCONSCIOUS

I assume, then, that when information is pressed (or kept) out of the conscious (by selection), this happens either because it is likely to be more useful in the subconscious or because it is not useful enough to warrant saving.

Evidently, some information moves into the subconscious as a result of repetition because it no longer requires conscious effort (e.g. playing a musical instrument, typing, or language). Some such information may be lost from retrieval to the conscious (even from the memory bank entirely) as a result of disuse (e.g. a little-used language). Still other information may never have been conscious, but may remain in the subconscious (and be available to the conscious under particular circumstances). It may have gotten into the subconscious by a process evolved to deal with useful information that was never conscious either (1) in the individual or (2) in the species. Or it may have been kept there by selection, in which case the analogy by Jaynes (1977)—mentioned earlier—between consciousness and a flashlight, would have to be modified to include that the flashlight cannot see into every corner but does not know it.

Deception, as with any life goal, can be deliberate or conscious or not. The conscious motive can be: (1) correct; (2) wrong because the real motive is inaccessible to the senses (e.g. the furthering of genic survival); or (3) wrong because the real motive is concealed from consciousness—either it was pressed out of the conscious or prevented from getting there.

The last must involve deception by self-deception. But why? Perhaps because conscious intent is not needed or conscious intent would interfere. Are we, then, ignorant of the self (genetic)-serving nature of our behaviour because (1) the true nature of our striving is not accessible to our senses; (2) there is no value in using the conscious; or (3) keeping our goals nonconscious aids us in serving them in social circumstances? In each case self-deception that effects deception of others is the aspect that I think is by far the most important in understanding the human psyche, and the one that I will develop here. There would be no premium on self-deception in nonsocial circumstances if this is the case, and this suggests the beginnings of a test. The concealment of ovulation in human females is an excellent case to analyse in terms of conflicts and confluences of interests in the parties involved (see Alexander and Noonan (1979) and Daniels (1983)—the latter especially for a review of published discussions following Alexander and Noonan). Mitchell (1986) criticizes Alexander and Noonan's discussion of deception and self-deception in connection with concealment of ovulation as "a confusion of a lack of information with deception". He does not, however, seem to grasp our argument that, if an event as central to reproduction and as physiologically profound as ovulation—and as available to both sexes as it is in perhaps all other mammals—is not conscious (and cannot be made conscious), there is a strong implication that it has been kept out of the conscious by selection. Alexander and Noonan noted that imperfect concealment—from either self or others—does not necessarily deny that selection has favoured concealment: we simply argued that less information about ovulation is available to the consciousness of either women or men than would be expected if selection had not favoured its exclusion from consciousness. If this conclusion is incorrect, then in this age of strong desires to control pregnancy there would surely be considerably less of a market for contraceptives.

Bernard Crespi (pers. comm.) has noted that males may react negatively to indications that their mates are aware of ovulation (implying control of fertilization and the possibility of cuckoldry), so that human females may be evolved primarily to avoid giving any such indication (for example, by maintaining a more or less unchanging interest in copulation during the ovulatory cycle) rather than to be entirely ignorant (unconsciously as well as consciously) of ovulation. This argument actually seems to predict the precise condition that exists in modern women— some ability to predict ovulation, but a general failure of such abilities to be acutely conscious, or even possible, sometimes, without modern medical information.

SELF-DECEPTION, DECEPTION, AND INTERGROUP CONFLICT

Arguments that the complex cooperativeness of human social life has been driven by intergroup competition and conflict call for all of what has just been said to be cast in terms of intergroup interactions. I have already argued that self-deception is a social phenomenon related to deception of others. Now I will argue further that self-deception explicitly plays a role in fostering and maintaining group unity, and that this role is intricated with the practice and prominence of familial, tribal, ethnic, racial, or regional myths, including organized religion. Indeed, I speculate that self-deception is a central factor in the group-unifying effects of patriotism, organized religion, and similar phenomena because it leads to acceptance of dogmas and myths that impart, at least temporarily, unity of purpose, interests, and striving. Myths need not represent the truth if their only significance is group unity: they need only be accepted. Acceptance, in turn, is expected to depend not on plausibility *per se* but on conviction, or acceptance, of a myth's value in group unification; scientific arguments about humans, for example, may be rejected because they do not have unifying effects, yet are seen as myths (as world views, ideologies, or even religious). Similarly, social, political, or religious leaders may find even otherwise highly laudable goals rejected by the populace if the effect of exhortations concerning them is divisive, restrictive, or self-deprecatory (for example, emphasizing misuse or over-use of environmental resources or projecting guilt rather than pride). Acceptance of unifying myths or information or goals depends on the individual's acceptance of the value of group unity, including the position or status of himself that will result, or other effects on himself and his intimates (children, spouse, relatives, reciprocants). Even myths widely regarded as counterfactual may be accepted, repeated, and elaborated if their effect is seen as unifying. The extent to which directly group-unifying effects of self-deception followed or paved the way for self-deception as a means of deceiving others within one's group seems moot (even self-deception with respect to one's own likelihood of recovering from pain or illness—in, say, a terminal illness—probably has as its primary function deception of others).

What about self-deception in cases of maladaptively extreme risk-taking, as in compulsive gambling or extreme heroism? In part the question seems always to be which is (1) extreme risk-taking as a result of self-deception either (a) about the risks directly or (b) as a result of coercion or gullible acceptance of exhortations from others; and which is (2) inaccurate assessment of risks owing to (a) incomplete information (and failure to assess properly the likelihood of its incompleteness), (b) inaccurate information, or (c) imperfect internal cost-benefit assessment machinery (including developmental misinformation and pathologies such as obsessiveness or addiction). Careful analysis of risk-taking dissected into some such categories seems necessary to resolve this question.

EVOLUTION OF THE EMOTIONS

The emotions can be defined as various complex reactions with both psychical and physical manifestations, as love, hate, anger, fear, grief, etc. (Webster's Unabridged Dictionary 1977; see also Panskepp 1982, and its following commentaries). Students of human behaviour are apt to regard the emotions as one of the principal features of the human psyche, along with consciousness, cognition, linguistic ability, and personality traits.

From either logic or the above definition, one can consider the emotions as comprising three more or less separate aspects:

1. The *expression* of the emotions (e.g. blushing, smiling, crying, frowning, screaming);
2. The *feelings* we associate with their expression (also sometimes used in definition, without mention of expression, as with "strong, generalized feeling; psychical excitement" or "any specific feeling");
3. The *underlying physiological activities or changes*.

Both the *expressions* of the emotions and the *feelings* associated with them can be significant to us either when they occur in ourselves or when they occur in others. Presumably, some of the underlying physiological activities or changes can occur without extrinsic expression or even feelings that we might term emotional (I emphasize the assumption that the first two aspects of the emotions above—at least as we know and experience them—would not be necessary for appropriate actions outside social contexts—i.e., in more or less completely solitary-living organisms). Also, presumably, the underlying physiological activities or changes have been modified (elaborated, altered) by the evolution of the expressions of the emotions and of the feelings we associate with the term.

How did emotions evolve to assume their current form and degree of expression in humans?

Stage 1: It seems reasonable to assume that there was a time, in the ancestry of humans, when the emotions were still incidental effects of physiological events that cause appropriate behaviour in specific circumstances, unnoticed by other individuals; this condition probably exists now in many or most nonsocial organisms. Such physiological activities or changes, which prime or adjust the organism to respond in ways favourable to its own survival or reproduction, could have (and must have) sometimes yielded, *strictly incidentally,* both extrinsic effects and internal feelings in the organism experiencing the physiological changes.

Stage 2: Incidental extrinsic changes reflecting physiological changes must have been noticed and eventually used by other organisms. Presumably, such observers would use the evidence of emotions in other organisms to their own advantage rather than to the advantage of the organism showing emotions, when there were differences in their interests. Such uses could include fleeing if a stronger individual showed evidence of anger or likelihood of attacking; taking advantage when another individual showed evidence of indecision or fear; searching for evidence of danger suggested by the emotional state of another so as to place one's self in a more advantageous position or condition, perhaps with respect to the individual showing the emotion; etc. External evidence of changes in the emotions could also be used by other individuals to assist them in inducing changes in the emotions of another, presumably in directions that benefited the individual inducing the changes.

Stage 3: Once individuals became capable of recognizing emotional states in other individuals, then it seems virtually certain that selection would alter both this ability and the emotional states themselves, or the external evidence of them, in ways that would be called communicative. In other words, as Darwin (1898) knew, at this point, the external expression of the emotions would surely be poised to become a major source of communication, especially in social species, and most especially in species with complex sociality in which the flow of social interactions tended to involve multiple and rapid emotional changes among many different states. This is the point (in evolution) at which it would become important for us to know about and assess our own 'feelings' or emotions—because we could then manipulate them to affect use by others of evidence about them.

Presumably, any organism that altered its emotional expressions under the influence of natural selection would do so in a way that affected its own interests *positively*. If the selection occurred because other organisms were already evolving to use the incidental expressions of the emotions to *their* advantage, then we can see that the organisms would tend to evolve to alter external expressions of their own emotions in such fashions as to thwart their use by others, at least when the others were using them to serve interests that differed from those of the individual showing them. This means that, at least most of the time, organisms would evolve to change their emotions in one or more of at least four ways: (a) to conceal some emotion being experienced; (b) to suggest an emotion not felt; (c) to indicate one emotion when actually experiencing a different one; or (d) to suggest either more or less intensity of emotion than felt.

All of these changes imply deception or manipulation of others. But scarcely anyone is likely to believe that all communication involves solely manipulation and deception. One wishes to explore the question whether or not expressions of the emotions have ever been altered during evolution in such ways as to convey *true* feelings—to tell the truth, so to speak, about one's emotions. Presumably, this could happen if social partners or companions were using the expressions of each other's emotions to help themselves because their interests were (at least temporarily) coincident. I presume that if two individuals sharing (at least temporarily) the *same* interests were to detect changes in one another's emotional states, in each case the detecting individual would use the information in its own interests, although such use should be imagined to include assisting either itself *(directly)* or the other individual (hence, in this case, itself *indirectly*). Such uses might include calming the other individual if its emotional state were placing it (or both individuals) in danger; trying to determine what was responsible for the emotional state of the other individual so as to respond to that environmental factor appropriately too; making some effort directly to attain an emotional state similar to that of the other individual if it seemed likely that the situation would call for cooperative effort; etc. On the other hand, if the interests of two individuals differed even slightly, we should expect each individual showing emotions to alter their expression so as to cause the responding individual to give a slightly different response than it might if following strictly its own interest. Situations may be rare in which two individuals share the same interests in such ways or to such degrees that neither can gain by deceiving the other into a little more assistance than it would give if it were acting according to complete and truthful information about the situation or the other individual's motivations.

The above arguments imply that expressions of the emotions are either incidental effects or else communicative, largely in the context of manipulation and deception. They also imply that virtually any extrinsic expression of the emotions, in an organism as complexly and continuously social as humans, is likely to have been noticed and used enough that some evolutionary modification has occurred in the context of communication (even non-noticeable, non-extrinsic expressions of the emotions are so used now, in polygraphs, or so-called lie detectors). It seems likely that a significant effect has been caused on how we feel about what we call our emotions. In other words, some, much, or perhaps virtually all of the ways that we feel, consciously, when we experience what we think of as changes in our emotional states, are results of emotions having evolved to be communicative. In all likelihood, selection on the expression of the emotions has modified not only the way we feel about our emotions, but the actual physiological events that underlie the emotions as well. There must have been considerable feedback among these three aspects of the emotions all during human evolution. Paradoxically, because so much of communication, especially that involving expressions of the emotions, may be non-conscious or even self-deceptive (as use of polygraphs suggests), it is difficult for us to accept that physiological changes resulting in appropriate behaviours can occur without the feelings that we associate with expressions of the emotions. This is so because the emotions have actually evolved to be communicative and presumably would not be experienced by us in the way they are if they had not evolved such a function. Again, the potential for confusing primary and secondary effects is evident.

In turn, it is difficult to argue that *because* we blush or smile or laugh or frown or grieve when *alone* (*as well as when with others*), this means that the emotions, as we experience them now, are often simply ways of changing ourselves physiologically to meet *nonsocial* eventualities, and may not be social or communicative at all. Presumably, however, if expressions of the emotions have evolved to be communicative, they may occur (secondarily) when we are alone either (1) because we cannot easily eliminate such nonsocial expressions, owing to the insignificance of their expense and the expense of eliminating them while retaining appropriate expressions in social situations; or (2) because we have evolved to use them in social scenario-building or planning when we are alone. It may be noticed that, to the extent that the latter is true, expressions of the emotions when one is alone should be honest and true reflections of at least the emotions that would be felt in the real situation that is being modelled in a mental scenario. In other words, truth in emotions may sometimes be expected when we are communicating with ourselves, if in no other situation.

As noted earlier, it is significant that the communicative function of expression of the emotions has not become entirely conscious and deliberate. Humans obviously do not have complete control of their emotions, including sexual excitement. If the function of expression of the emotions is, as I have suggested above, communicative, then why this should be true becomes a significant question. I believe that the answer is that, first, evidence of complete control of the emotions would indicate to others that there is no reliable way of assessing the effects of social events, or their own or others' presence and actions, on others. Accordingly, it seems likely to be a disadvantage in social matters to give the impression of such control over one's own emotions. We are in awe of actors and actresses who can produce emotions at will, but we are suspicious and negative toward individuals who do the same thing in our social interactions with them (consider such derogatory remarks as "I think she is just turning on the tears!"). Anyone engaged in establishing an important social interaction (such as seeking a long-term or lifetime mate) is bound to respond negatively to actions making it appear that the prospective partner can control at will its reactions to social, emotional, or sexual intimacy with us. We expect social interactants sometimes to behave in a certain fashion despite any conscious intentions. We look for evidence of such effects, we try with increasing effort to cause them to occur, and we are likely to regard their absence as evidence that the other party is less interested in us than we would like or may require.

Accordingly, in the degree of consciousness of expression of the emotions, we humans tread a fine line that can exemplify the aspects of consciousness that are most difficult to understand, and that cause consciousness—which represents the means by which we examine ourselves in the first place—be to quite poorly suited to self-analysis, even if, paradoxically, it is the only analytical device available to us. For it seems apparent that none of the attributes we need most to understand if we are to comprehend our psychical nature is likely to be more completely available to the conscious than are the emotions. The reason is evidently that humans have evolved to be so adept at identifying falseness in deliberate (or conscious) actions and motivations that they have also evolved to deceive by keeping many aspects of motivation out of the conscious. Even more paradoxically, this complexity would not have arisen without the evolution of consciousness in the first place. To use Humphrey's (1986) analogy of the inner eye, there is no inner eye eyeing the inner eye (of consciousness), so that we are left to analyse these problems (create such an eye) by using the procedures of science. This is, to some extent, a procedure advanced by Sigmund Freud. It would appear, however, that to continue the process—to understand motivations ever more deeply so as to understand the human psyche and all our mental activities and tendencies more deeply as well—we will be required to refer continually to the best available understanding of natural selection, because that is the ultimate designer of motivation.

GROUP-COORDINATION AND THE EMOTIONS

Scenarios constructed earlier in this essay for the early evolution of hominids included three major stages (see also Alexander 1979):

1. Group-living because of predation (probably most group-living primates and early hominids);
2. Group-living that includes cooperative group-hunting (chimpanzees, humans);
3. Group-living that includes direct intergroup competition or aggression (chimpanzees, humans).

Beginning with this sequence, a major question is what kinds of mechanisms enabled group-cooperative humans to conduct intergroup aggression cooperatively. How did humans manage the coordination necessary to carry out raids efficiently, especially against enemies belonging to their own species and possessing the same general abilities and tendencies? What kinds of evolutionary change elaborated and perfected the ability to coordinate cooperative efforts of individuals in complex fashions? Although group hunting may have been the initial circumstance in which mechanisms of cooperation evolved, the greatest challenges would obviously have been in connection with intergroup conflicts and raids involving conspecifics.

Coordination of the emotions almost certainly plays a central role in group cooperation during intergroup aggression, as it does in group hunting. Demagogues can coordinate group emotions, and recognition of the value of leaders in such contexts could lead to acceptance of despotism as group sizes increase. Ritual, myth, religion, patriotism, xenophobia, ceremonies, cheerleading, and pep rallies can all be seen as related to the coordination (and testing) of emotions in connection with specific cooperative tasks. One can hardly fail to see parallels between the elaborate and ceremony-like expressions of excitement among diverse organisms such as African wild dogs about to depart on a hunt (Lawick and Lawick-Goodall 1971), and humans engaged in stirring their fellows to participation in risky activities.

Robert Hinde has suggested (in a lecture at the University of Michigan, April 1987) that emotions in modern wars (as opposed to the raids of bands or tribal groups on neighbours) are tuned not to developing and showing anger and aggressive tendencies and passions but to the support of an institution (or institutions); that patriotism, and economic, political, and religious responsibility, are called upon by the orators and demagogues and leaders; that modern soldiers fight out of responses to these kinds of exhortations; and that we must understand the genesis of iinstitutions to understand modern war. This argument is probably slightly over-simplified. Thus, when I was in the US Army, we were exhorted by being told, first, that we should hate 'gooks' (the enemy); second, that we were, ultimately, defending our sweethearts, sisters, wives, mothers, and children; third, that we were, again, ultimately, defending our homes and land. (These last two exhortations were especially clear during World War II, when there were also emotional songs about home and family, such as "This is worth fighting for!" But both were also used when the US was fighting in Korea: "If we don't fight them there, we'll be fighting them here!"); and, fourth, that we were defending democracy and all the good institutions that are America (I also entered the Army with these final, electrifying words from my own mother: "Although I did not raise you to be a soldier, I *know you will be a good one!*'). But the idea is worth considering that— together with the emotions *per se*—Hinde's proposition can be related to the difficult problem of why despotism rises then wanes as social systems change so as to allow or cause larger and larger groups to be unified (Alexander 1979; Betzig 1986). Presumably, with very small groups, the emotions of the moment determine the efficacy of a raid. Perhaps the exhortations of leaders and others, through shows and exaggerations of their own emotions, cause everyone else to become aroused enough to carry out a raid and do it well. Maybe increasing extremes of despotism work similarly as group sizes enlarge—up to a point, but problems arise in managing very large groups through despotism (although multiple episodes in recent history show that in specific situations individual demagogues can be appallingly effective). Perhaps such problems provide part of the explanation for the rise of democracies and what I have previously called reproductive opportunity levelling (Alexander 1987).

Surely one of the most consequential uses of linguistic ability must have been in coordinating group efforts. And as it became significant, language would have become a vehicle for expression of the emotions, and as well would surely have altered their expression as a communicative device (Burling 1986).

LANGUAGE AND SCENARIO-BUILDING

Laura Betzig (pers. comm.) has reminded me of the relationship between (1) what Hockett (1960) called "displacement" in human linguistic communication, and (2) scenario-building as a modelling and testing activity with respect to possible later events. Displacement refers to the human capability of communicating linguistically about events removed in time and space from the act of communication—the use of past and future tenses, and the discussion of events involving some different spatial location. Although many species may have evolved some capability of building and testing mental scenarios that involve displacement, communication between individuals about such displaced events would be exceedingly difficult without language, and, especially, the use of past and future tenses (functionally, consideration of past events would seem always to be in the interest of learning more about possible future events). Displacement, so defined, is in some sense not limited to human language, as Hockett knew: honeybee "dance language", in which distance and direction of sources of food or hive sites are communicated with precision and detail, is one of the phenomena of animal communication that has most intrigued and baffled students of human behaviour (Gould 1975). Leaving aside for the moment the problem of comparing adequately the physiological mechanisms of honeybee and human communication, the relationship between the rise of scenario-building in human mental activities and the value of evolving abilities to communicate about them is obviously worth the attention of those wishing to understand linguistic ability and the evolution of human mentality. Lieberman (1984: 248) suggests, from the work of Fouts, that both temporal and spatial displacement occur in the communication of chimpanzees, through signing, with humans.

Testing the General Hypothesis

WHAT THINGS SEEM RIGHT WITH THE HYPOTHESIS?

1. It has the potential to account for any and all sizes of socially complex groups. Critics of hypotheses giving 'war' a central role in human evolution sometimes have asserted that war cannot account for the rise of nations because different social or political groups have been at war more or less continually without having turned their social systems into nation-states. But this criticism misses the significance of particular kinds of expenses in increasing group sizes in some localities—such as uncrossable mountain ranges or rivers. An intergroup competition hypothesis includes the condition that suitable adversaries must exist to account for continual increases in group size and complexity (see Carneiro 1970; Alexander 1979, and references cited therein).

2. It accords with recorded history with respect to prevalence of intergroup competition. This fact seems to me at least to shift the burden of proof to those who would claim that prehistoric humans did not live in situations that would have caused them to evolve tendencies and abilities to be aggressive when circumstances demanded.

3. It accords with the unique human attribute of group-against-group competition in play, and with the centrality of such play in human sociality. I repeat my acceptance of the general theory that play represents practice and low-cost testing.

4. It accords with the ecological dominance of the human species. This, of course, is just a modification of the widespread anthropological description of humans as the species that, more than any other, creates its own environment. As remarked earlier, I am not referring to ecological dominance of a sort that could only postdate agriculture, but rather a kind that, except for the presence of humans, is probably possessed even by chimpanzees.

5. It accords with the disappearance of all close human relatives despite our rapid evolution (and probably *requires* it). This requirement thus approaches becoming a falsifying proposition (that fails: see also, comments below on apes and dolphins).

6. It accords with the 'autocatalytic' model of brain size increase (Stringer 1984). Of course, internal changes in the brain that increased intellectual capacity were surely occurring simultaneously with increases in size of the brain itself, or of the brain cavity; and, as already noted, it is possible for social structures to be achieved in which strong selection for increases in brain size might taper off and disappear, as the fossil record suggests.

WHAT THINGS SEEM WRONG WITH THE HYPOTHESIS?

1. Early humans and pre-humans are generally assumed to have very low population densities—too low for intergroup competition to have been significant (e.g. Martin 1981). We must, however, wonder how much of this assumption is due to inadequate information, and to the tendency to assume *a priori* that weather, climate, and food were early humans' greatest problems. To maintain my argument I have to conclude that densities *per se* were not critical, or else that estimates of densities were wrong. It is probably more important to know what kinds of social groups people lived in, and why, than to know densities *per se*. We must consider the possibility, I think, that social groups may have been in intense competition with one another for scarce or localized resources even if overall population density was low (e.g. Ember 1978). Hamilton (1975) has stressed that life in small kin groups—such as presumably would occur under low population densities—could well exacerbate the tendency to be ethnocentric and xenophobic.

The question of densities also seems to bear on hypotheses about rates of movement between geographic regions, and thus would appear to bear on the alternatives of (1) a multi-regional hypothesis involving a single species in which genetic (or cultural) changes appearing in a few or many separate localities are repeatedly spread throughout the species; or (2) a single-locality hypothesis involving appearance of either a separate species or a strikingly different form in one locality, spreading without much (or any) interbreeding to cover eventually the entire range of modern humans and replacing the forms previously living there (e.g. Wolpoff, this volume). Either of these alternative hypotheses is compatible with the hypothesis advanced here. Intergroup competition could have been (and probably was) between both conspecific social groups and similar species.

That humans reached Australia and New Guinea about 40 000 years ago—apparently across a sizeable stretch of water—and penetrated to southern South America after crossing the Bering Strait 10 000 or 15 000 years later, and also the broad overall distribution of evolving hominids for several million years, implies considerable ability of hominids to move great distances—hence, to interbreed, and to interact with strange groups. It also implies a strong likelihood of evolution according to the amity-enmity polarity of intergroup competition as outlined here, regardless of actual densities or general way of life. Only a close intragroup cooperativeness, leading almost automatically to intergroup hostility, is necessary. The entire recorded history of humanity, and our direct knowledge of ethnocentricity and xenophobia (see Reynolds et al. 1987), is also consistent with these implications, hence, in this sense, with the general hypothesis developed here.

2. Early humans and pre-humans are generally assumed to have been under great food stress. There is, however, more than one reason for food stress. As others have pointed out, a dominant male in a highly polygynous

species, at the height of his breeding performance, is almost certain to be under 'food stress'. So may be a subordinant male, ostracized from his social group by a dominant male or for any other reason, or a subordinate *group*.

3. War and large groups are both recent; as discussed below, however, there is no unambiguous evidence of either early aggression or its absence.

4. Great apes have some of our mental attributes, including a kind of self-awareness (Suarez and Gallup 1981), but perhaps only chimpanzees currently have an appropriate social structure. It might be possible to argue that orangutans and gorillas do not have intellects that are sufficiently like those of humans to require (by my hypothesis) that they have lived in social groups that would have caused the kind of runaway intellectual evolution I have been describing. But I cannot easily reject the notion that the intellects of the great apes may in fact demand explanation in terms of my hypothesis. I am led, then, to wonder—entirely without empirical evidence—whether or not orangutans and gorillas once lived in social groups more like those of chimpanzees and humans than is presently the case: in other words, multi-male groups, and perhaps multi-male groups in which the males were cooperative in hunting, or even in intergroup aggression (see Wilson 1975: 568, and Wrangham 1987: 60, for comparisons of some relevant attributes). Wilson (1975: 36) suggests that orangutans may have once been more social than they are now, and that conditions leading to extreme sexual dimorphism (exaggerated in both orangutans and gorillas) may have reduced sociality (Ciochon 1987) summarizes information on ape and hominid phylogeny, and Robert Smuts has suggested to me that juvenile orangutans may be more social than expected from the current social life of the species.

5. The problem of explaining the size and complexity of dolphin brains, and their apparently remarkable learning abilities, parallels that of understanding the great apes (Connor and Norris 1982; Schusterman et al. 1986; Herman 1980; Connor, pers. comm.). Dolphins may be as unusual in these respects compared to other inhabitants of the sea as humans (or humans and apes) are compared to other inhabitants of terrestrial habitats (e.g. Worthy and Hickie 1986). Dolphins do not construct tools or other complex artifacts. They are evidently subject to severe predation from sharks (Krushinskaya 1987), and are remarkable navigators (Kellogg 1958; Norris et al. 1961; Klinowska 1986). If these features of their environment are not responsible for their large brains and apparently complex mental abilities we are left with the complexity of their social life by default. Unfortunately, because of rudimentary knowledge of the details of everyday dolphin sociality (Norris and Dohl 1980; Krushinskaya 1986), we can make little further comment on this topic. This ignorance, coupled with the unusual brain and learning abilities of dolphins, has caused a great deal of public interest and considerable speculation among scientists (for example, the highly publicized speculations about dolphin communication, and hypotheses

that dolphins may engage in reciprocity (Connor and Norris 1982) and that odontocetes may stun their prey with high-intensity sounds (Norris and Mohl 1983; Morris 1986)). The importance of examining both dolphin and ape sociality more intensively seems apparent.

HOW CAN THE HYPOTHESIS BE FALSIFIED (BEYOND THE ABOVE DIFFICULTIES)?

1. One obvious possibility of falsifying the general argument that humans have evolved largely through a process of runaway social competition and imbalances of power between competing groups would be to show that such activities are too recent in human history to account for evolution of humans from nonhumans or for evolution of modern humans from archaic humans. I think that most accounts of human social evolution (e.g. see Mann 1986, and references cited therein) imply that this is in fact the case. If, however, chimpanzees have already embarked upon a path involving significant inter-group aggression—a proposition developed independently of the argument generated here and earlier (compare Alexander 1967–1988, with King 1976; Goodall 1986; Wrangham 1987; and Manson and Wrangham, n.d.), then this potential falsifier, it seems to me, is itself falsified. Nevertheless, it seems necessary to understand how it might be that current accounts of the evolution of civilization do not always seem to support the ideas I am espousing. Part of the reason may result from views about hunter-gatherers that may be untenable (e.g. see Ember 1978; Alexander 1979). A second part involves the nature of evidence about intergroup aggression.

2. Absence of indicators of significant intergroup aggression is a second possible falsifier of my arguments. My previous comments on this question (Alexander 1979: 227–228) are here paraphrased and supplemented:

Two kinds of evidence bear on the question whether or not intergroup competition and aggression have played a central role in human evolution. One kind is physical evidence of aggression, including fossils. Little or none of this evidence is unequivocal: spear points, arrowheads, and stone axes all have been called 'tools' or 'weapons', depending on one's bias, and they could have been either or both; skulls could have been crushed by predators or damaged after death; evidence of cannibalism could have been interpreted differently if it came from ceremonial affairs within groups rather than from wars; etc. For example, as suggested by Darwin (1871), Pilbeam (1966), Wolpoff (1971), Lovejoy (1981), and others, in the light of the evidence that humans are willing and adept at complex inter- and intragroup competition, stone 'tools' (weapons) could have lowered the usefulness of teeth *as weapons* (of defence or offence, and against predators as well as conspecifics) as much as (or rather than) "removed the need for use of the anterior teeth in aiding the hands to hold various utensils or materials such as skin or wood" (Howells 1976).

Even continuous intergroup hostility and aggression do not necessarily leave a record for archaeologists to trace. If there were no written records, what evidence would there be to tell us what happened to the Tasmanians and the Tierra del Fuegians? Without written records could we have been unequivocal thousands of years later about what the invading Europeans did to the Native Americans on both continents of the New World? Consider the most monstrous cases of genocide in recorded history: can we even be sure, again without written words, that what happened in the twentieth century at Buchenwald and Auschwitz, and in Nigeria and Cambodia, would be properly interpreted, say, a million years from now? Yet more people may have been killed in these places than existed in all of the time before recorded history. Such questions, it seems to me, cast doubt on the interpretation that equivocal evidence of human aggression, not to say the milder yet potentially continual and crucial forms of intergroup competition, must automatically be discarded.

The second kind of evidence comes from interpreting recent history and the behaviour of modern humans, and then asking about the legitimacy of extrapolating backward in time, both to postulate what happened and to interpret the otherwise equivocal evidence from archaeology and palaeontology. We know that intergroup competition and aggression have been continuous across nearly the whole face of the earth throughout recorded history. We know that cooperativeness on the grandest scale, and the greatest of all the alliances of history, were in response to upsets in balances of power and the aggression of one nation against another. We know that competition is continuous among the various kinds of political groups, large and small, that exist across the whole earth. We know that atomic fission, space travel, and probably most of the remarkable modern advances in science and technology occurred or were accelerated as a consequence of intergroup competition or outright war.

Not only are there two kinds of evidence with respect to intergroup aggression, but the nature or effects of intergroup aggression on human evolution may be better understood if it is considered in at least two major stages. The stage that is more understandable to us, and better represented by evidence, is the later stage, involving organized military interactions, extensive weaponry and strategizing, armies, and sometimes very large scale operations. This is the kind of intergroup aggression that is typically called 'war'. It extends from the beginnings of recorded history to the present, virtually continuously, is often highly organized and complex, and was evidently instrumental in the development and maintenance of the kinds of social systems that have prevailed across recorded history. As already mentioned, these facts place a certain burden of proof on those who would have intergroup aggression disappear as one moves back into prehistory.

Intergroup aggression prior to recorded history is more difficult to substantiate directly. As already suggested, tools may have been weapons, and most evidence is equivocal. It seems legitimate to consider chimpanzees as approximating a model of some early stage of such aggression. The time between such a stage and the appearance of full-fledged 'war' could have involved several hundred thousands of generations and, at times, more than a single species. It would have been during this period that the transition from the pre-human to the human condition would have occurred.

3. Show that, historically, the *topics, timing,* or *sequences* of representational ability (evidenced by tools, weapons, art, ceremony) are wrong; or, by other means, that the complexity of the psyche is not tied primarily to success in social matters.

Any interpretation of the function or operations of the modern human psyche, and of its manner of evolution, must be compatible with the fragments of evidence that exist in fossil or other forms (tools, sculptures, paintings, evidence of planning and social structure, items associated with burials, and what is known about the behaviour of nonhuman primates). Meanings appropriate to the rest of the arguments must eventually be derivable from the nature of the artifacts, their ages, and their sequence. One wishes to ask whether or not evidence of appropriate representational ability appears at the right times and in the right sequences during human history to support the arguments here generated. What are the (1) real and (2) expected sequences of changes in the fossil and other evidence of scenario-building (as well as tool- and weapon-use and social structure) across the relevant periods of human evolution? When and how did scenario-building ability become a means of acquiring status—as in sexual selection or leadership? What kinds of findings would support or falsify the ideas expressed here? Two kinds of evidence are available to us on this question: (1) indications of changes in sizes, compositions, and interactions of human social groupings leading toward the ranges of variation found in modern humans; and (2) artifacts, fossils, and remains relevant to psychical abilities of the sort found in modern humans. How can they be used to test the arguments advanced here? The seminal contributions to this projected test appear to be those of Wynn (1979, 1981) and Gowlett (1984). Using the artifactual record, these authors begin tracing the gradual appearance of consciousness, self-awareness, foresight, and the internal representational abilities of modern humans. Although the data are sparse, they are at least able to suggest that *Homo erectus,* as we would also imagine from its phylogenetic position, possessed an elaborated kind of 'great apes' mentality appropriate to the predecessor of modern humans. So far as I can tell, the meagre evidence from this kind of analysis as yet casts no doubt on the hypotheses advanced here.

4. Show that the greatest increases in intellectual or mental capacities (brain size?) occurred when nonhuman or nonbiotic hostile forces were most severe *rather than vice versa.* To me this test seems the most unequivocal, and the most likely to be useful (see Stringer 1984). It requires evidence as to whether or not the greatest changes in intellect

occurred during maximum extent of glaciation (and near the glaciers or in otherwise severe—perhaps xeric—climates), and when population densities were lowest as a result of such extrinsic hostile forces; or on the other hand under mild climatic conditions when food was relatively abundant and population densities were highest.

Additional tests may be possible from psychological studies that bear on the use of the human intellect in social matters *versus* other circumstances. No such test is obvious to me now, since it would appear that usefulness of the brain in other circumstances may have evolved concomitantly in such fashion as to render inextricable the two aspects of brain function. I suspect, however, that continued confirmation of the significance of the emotions and personality traits in social and communicative matters, and especially in manipulating and deceiving others, would provide strong support for the arguments advanced here. This will be especially true if such features of the psyche are also related to linguistic ability and the various aspects of consciousness and cognition in ways that reinforce the argument for social significance.

When I described to a friend the problem of titling this essay about the human psyche, he suggested with a sly smile that I might call it "Psychology". Having completed the essay, I discover that, indeed, I have argued, in agreement with Humphrey (1976–1986) and using his phrase, that the function of the human psyche is to "do psychology"—that is, to study itself as a phenomenon, in ourselves and other conspecific individuals, and to manipulate, in particular, the versions of itself found in those other individuals. When I read this statement back to the same friend, he nodded and added, "Unconsciously".

Acknowledgements

I thank Theodore H. Hubbell, Richard C. Connor, Robert W. Smuts, Donald Symons, Pat Overby, Lars Rodseth, Laura Betzig, Martin Daly, Margo Wilson, George C. Williams, William D. Hamilton, Daniel Otte, Robert Foley, Katharine M. Noonan, Mark V. Flinn, Paul Turke, Cynthia K. Sherman, Paul W. Sherman, John Speth, Bernie Crespi, Kyle Summers, Randolph M. Nesse, Milford Wolpoff, and Frank Livingstone for stimulating discussions on this and related topics, and for help with the manuscript. Joseph Manson and Richard Wrangham allowed me to discuss their unpublished manuscript. Bernie Crespi and Aina Bernier helped immensely with the literature search. Richard C. Connor, in particular, assisted in developing ideas about reciprocity and the stratification of coalitions. Financial support is acknowledged from the Frank Ammerman Fund of the Insect Division of the University of Michigan Museum of Zoology, and the Evolution and Human Behavior Program of The University of Michigan College of Literature, Science, and the Arts.

References

Alexander, R. D. 1967. Comparative animal behavior and systematics. In (Anonymous) *Systematic Biology*. National Academy of Science Publication 1692: 494–517.

Alexander, R. D. 1971. The search for an evolutionary philosophy of man. *Proceedings of the Royal Society of Victoria* (Melbourne) 84: 99–120.

Alexander, R. D. 1974. The evolution of social behavior. *Annual Review of Ecology and Systematics* 5: 325–383.

Alexander, R. D. 1975. Natural selection and specialized chorusing behavior in acoustical insects. In D. Pimentel (ed.) *Insects, Science and Society*. New York: Academic Press: 35–77.

Alexander, R. D. 1979. *Darwinism and Human Affairs*. Seattle: University of Washington Press.

Alexander, R. D. 1986. Ostracism and indirect reciprocity: the reproductive significance of humor. *Ethology and Sociobiology* 7: 253–270.

Alexander, R. D. 1987. *The Biology of Moral Systems*. Hawthorne (NY): Aldine.

Alexander, R. D. 1988. The evolutionary approach to human behavior: what does the future hold? In L. L. Betzig, M. Borgerhoff Mulder and P. W. Turke (eds) *Human Reproductive Behavior: a Darwinian Perspective*. Cambridge: CambridgeUniversity Press: 317–341.

Alexander, R. D. (in press). Über die Interessen der Menschen und die Evolution von Lebensabläufen. In H. Meir (ed.) *Die Herausforderung der Evolutionsbiologie*. München: Piper: 129–171.

Alexander, R. D. n.d. Why human babies are helpless: a general theory of altriciality. Unpublished manuscript.

Alexander, R. D. 1990. How did humans evolve? Reflections on a uniquely unique species. *University of Michigan Museum of Zoology, Special Publication* 1: 1–38.

Alexander, R. D. and Borgia, G. 1978. Group selection, altruism and the levels of organization of life. *Annual Review of Ecology and Systematics* 9: 449–474.

Alexander, R. D. and Noonan, K. M. 1979. Concealment of ovulation, parental care and human social evolution. In N. A. Chagnon and W. G. Irons (eds) *Evolutionary Biology and Human Social Organization: an Anthropological Perspective*. North Scituate (Mass): Duxbury: 436–453.

Alexander, R. D. and Tinkle, D. W. 1968. Review of K. Lorenz: *On Aggression* and R. Ardrey: *The Territorial Imperative*. *Bioscience* 18: 245–248.

Alcock, J. 1984. *Animal Behavior: an Evolutionary Approach*. Sunderland (Mass): Sinauer. Third Edition.

Axelrod, R. and Hamilton, W.D. 1981. The evolution of cooperation. *Science* 211: 1390–1396.

Betzig, L. L. 1986. *Despotism and Differential Reproduction: a Darwinian View of History*. Hawthorne (NY): Aldine.

Bigelow, R. S. 1969. *The Dawn Warriors: Man's Evolution toward Peace*. Boston (Mass): Little, Brown.

Box, H. O. and Fragaszy, D. M. 1986. The development of social behaviour and cognitive abilities. In J. G. Else and P. C. Lee (eds) *Primate Ontogeny, Cognition and Social Behaviour*. Cambridge: Cambridge University Press: 119–128.

Burling, R. 1986. The selective advantage of complex language. *Ethology and Sociobiology* 7: 1–16.

Campbell, D. T. 1975. Conflicts between biological and social evolution and between psychology and moral tradition. *American Psychologist* 30: 1103–1126.

Carneiro, R. L. 1970. A theory of the origin of the state. *Science* 169: 733–738.

Cheney, D. 1987. Interactions and relationships between groups. In B. B. Smuts, D. Cheney, R. M. Seyfarth, R. W. Wrangham and T. T. Struhsaker (eds) *Primate Societies*. Chicago: University of Chicago Press: 267–281.

Cheney D. and Wrangham R. W. 1987. Predation. In B. B. Smuts, D. L. Cheney, R. M. Seyfarth, R. W. Wrangham and T. T. Struhsaker (eds) *Primate Societies*. Chicago: University of Chicago Press: 227–239.

Chomsky, N. 1980. Rules and representations. *Behavioral and Brain Sciences* 3: 1–61.

Ciochon, R. L. 1987. Hominid cladistics and the ancestry of modern apes and humans. In R. L. Ciochon and J. G. Fleagle (eds). *Primate Evolution and Human Origins*. Hawthorne (NY): Aldine.

Connor, R. C. and Norris, K. S. 1982. Are dolphins reciprocal altruists? *American Naturalist* 119: 358–374.

Daniels, D. 1983. The evolution of concealed ovulation and self-deception. *Ethology and Sociobiology* 4: 69–87.

Dart, R. 1949. The predatory implemental technique of *Australopithecus*. *American Journal of Physical Anthropology* 7: 11–38.

Dart, R. 1954. The predatory transition from ape to man. *International Anthropological and Linguistic Review* 1: 201–213.

Darwin, C. R. 1859. *On the Origin of Species*. Facsimile of the first edition with an Introduction by Ernst Mayr. Cambridge (Mass): HarvardUniversity Press, 1967.

Darwin, C. R. 1871. *The Descent of Man and Selection in Relation to Sex* (2 Vols). New York: Appleton.

Darwin, C. R. 1898. *The Expression of the Emotions in Man and Animals*. New York: Appleton.

Dawkins, R. 1976. *The Selfish Gene*. Oxford: Oxford University Press.

Dawkins, R. 1982. *The Extended Phenotype: the Gene as the Unit of Selection*. San Francisco: Freeman.

Dawkins, R. 1986. *The Blind Watchmaker*. New York: Norton.

Ember, C.R. 1978. Myths about hunter-gatherers. *Ethnology* 17: 439–448.

Fagen, R. 1981. *Animal Play Behaviour*. New York: Oxford University Press.

Fagen, 1982. Evolutionary issues in development of behavioral flexibility. In P. P. G. Bateson and P. H. Klopfer (eds) *Perspectives in Ethology*. New York: Plenum Press: 365–383.

Fisher, R. A. 1958. *The Genetical Theory of Natural Selection*. New York: Dover. Second Edition.

Flinn, M. and Low, B. S. 1986. Resource distribution, social competition, and mating patterns in human societies. In D. I. Rubenstein and R. W. Wrangham (eds) *Ecological Aspects of Social Evolution*. Princeton (NJ): Princeton University Press: 217–243.

Fox, R. 1980. *The Red Lamp of Incest*. New York: Dutton.

Gallup, G. G. 1970. Chimpanzees: self-recognition. *Science* 167: 86–87.

Ghiglieri, M. P. 1987. Toward a strategic model of hominid social evolution. In *Understanding Chimpanzees*. Chicago: Chicago Academy of Sciences.

Ghiglieri, M. P. 1988. *East of the Mountains of the Moon: Chimpanzee Society in the African Rain Forest*. New York: Free Press.

Godfrey, L. and Jacobs, K. H. 1981. Gradual, autocatalytic and punctuational models of hominid brain evolution: a cautionary tale. *Journal of Human Evolution* 10: 255–272.

Goodall, J. 1986. *The Chimpanzees of Gombe: Patterns of Behavior.* Cambridge (Mass): Belknap Press.

Gould, J. 1975. Honeybee communication: the dance-language controversy. *Science* 189: 685–693.

Gowlett, J. A. J. 1984. Mental abilities of early man: a look at some hard evidence. In R. Foley (ed.) *Hominid Evolution and Community Ecology: Prehistoric Human Adaptation in Biological Perspective.* New York: Academic Press: 167–192.

Griffin, D. R. 1978. Prospects for a cognitive ethology. *Behavioural and Brain Sciences* 1: 527–538.

Hamilton, W. D. 1964. The genetical evolution of social behaviour I, II. *Journal of Theoretical Biology* 7: 1–52.

Hamilton, W. D. 1971. Geometry for the selfish herd. *Journal of Theoretical Biology* 31: 295–311.

Hamilton, W. D. 1975. Innate social aptitudes of man: an approach from evolutionary genetics. In R. Fox (ed.) *Biosocial Anthropology.* New York: Wiley: 133–155.

Harpending, H. 1986. Review of L. Betzig: Despotism and Differential Reproduction: a Darwinian View of History. *American Scientist* 75: 87.

Herman, L. M. (ed.) 1980. *Cetacean Behavior: Mechanisms and Functions.* New York: Wiley.

Hockett, C. F. 1960. Logical considerations in the study of animal communication. In W. E. Lanyon and W. N. Tavolga (eds) *Animal Communication.* Washington (DC): American Institute of Biological Sciences: 392–430.

Howells, W. W. 1976. Explaining modern man: evolutionists *versus* migrationists. *Journal of Human Evolution* 5:477–495.

Humphrey, N. K. 1976. The social function of intellect. In P. P. G. Bateson and R. A. Hinde (eds) *Growing Points in Ethology.* Cambridge: CambridgeUniversity Press: 303–317.

Humphrey, N. K. 1978. Nature's psychologists. *New Scientist* 78: 900–903.

Humphrey, N. K. 1979. Nature's psychologists. In B. Josephson and B. S. Ramchandra (eds) *Consciousness and the Physical World.* New York: Pergamon: 57–75.

Humphrey, N. K. 1983. *Consciousness Regained: Chapters in the Development of Mind.* Oxford: Oxford University Press.

Humphrey, N. K. 1986. *The Inner Eye.* London: Faber and Faber.

Hutchinson, G. E. 1965. *The Ecological Theatre and the Evolutionary Play.* New Haven: Yale University Press.

Irons, W. 1981. Why lineage exogamy? In R. D. Alexander and D. W. Tinkle (eds) *Natural Selection and Social Behavior: Recent Research and New Theory.* New York: Chiron Press: 476–489.

Isaac, G. Ll. 1978. Food-sharing and human evolution: archaeological evidence from the Plio-Pleistocene of East Africa. *Journal of Anthropological Research* 34: 311–325.

Isaac, G. Ll. 1979. Evolutionary hypotheses. *Behavioral and Brain Sciences* 2: 388.

Jaynes, J. 1977. *The Origin of Consciousness in the Breakdown of the Bicameral Mind.* Boston: Houghton Mifflin.

Jolly, C. 1970. The seedeaters: a new model of hominid differentiation based on a baboon analogy. *Man* 5: 5–26.

Keith, A. 1949. *A New Theory of Human Evolution.* New York: Philosophy Library.

Kellogg, W. N. 1958. Echo ranging in the porpoise. *Science* 128: 982–988.

King, G. E. 1976. Society and territory in human evolution. *Journal of Human Evolution* 5: 323–332.

Klinger, E. 1971. *Structure and Functions of Fantasy.* New York: Wiley-Interscience.

Klinowska, M. 1986. The cetacean magnetic sense: evidence from strandings. In M. M. Bryden and R. Harrison (eds) *Research on Dolphins.* Oxford: Clarendon Press: 401–432.

Klopfer, P. H. 1970. Sensory physiology and esthetics. *American Scientist* 58: 399–403.

Krushinskaya, 1986. The behaviour of cetaceans. In G. Pilleri (ed.) *Investigations of Cetacea* 19: 115–273. Berne (Switzerland): Brain Anatomy Institute.

Kummer, H. 1971. *Primate Societies: Group Techniques of Ecological Adaptation.* Chicago: Aldine-Atherton.

Kurland, J. A. and Beckerman, S. J. 1985. Optimal foraging and hominid evolution: labor and reciprocity. *American Anthropologist* 87: 73–93.

Lawick H. van and Lawick-Goodall J. van. 1971. *Innocent Killers.* Boston: Houghton-Mifflin.

Lee, R. B. 1968. What hunters do for a living, or how to make out on scarce resources. In R. B. Lee and I. DeVore (eds) *Man the Hunter.* Chicago: Aldine: 30–48.

Lenneberg, E. 1971. Of language knowledge, apes and brains. *Journal of Psycholinguistic Research* 1: 1–29.

Levi-Strauss, C. 1949. *Les Structures Elémentaires de la Parenté.* Paris: Plon.

Lewontin, R. C. 1966. Review of G. C. Williams: Adaptation and Natural Selection. *Science* 52: 338–339.

Lewontin, R. C. 1970. The units of selection. *Annual Review of Ecology and Systematics* 52: 1–18.

Lieberman, P. 1984. *The Biology and Evolution of Language.* Cambridge (Mass): Harvard University Press.

Loizos, C. 1967. Play behaviour in higher primates: a review. In D. Morris (ed.) *Primate Ethology.* Chicago: Aldine: 226–282.

Lovejoy, C. O. 1981. The origin of man. *Science* 211: 341–350.

Mann, M. 1986. *The Sources of Social Power. Vol. 1: A History of Power from the Beginning to AD 1760.* Cambridge: Cambridge University Press.

Manson, J. and Wrangham, R. W. n.d. The evolution of hominoid intergroup aggression. Unpublished manuscript.

Martin, P. 1984. The time and energy costs of play behaviour in the cat. *Zeitschrift für Tierpsychologie* 64: 298–312.

Martin, P. and Caro, T. 1985. On the functions of play and its role in behavioral development. *Advances in the Studies of Behavior* 15: 59–103.

Martin, R. A. 1981. On extinct hominid population densities. *Journal of Human Evolution* 10: 427–428.

Mitchell, R. W. 1986. A framework for discussing deception. In R. W. Mitchell and N. S. Thompson (eds) *Deception: Perspectives on Human and Nonhuman Deceit.* Albany: State University of New York Press: 3–40.

Morris, I. 1975. *The Nobility of Failure: Tragic Heroes in the History of Japan.* New York: Holt, Rinehart and Winston.

Morris, R.J. 1986. The acoustic faculty of dolphins. In M. M. Bryden and R. Harrison (eds) *Research on Dolphins.* Oxford: Clarendon Press: 369–399.

Neisser, U. 1976. *Cognition and Reality: Principles and Implications of Cognitive Psychology*. New York: Freeman.

Norris, K. S. and Dohl, T. P. 1980. The structure and functions of cetacean schools. In M. M. Bryden and R. Harrison (eds) *Research on Dolphins*. Oxford: Clarendon Press: 369–399.

Norris, K. S. and Mohl, B. 1983. Can odontocetes debilitate prey with sound? *American Naturalist* 122: 85–104.

Norris, K. Prescott, J. H., Asa-Dorian, P. V. and Perkins, P. 1961. An experimental demonstration of echolocation behavior in the porpoise, *Tursiops truncatus* (Montagu). *Biological Bulletin* 20: 163–176.

Nunney, L. 1985. Group selection, altruism and structured-deme models. *American Naturalist* 126: 212–230.

Panskepp, J. 1982. Toward a general psychobiological theory of emotions. *Behavioral and Brain Sciences* 5: 407–467.

Parker, G. A. 1974. Assessment strategy and the evolution of fighting behavior. *Journal of Theoretical Biology* 47: 223–243.

Parker, S. T. 1984. Playing for keeps: an evolutionary perspective on human games. In P. K. Smith (ed.) *Play in Animals and Humans*. Oxford: Blackwell: 271–293.

Parker, S. T. and Gibson, K. R. 1979. A developmental model for the evolution of language and intelligence in early hominids. *Behavioral and Brain Sciences* 2: 367–408.

Piaget, J. 1962. *Play, Dreams and Imitation in Childhood*. New York: Norton. Revised Edition.

Pilbeam, D. R. 1966. Notes on *Ramapithecus*, the earliest known hominid, and Dryopithecus. *American Journal of Physical Anthropology* 25: 1–6.

Pitt, R. 1978. Warfare and hominid brain evolution. *Journal of Theoretical Biology* 72: 551–575.

Premack, D. and Woodruff, G. 1978. Does the chimpanzee have a theory of mind? *Behavioral and Brain Sciences* 1: 515–526.

Pusey, A. E. and Packer, C. 1987. Dispersal and philopatry. In B. B. Smuts, D. Cheney, R. M. Seyfarth, R. W. Wrangham and T. T. Struhsaker (eds) *Primate Societies*. Chicago: University of Chicago Press: 250–266.

Rawls, J. 1971. *A Theory of Justice*. Cambridge(Mass): Harvard University Press.

Reynolds, V. 1965. Some behavioural comparisons between the chimpanzee and the mountain gorilla in the wild. *American Anthropologist* 67: 691–706.

Reynolds, V., Falger, V. S. E. and Vine, I. (eds) 1987. *The Sociobiology of Ethnocentrism*. London: Croom Helm.

Savage-Rumbaugh, E. S., Rumbaugh, D. M. and Boysen, S. 1978. Linguistically mediated tool use and exchange by chimpanzees (*Pantroglodytes*). *Behavioral and Brain Sciences* 1: 539–554.

Schusterman, R. J., Thomas, J. A. and Wood, F. G. (eds) 1986. *Dolphin Cognition and Behavior: a Comparative Approach*. New York: Erlbaum.

Singer, J. L. 1966. *Daydreaming: an Introduction to the Experimental Study of Inner Experience*. New York: Wiley.

Smith, P. K. 1982. Does play matter? Functional and evolutionary aspects of animal and human play. *Behavioral and Brain Sciences* 5: 139–184.

Smuts, B. B., Cheney, D. L., Seyfarth, R. M., Wrangham, R. W. and Struhsaker, T. T. (eds) 1987. *Primate Societies*. Chicago: University of Chicago Press.

Speth, J. Early hominid subsistence strategies in seasonal habitats. *Journal of Archaeological Science* 14: 13–29.

Strate, J. M. 1982. *An Evolutionary View of Political Culture.* Unpublished Ph.D. Thesis, University of Michigan, Ann Arbor.

Stringer, C. 1984. Human evolution and biological adaptation in the Pleistocene. In R. Foley (ed.) *Hominid Evolution and Community Ecology: Prehistoric Human Adaptation in Biological Perspective.* New York: Academic Press: 55–83.

Suarez, S. D. and Gallup, G. G. 1981. Self-recognition in chimpanzees and orangutans, but not in gorillas. *Journal of Human Evolution* 10: 175–188.

Symonds, P. M. 1949. *Adolescent Fantasy.* New York: Columbia University Press.

Symons, D. 1978a. *Play and Aggression: a Study of Rhesus Monkeys.* New York: Columbia University Press.

Symons, D. 1978b. The question of function: dominance and play. In E.O. Smith (ed.) *Social Play in Primates.* New York: Academic Press: 193–230.

Tanner, N. and Zihlman, A. 1976. Women in evolution. Part 1: innovation and selection in human origins. *Signs: Journal of Women in Culture and Society* 1: 585–608.

Tenaza, R. and Tilson, R. 1985. Human predation and Kloss's gibbon (*Hylobates klossii*) sleeping trees in Siberut Island, Indonesia. *American Journal of Primatology* 8: 299–308.

Tooby, J. and Devore, I. 1987. The reconstruction of hominid behavioral evolution through strategic modelling. In W. G. Kinzey (ed.) *The Evolution of Human Behavior: Primate Models.* Albany (NY): State University of New York Press: 183–237.

Trivers, R. L. 1971. The evolution of reciprocal altruism. *Quarterly Review of Biology* 46: 35–57.

Trivers, R. L. 1972. Parental investment and sexual selection. In B. Campbell (ed.) *Sexual Selection and the Descent of Man.* Chicago: Aldine: 136–179.

Trivers, R. L. 1985. *Social Behavior.* Menlo Park: Benjamin/Cummins.

Vining, D. R. 1986. Social *versus* reproductive success: the central theoretical problem of human sociobiology. *Behavioral and Brain Sciences* 9: 167–187.

Wade, M. J. 1976. Group selection among laboratory populations of *Tribolium. Proceedings of the National Academy of Science (USA)* 173: 4604–4607.

Wade, M. J. 1978. A critical review of the models of group selection. *Quarterly Review of Biology* 53: 101–114.

West Eberhard, M. J. 1979. Sexual selection, social competition and evolution. *Proceedings of the American Philosophical Society* 123: 222–234.

West Eberhard, M. J. 1983. Sexual selection, social competition and speciation. *Quarterly Review of Biology* 58: 155–183.

Williams, G. C. 1966. *Adaptation and Natural Selection.* Princeton (NJ): Princeton University Press.

Williams, G. C. 1985. In defense of reductionism in evolution. *Oxford Surveys in Biology* 2: 1–27.

Wilson, D. S. 1975. New model for group selection. *Science* 189: 8701.

Wilson, D. S. 1980. *The Natural Selection of Populations and Communities.* Menlo Park: Benjamin/Cummins.

Wilson, E. O. 1971. *The Insect Societies.* Cambridge (Mass): Harvard University Press.

Wilson, E. O. 1973. The queerness of social evolution. *Bulletin of the Entomological Society of America* 19: 20–22.

Wilson, E. O. 1975. *Sociobiology: the New Synthesis.* Cambridge (Mass): Belknap Press.

Wolpoff, M. H. 1971. Competitive exclusion among Lower Pleistocene hominids: the single species hypothesis. *Man* 6: 601–614.

Worthy, G. A. J. and Hickie, J. P. 1986. Relative brain size in marine mammals. *American Naturalist* 128: 445–459.

Wrangham, R. W. 1979. On the evolution of ape social systems. *Social Science Information* 18: 334–368.

Wrangham, R. W. 1980. An ecological model of female-bonded primate groups. *Behaviour* 75: 262–299.

Wrangham, R. W. 1987. The significance of African apes for reconstructing human social evolution. In W. G. Kinzey (ed.) *The Evolution of Human Behavior: Primate Models.* Albany (NY): State University of New York Press: 51–71.

Wynn, T. 1979. The intelligence of later Achulean hominids. *Man* 124: 371–391.

Wynn, T. 1981. The intelligence of Oldowan hominids. *Journal of Human Evolution* 10: 529–541.

Wynne-Edwards, V. C. 1962. *Animal Dispersion in Relation to Social Behaviour.* Edinburgh: Oliver and Boyd.

Zahavi, A. 1975. Mate selection—a selection for a handicap. *Journal of Theoretical Biology* 53: 205–214.

Evolution and Morality

Injustice

and the continuing weight of conscience

I, for one, cannot deny having used,
if only long ago, and senselessly cued,
places of origin, shapes of eyes, colors of skin,
to set myself apart from other men

those whose greatest breaches with me
were fashioned through a history
of bullying and slavery and penury

that in some cruder, crueler times
were started by those self-same signs.
Alexander 2011, p. 25

INTRODUCTION

Twelve (More) Things about the Evolution of Morality Make People Nauseous
David C. Lahti

"Over there they believe society is based on lies, but here we believe in justice."
That is essentially what I was told by an experienced graduate student at the
University of Michigan when I mentioned that I was going to visit Dick Alexander.
When I arrived in the orthopteran range at the Museum of Zoology, eight or nine
backs were hunched over a new *Nature* paper looking for a fatal error (Keller &
Ross 1998). It seems that a "green beard"—a previously hypothetical trait that is
claimed to indicate relatedness accurately and that individuals can recognize in
each other and use as a signal for preferential treatment—had been discovered in
the red fire ant *Solenopsis invicta*. Perhaps the experimental odor transfer control
hadn't been done perfectly. Perhaps they used the term "outlaw gene" (Alexander
& Borgia 1978) too loosely. But all the elements were there! Queen ants that
reproduce have a genotype *Bb*. Queens with a *BB* genotype attempting to reproduce get killed, primarily by those with the *Bb* genotype, who can tell the alleles
apart via cuticular hydrocarbons. Fine, it's probably a real green beard in nature.
But theorists have warned that such a system is vulnerable to cheating. A *BB* ant
might eventually masquerade with the *Bb* signal, if it is possible to separate those
two effects of the gene. Deceit would be adaptive for *BB* ants. Later I hesitantly
raised the objection that any deceit, or even concealment for that matter, might be
hard to maintain long term in the face of others' interests in uncovering the truth.
"Does your wife know when she is ovulating?" retorted one graduate student.
No, I admitted. Of course she can learn to do so, but only because of modern
medical and physiological advancements that have revealed the very existence of
ovulation. And yet how important ovulation is to reproduction! And how flam-
boyantly some other female apes advertise it! So why do we conceal it? Very interesting. Then I read the so-called Uniquely unique paper (Alexander 1990), which
had been published as an unprecedented Special Publication of the Museum of
Zoology because it was deemed too controversial for the ordinary Occasional
Publications (*You have got to be kidding me*, I thought—but I suppose I'd rather a
paper of mine be considered Special rather than merely Occasional). I was shortly
convinced that hiding the truth, not only from mates but even from oneself, was
adaptive for female humans in the case of ovulation, as it made males more likely
to pair-bond and provide biparental care, features so important in our socially

competitive environment. Adaptive deceit in ants, adaptive evasion in humans: *We're already getting close to a society built on lies and I've only been here one day. No wonder the ecologists for social welfare across the street warned me. But it's too late now.*

Why do we see resistance to understanding ourselves evolutionarily, even among those (like those ecologists) who broadly understand and accept evolution? Alexander, whose synthetic evolutionary approach to human culture is the most extensive, careful, and consistent around, has returned to this question repeatedly, especially in recent years (Alexander 1978; Alexander 1988; Alexander 2005; Alexander 2009). I'd like to focus this question further here and apply it specifically to our understanding of the origins of morality, especially the understanding that has been promoted by Alexander's own work. The paper "Biology and the Moral Paradoxes" (this volume, and Alexander 1982) is a succinct and effective summary of the aspects of Alexander's thought that can lead us into the resulting controversy and resistance.

One fascinating and frustrating phenomenon is that those who study morality professionally are among the least likely to understand or accept our best theory of the evolution of morality. I would argue that this rejection is not primarily because of anything they know about morality that evolutionary biologists do not (although that is certainly the case, and a problem that should be redressed). A more pervasive and general reason is simply that most philosophers and social scientists still practice their arts as though humans are not a product of evolution. This is astonishing and becomes more so with each passing decade. However, aside from their attitude that evolution in general is irrelevant, their specific loathing and dismissal of the emerging picture of the evolution of morality needs special explanation; and I argue that *this* attitude is not at all surprising and is likely to continue for a long time. Even many biologists who do not think much about human evolution have a similarly negative response to evolutionary moral theory, almost by default. I think the main reason for the nausea is that our best theory of the evolution of morality looks as though it were concocted precisely for the purpose of slapping moralists in the face, making fun of goodness, and dragging through the mud everything pure and decent in the world.

Before we get to this, however, there are several arguments the professional morality folks use to stop evolutionary analysis before it even starts. Most of these have already been discussed to death and I will not deal with them again here: a group of doubts about the explanatory efficacy of science in general or evolutionary science in particular, including concerns about the nature of history, social and other influences that compromise objectivity, and the cute postmodern idea that science might simply be one of many possible narratives; another family of concerns encompasses metaphysical or religious (including political and social theoretical) taboos about "human nature," including a preference for absolute freedom of the will, worries about genetic determinism, throwback Marxist hopes, or just a general distaste for inherited influences on the human mind. These venerable arguments could be parsed into a dozen perceived problems with an evolution-ary approach to morality. I will leave those aside here in favor of another dozen. Actually, the first two remain in this general show-stopper category, often surfac-ing as peremptory blurts. I will consider them only briefly: That evolution may well happen but it is irrelevant to philosophy, and that a lack of consensus is an excuse not to think about the issue.

Philosophers frequently sniff at **(1) the outrageous suggestion that evolu-tionary biology has anything at all to say about morality**. Such a radical sepa-ration of science and philosophy would have been alien to nearly all the great philosophers of old, including Aristotle, Descartes, Hume, and Kant, who were inspired by the latest empirical knowledge and stretched its implications as far as they could into a springboard from which to leap into realms that science had not yet fathomed. A few philosophers still retain this idea of science as handmaiden to philosophy. How different Wittgenstein, who was so content with a narrow conception of philosophy that he could write (and I am so sick of seeing this quoted) that "Darwin's theory has no more to do with philosophy than any other hypothesis in natural science" (Wittgenstein 1921). I respect Wittgenstein's exper-iment in philosophical minimalism, in the spirit of the beginning of Descartes' *A Discourse on Method* (1637). But should philosophy really be nothing but con-ceptual clarification? No. Philosophy, φιλοσοφια, is literally the love of wisdom, it is the search for answers to the big questions. And it has a responsibility to take all human observation into account in order to do this. Your concepts them-selves, philosophers, depend on your psychology, that is to say your minds, which are a product of history and have evolved genetically and culturally according to particular rules for particular reasons in particular environments. Your concepts would be different if any of those things were different. To practice philosophy as if conceptual clarification can be done in ignorance of human natural history is to assume an idiosyncratic human natural history of your own, one that prob-ably dates back to Hume or even to Plato, and represents the best of what they knew about humans. To engage in metaphysics without understanding human evolution is to assume a false preevolutionary anthropology. To engage in moral philosophy without understanding human evolution is to assume a false preevo-lutionary psychology and sociology.

Alexander, building on the progress of other evolutionary biologists such as Darwin, Williams, Hamilton, and Trivers, has birthed an expansive and vibrant philosophy, representing the first several steps in a reevaluation and recasting of the world's big questions in the light of evolution, including those relating to morality (e.g., Alexander 1979; 1982; 1987; 1989). Of course Alexander, like many scientists, considers it a slight for his work to be referred to as *philosophy*, which everybody knows is a code word for a pompous and meticulous web of utter speculation. This is not philosophy at its best, however. Philosophy is an art of precise and internally consistent hypothesis generation. And the more there is to be explained, the more good philosophizing is necessary in order to develop a hypothesis that is broad and detailed enough to explain it. Philosophizing about something as formidable as human nature must amount to the generation of a large ball of interacting hypotheses, all of which are consistent with each other and explain all observations to date.

Another case rather shamelessly proffered by professional thinkers about morality is that **(2) the difference of opinion among evolutionary biologists** as to the explanatory power of their field with respect to morality gives the rest of the world an excuse to let the matter sit. Certainly there are many theories out there, and everyone has a twist: cognitive scientists, psychologists, a diverse array of anthropologists, primatologists, economists, journalists, and yes zoologists. Some treatments are better than others, most are rhetorically overblown and oversimplified, and (to be frank) few of them deserve to have seen the light of day (Lahti 2003). If the students of evolution can't even get it straight among themselves, why should philosophers and social scientists be concerned? Of course, this logic is especially dangerous for philosophers to use because there is not a single important matter in the world on which there is a reasonable consensus among their own ranks. The fact is, although science does proceed socially by consensus, no philosopher nor scientist can get any sort of answer to any big question by depending on, much less waiting for, consensus. The bigger the question, the more difficult wrapping one's head around it becomes, the more axes are out there to grind, the more diverse is the field of ideas, and so the poorer the consensus. Humans will always fret and explore most intensely about issues that have to do with humans—that's almost a truism. Rather than letting the matter sit, we must compare the extant hypotheses, winnowing them for consistency, explanatory power, and their ability to pass tests having the potential to falsify them.

Provided that evolutionary analysis of morality is actually allowed to begin, any serious work might very well meet with a turned-up nose regardless of what the final picture looks like, because of a three-pronged arsenal of arguments, almost like an immune defense. First we have the lingering effects of a prior infection: **(3) guilt by association with "social Darwinism."** In the early days, some theorists attempted to derive morals from the evolutionary process, some of which claimed some humans to be better than others and justified power disparities and social injustices. Such detestable ideas had the effect of virtually inoculating the academy against susceptibility to any future outbreaks of evolutionary thinking about morality. Even today, the mere mention of the term "social Darwinism" exerts great rhetorical power, whether it is used appropriately or not. If for some reason that prophylactic doesn't do the trick, the second line of defense against evolutionary ethics is to take two fallacies and call a philosopher in the morning. The two fallacies in question do highlight actual errors of thought when used properly, but with regard to the connection between evolution and morality they can be portrayed as having mighty accusatory powers far beyond their logical reach. One is **(4) the *genetic* fallacy**, where the origin of an idea is used as an argument about its validity or truth. All you have to do (some think) is accuse someone of this, and they will be prevented from considering the origin of morality to have any relevance for...morality. Wielding **(5) the naturalistic fallacy** can be just as vital. This fallacy is committed whenever (depending on its formulation) someone claims that moral values are implied by, are defined as, are really the same thing as, or somehow arise out of, facts of the sort that natural science can countenance. Some take this fallacy to absolutely debilitate any attempt to make morality natural, or rooted in things that are accessible to natural science, which presumably includes anything that has evolved. With this arsenal in place, the professional students of morality might not have needed much else in order to protect their subject from evolution. (Someone should name a fallacy for the reckless wielding of fallacies.) However, strange as it may seem, these defenses against evolutionary biology's intrusion were probably not even necessary: Evolutionary biology became its own worst possible publicist.

To start with, although evolutionary biology does not preach a morality, it describes the human condition in the context of that of all life, where traits persist and spread insofar as they benefit their individual bearers. Thus, any consistent explanation of human action from evolutionary biology will look like **(6) egoism**, which is often considered "knockabout philosophy" as my philosophy advisor used to use the term—a view you toss around to whet the critical abilities of young thinkers, but eventually discard for more serious contenders. Still, there are always plenty of "enlightened" egoists around, people who believe that in some way what is good is so because (or consists in the fact that) it is good for us. The horrifying part is the particular meaning of "good for us" that evolution brings to the table. This is where evolutionary biology irreparably ruins its image. What we humans consider good, we came to consider such because...it is good for...**(7) *reproductive success***. This means, for starters, *sex*. After thousands of years of morality being a vanguard against...well, immorality if you know what I mean, we are asked to believe that the two have always been in bed with each other! The idea that moral goodness or rightness has any sort of basis at all in making babies—no, worse, in making *more babies than other people do*—is probably the most odious thing any thinker could suggest about the nature of morality. Or, if there were a more repulsive connection, it would have to be the other thing we protect our children from in the name of goodness: **(8) *violence***, and especially *war*. And lo and behold, this becomes in the evolutionary picture the *sine qua non* of morality! All of nature struggles to reproduce, but only humans have morality. Why? Because we humans have lived in social groups that have competed so fiercely with each other that it led to arms races, not only of literal weapons but also, and more crucially, of minds. We outsmarted other hostile forces to become each other's worst natural enemy. And the group competition was so unrelenting that the social group had to be unified or else disintegrate. Any individual's successful reproduction would depend on the persistence of the group, necessitating cooperation with other group members, and standards for such cooperation, including to some extent the sacrifice of immediate individual interests for the greater good. Thus, paradoxically, war led to goodness, and without war we would never have evolved the intellectual ability to decry war, nor to hope for something better. Moreover, the mechanism by which this happened was a "selfish" maximization of individual reproductive success.

What should be our response to repugnance at this history? In my opinion, our response should be to encourage quality science education so that the repugnance declines with time. As morally ambiguous as this proposed history is, it is our most powerful explanatory framework for human nature. Those who understand natural selection and appreciate its implications, and who are aware of the nature of social behavior in other species, are not likely to encounter a tremendous barrier to realizing or accepting this picture. And at this point, despite evolution-inspired nihilistic popularizations, those who respect morality and the idea of goodness might still rescue them in dignity from the rubble of their history. After all, egoism in an evolutionary sense includes concern for one's kin and group, reproductive success for humans includes not only sex but pair-bonding and parental care, and fighting only sowed the seeds for morality insofar as it protected one's social group and became the wellspring of cooperation. More generally, many human traits derive from precursors that were humbler in some (anthropocentric) sense. Just as many of us have risen above a Wilberforcian indignance at the idea of having evolved from animals that don't wear clothes or speak very well, maybe we can stave off moral discomfort at the centrality of selfishness, sex, and violence in our moral evolution.

This is not the end of the story, however. The foregoing is prudish stuff compared to the next level at which moral ambiguity enters the evolutionary picture, the full extent of which many philosophers and social scientists are not even aware, so poor is the diffusion of ideas across disciplinary borders. This is the level of the individual human psyche, and of the motivations or intentions that guide our actions. For morality to have fulfilled its ancient and continuing evolutionary function (thus explaining its persistence in the repertoire of the human species) we can expect (9) **a correlation between the interests of the genes and the goals driving human action**. To the extent that reproductive success has been the currency of natural selection, it will accordingly have evolved to be a predominant and universal human aim. And to the extent that group competition has been the engine of cooperation, corresponding desires and motivations will have evolved in our psyche as well. Moreover, any exceptions or supplements to this picture are not expected to be universal or particularly lasting. Views and people going against this trend will be statistical residuals in the long view. What this means for human psychology and personality is morally complex: We are expected to act primarily to benefit ourselves individually, as well as our mates and offspring, and (because of indirect fitness benefits) our nondescendant kin. We will tend to value members of our own various social groups over nonmembers, and we will tend to withhold beneficence and be less caring the more distantly related people are to us. Some of these tendencies and the implications from them we might consider morally praiseworthy, others acceptable, and many we would consider immoral. There is no consistent explanation for the evolution of moral behavior—rather, there is the explanation for the evolution of *human* behavior, including the moral and immoral mixed up together.

To such standard material from the Alexandrine picture of human behavior we must add still more troubling consequences from the interaction between group competition and indirect reciprocity. Indirect reciprocity, the mechanism by which rewards and punishments can return to someone through any individual in society or even from society as a whole, is the most pervasive and explanatory element in the evolutionary analysis of morality (see Sigmund, this volume). While still yielding falsifiable predictions, indirect reciprocity renders unproblematic a majority of the scenarios that are typically offered to challenge to the evolutionary picture, from blood donation and philanthropy to monasticism and celibacy. Considering the benefit of group unity, indirect reciprocity becomes a vehicle for widespread social pressure to contribute service to the group. Not surprisingly, much of what we consider moral or good tends to be consistent with this rubric of service to the group. Insofar as group cohesion preserves the group, it benefits the reproductive success of its members; consequently these members will tend to benefit by serving the group, encouraging this group service in others, and socially enforcing their commitment. However, more broadly, each individual will also gain in reproductive success within the group by seeking benefits for oneself at the expense of others in the group. The result of this analysis is that each individual should encourage in others a greater level of commitment to group service than one's own optimal level of commitment. Moreover, by the mechanism of indirect reciprocity it is less the actual good behavior or intentions that lead to reproductive success, and more the *perception* by others that one is good. Thus we are likely adapted to inflate the impression we lend to others of our goodness. The evolutionary prediction in short is that **(10) hypocrisy is inherent in human nature and an essential part of adaptive moralizing**. Many adaptive strategies can be drawn from this prediction, including ways to discriminate in our dispensing of benefits, to play upon the altruism of others, and to anticipate and manipulate others' impressions of us for our own gain. The discussion of morality has turned into a discussion of how we get around morality, and the evolution of goodness has turned into the evolution of the best tool in the con man's kit. Thoroughgoing goodness doesn't exist except in the sucker, and so those of us in the know have every reason to exchange evil grins and high fives (or perhaps more prudently, discreet winks) when we encounter idealistic people who really appear to have some respect for the concept.

Actually this is not quite right. According to the emerging evolutionary picture of morality, the main reason we do not see so many evil grins and high fives, or even winks, in response to others' moralizing or greater group service than our own, is more than just prudence. It is possible that some or even most of us do not generally stifle obvious feelings. Alexander's more likely alternative is

(11) *self-deception.* We are not perfect at hiding our conscious motives, and others are evolved to detect them. Therefore the mechanisms just described work better when we are just as clueless as our audience is, to the reasons why we have come to encourage moral adherence and laud heroism. (Hence, Alexander adds, our resistance to accepting those mechanisms [Alexander 1987].) We tend to be conscious of things when being conscious of them is advantageous, and not otherwise. (Note that modern psychology has it backward when it concentrates on why we sublimate or render certain events or motives unconscious, when this is in fact the primitive and much more widespread state of affairs in nature. The real question is why our ancestors became conscious of certain things.) This incomplete and even false knowledge about ourselves allows most of us to have the comfortable sensation that we are unitary and consistent in the causes of our actions. Many lines of research have since converged on this idea that we humans are inherently inconsistent and either cognitively fragmented or self-deceived. The idea has an array of explanations in the literature besides the facilitation of moralizing: is it also because we have old and young parts of our brains, or two lateral halves (Haidt 2006)? Or is it because we have some genes inherited from dad and some from mom and these can predispose us differently (Burt & Trivers 2006)? Or is it because our brain is comprised of modular neural networks that evolved for different and often contradictory functions (Kurzban 2010)? Whatever factors besides selection for moral manipulation contribute to our "impurity of heart" (Kierkegaard 1846), the upshot for our psyche is that we are far from the internally consistent and honest wills that we generally consider ourselves to be. Worse, in a sense our hypocrisy probably runs even deeper than a temptation or felt preference—it may run so deep that it is cognitively inaccessible to us, while its guidance of our action is nevertheless successfully operating in our day to day lives.

With this, even many who respect the role of evolution in human psychology and behavior have had enough. The paradox that egoism lies at the base of altruism, that self-regard somehow subsumes other-regard, is morally ambiguous enough at the level of evolutionary mechanism. For it to threaten much more seriously to undermine our integrity at the level of individual thought and action is just too much. Altruism is reduced to what we are self-deceived to believe we are being when we are actually—perhaps unconsciously—being egoistic. Any view of human thought and action that can cut through to causes beneath conscious motivation has teeth so long that an audience will be quick to conclude that this is a wolf and not grandma talking to them. The visceral responses to such a view can be potent and surprising. I know one reader of this theory who claims to have thrown *The Biology of Moral Systems* (Alexander 1987) across the room at this point. I myself admit to having written "PALTRY" in huge letters across a page of an article when I first saw this view (Alexander 1992), by which I meant that it destroyed morality and made it paltry. One respected speaker, whose lecture was entirely on the subject of Alexander's theory of self-deception, shook his head in disgust and said from the podium, "If you can't see what is wrong with that, I have nothing more to say to you". A mere description of the theory once prompted an otherwise kindly ex-president of a prestigious Cambridge college to shout "NO!" and pound his fist on the table of a fancy and quiet restaurant, shaking glasses and turning heads. Surely, if one accuses humanity of wholesale lies and posturing precisely in place of everything heroic and generous, one should expect an uneasy reception. Goodness is supposed to be a heavenly ideal, something to which we can aspire with all of our being, and its behavioral counterpart rightness is born of love, the best and purest thing this world has to offer. Yet our most explanatory evolutionary theory insists these wonderful things to have arisen out of social conflict and competition, and to be saturated with manipulation, self-deception, and the danger of being found out.

Is this the extent of the better angels of our nature? Where is the goodness that moral philosophies and religions advertize and that we are apparently evolved to display and encourage in others? Where does a deep-seated moral integrity fit in this theory? The answer appears to be that it fits nowhere, or at least does not demand a great deal of scientific explanation because we do not expect such a phenomenon to be very common. We are not *Homo bonus* or *beneficiens*, but *Homo sapiens*: thinking, clever, calculating, option-weighing humans. Evolutionary biology explains the existence of the moral ideals and also explains why our motivations and behaviors do not match those ideals. In contrast, perhaps most who discuss the biology of prosocial behavior leave us with the idea that niceness is the norm and the rare Machiavellians are the exceptions to explain and avoid (e.g., de Waal 1996; Oakley 2007; Hrdy 2009; Churchland 2011). Humans do get along remarkably well compared to a chaotic "state of nature" (Hobbes 1651), and so the nicer perspectives do have explanatory power. The less savory parts of the story should not be ignored, however. If Alexander and others (Batson et al. 1999; Trivers 2000) are correct, in terms of our adherence to the ideals of the moral point of view the goodies are the exception, if they exist at all. If there is any consistent, uncalculating, and deep devotion to the moral life in actual humans, its manifestations are odd points off the trendline, exceptions that are smoothed away by the averaging of statistical tendency. Perhaps the most nauseating thing about the evolutionary moral theory for the contemporary moralist is that when one steps away from the whole picture, we can't help but notice that despite our best hopes and cozy thoughts, **(12) we are not fundamentally good**.

The biggest and least controversial shortcoming of the evolutionary account of morality is that an explanation of morality leaves open the question, *What we are to do?*—the primary moral question of the ages and (if ancient literature is an indication) a main starting point for our curiosity about ourselves as humans. Evolutionary biology leaves this question not only unanswered but in a way less sensible than when we started, because its analysis has muddied the moral waters. In fact, one who understands the evolutionary account can be forgiven for suspecting that there may actually be no answer to this question in the deep sense in which it has historically been asked; that after an iconoclastic evolutionary analysis the things most vital to be said about morality and life-living have at last been said. We are left with ethical nihilism, and so although we still need to make practical decisions, to get too excited or evangelistic about which path we take would seem forced.

If this is the case, however, why do the ends of books and papers on the subject of the evolution of morality remain so strongly moralistic? Take the end of "Biology and the Moral Paradoxes," for instance: what is the impetus for the "goal of diminishing human problems through improving self-knowledge"? And on what basis do we consider "intergroup conflict" something that "we must supercede"? Why, given the evolutionary function of moralizing, should we give credence to claims that "we seek... world peace and world law" (all from Alexander 1982)? Moralizing tendrils twine through Alexander's writings, clinging to the very moral theory that has just revealed the sordid function of moralizing. In fact, Alexander's works deal in increasing detail with the future of religion and hope for humanity (Alexander 2009; 2012; this volume). Values poke through the evolutionary analysis, particularly at the beginnings and ends of papers, like bold weeds encroaching on a carefully tended garden: End mass violence and hatred of outgroups! Promote a ladder of affluence to enfranchise all! Demand honesty in our leaders!

What is going on in these statements and encouragements, in light of the evolutionary background? Trying to answer this question shows the difficulty of applying or testing the hypocrisy and self-deception hypotheses in any particular case. The morals could simply be the theory at work. They could benefit the author either by manipulating us into being more moral than the author so that he benefits disproportionately from our group service, or else from our lauding of his efforts. This is harsh, though exactly what the theory would claim. I do offer this possibility tongue-in-cheek, though, since such a strategy would seem comically inauspicious right on the heels of sensitizing us to such ploys. Alternatively, the morals promoted might approximate strategies that are actually reproductively advantageous for an individual in our current environment. In this case the author might be behaving altruistically by giving us a heads up, contrary to theoretical predictions—although more likely our collective appreciation of him far outweighs his cost in letting the cat out of the bag, such that our advantage and his own end up being in line with each other. If all of this is too cynical, a brighter alternative is that there is wiggle room in the system, such that goodness is not entirely something cast out as a carrot to lure donkeys into doing work for us; and that striving after an ideal other than individual reproductive success is actually something that some people, even after understanding the evolutionary account, are still willing to countenance, at least to a limited extent. But lest we be too proud, striving after such an ideal for its own sake might itself generate a pretty nice reputation, which points us again towards the possibility of hypocrisy and self-deception. Thus our interpretation is plagued by an endless cyclical regress between our behavior being captive to the evolutionary moral theory and rising above it. The solution to this regress may remain forever elusive. We have no comprehensive algorithm by which to divine the impact on reproductive success of every theory, belief, and action; and we have no intrusive psychoanalysis by which we can discern an agent's motivation underlying these same events. This is not a logical or conceptual problem with Alexander's view, as much as an inherent problem (if the view is correct) in the project of humans studying themselves.

However we might choose to interpret moralizing today, one sure fact is that morals are remarkably resilient to analysis—they survive terrible beatings. Apparently we can't kill the concept of goodness even when we riddle its evolution with scandal. The concept's very nature is to remain an unassailable ideal, regardless of its origins, history, or practice, and whatever the content we ascribe to it (Murdoch 1970). We may not live the ideal, but we can't fault it for that. Thus we can talk about the corruption inherent in moralizing and then turn right around and moralize with a straight face. Likewise we can bite the hand that fed us, using our cooperative value system that was forged in group competition to bash group competition.

There are good reasons why morals can survive the harsh treatment of an evolutionary analysis. Some of these are very basic—we are moral animals, after all, and regardless of how much we morally wander, it's not clear that most of us are capable of living and thinking in line with nihilism or even according to an ethic we invent from scratch, with all due respect (i.e., very little) to postmodernists and social constructionists. Here I will mention three other kinds of reasons—that is, rational reasons, or rationalizations—why morals don't completely lose face in light of evolutionary analysis; in other words, three reasons why the last four nauseating elements presented above do not do morality as much damage as they may seem to threaten, and as many people (such as those giving the visceral responses above) have feared they would. In brief they are (1) that there is indeed a place for the partially and even fully honest among alternative moral strategies, (2) that self-deception means precisely that the "hypocritical" might not be hypocritical, and (3) that the whole point of an ideal is that no amount of violation necessarily destroys it.

First, I suggest that the few adaptive strategies so far discussed in the area of the evolution of moral psychology need additional development. In particular, the hypocrisy and self-deception model works as far as it goes but is too simplistic by itself. Humans operate with varying personalities and in diverse environments and do not always employ a uniform strategy for reproductive success, especially when a particular strategy is risky and prone to backfire. In this area, evolutionary theorists about morality should learn from salamanders and fish. In several species, males can opt into sexual selection and compete for mates if they have the wherewithal, or else (if they do not) they can masquerade as females and gain matings by this alternative route (Gross 1996; Brockmann 2001). If these slippery creatures can opt for different strategies based on various individual and environmental variables, all functioning in the same general pursuit of individual reproductive success, one might suspect that humans can do the same in the area of cooperation. A strategy of self-deception and aggrandizing one's moral fiber, together with the attempt to encourage more moral commitment in others than one's own level, comprises but a portion of a range of activities that can be adaptive in a social context. It almost goes without saying that with our abilities of perception and discrimination, behaving cooperatively in actuality, especially in the low cost situations we face every day, can be a more viable alternative than attempting to deceive or manipulate others, especially considering the house of cards that reputation can be, where a lifetime of cooperation can be undone with a few or even one detrimental act (Alexander 1987). Surely not everyone has equal powers to dissemble, whether just to others or to oneself as well. Possibly emetics #10 and #11 above are better seen as two common (and complementary) tools of the human trade, than as necessarily universal and incessant practices. Of course, actually testing for the role and importance of these psychological or subpsychological tendencies would be more helpful than the speculation I am doing here.

Second, fish and salamanders aren't the only things that are slippery: so are words. When we use terms like "selfish," "altruism," "hypocrisy," and "self-deception" in an evolutionary context, we do not mean precisely the same thing that we mean by them in social usage, where these words have their ancestral currency. In typical uses these words imply a key role for intentionality. We would never denigrate as "selfish" a person whose actions merely tended to benefit oneself, if the person's intention was to benefit someone else or to accomplish something equally unrelated to self-benefit. Following the growth of evolutionary thought, however, we tend to repurpose existing terms for all the new conceptual spaces that have suddenly opened up. We now have various proximate (mechanistic, developmental) and ultimate (functional, historical) levels on which the identical terms can be used, and the uses at the different levels are not always consistent, nor do they always imply one another. For instance, we use "selfish" of genes as a heuristic tool, despite the fact that genes cannot possibly have the intentionality that is typically implied by that term. We are to understand that the word is being used in a role analogous to its typical one, despite some discontinuities, like when we say that skies are "threatening" or that a rattling screw "wants" to come loose from a machine. Therefore when we read "hypocrisy," and yet admit that this phenomenon might not be cognitively accessible to us because we are self-deceived, what we are really saying is that we might not really be hypocritical at all, but we might be doing something that is analogous to hypocrisy. Real hypocrisy is presenting an appearance of a virtuous intention that one does not really have. Hypocrisy in the evolutionary theory of morality has a range of possibilities, including this real hypocrisy as well as alternatives that do not actually qualify as such. One example is a situation where one's intention is entirely virtuous, but where that virtuous intentionality is based partly on a series of alleles that evolved by natural selection because that psychological state was beneficial to its bearer. Another is a situation where (at least some of) one's conscious motivations are virtuous, but they are minor in their effect on action compared to certain unconscious motivations that are self-interested and (in this case) are motivating the very same action. This situation too would not satisfy an ordinary definition of hypocrisy because, in this case again, one does actually have virtuous intentions. These arguments apply equally well to "self-deception" and "altruism" (Lahti 2003): these terms in the mouth of the evolutionary biologist are not really what they seem to be. Emetics 10 and 11 are not quite as nauseating when one takes them with this grain of salt.

Actually, the evolutionary sense of hypocrisy and self-deception leads to what I would consider to be one of the striking paradoxes in the evolution of morality: that "hypocrisy" and "self-deception," in their evolutionary senses, instead of making us worse, make us better—in fact they are our saving grace. Their importance in the theory arises because of the possibility for inconsistency within but especially between the levels on which human behavior is controlled, the level most relevant to morality being that of intentionality. Real hypocrisy occurs when our intentions are inconsistent with how we represent them to others; and nearly all of us would agree that this kind of inconsistency is generally (a bit of moralizing now) a bad thing. But the goal of human lifetimes, insofar as we are an evolved organism, is reproductive success. In light of that, if it were not for a similar kind of inconsistency between levels of behavioral control, we would have exclusively selfish intentions. In other words, if the evolutionary function of a behavior were always manifest in our intentions, we would only be able to help others by feeling or thinking selfishly, only care by faking it. Enter the ridiculous comment that "No hint of genuine charity ameliorates our vision of society, once sentimentalism has been laid aside.... Scratch an 'altruist' and watch a 'hypocrite' bleed" (Ghiselin 1974: 247). We can see the germ of truth in this, but it does not reflect how humans actually work. Precisely because of inconsistency between levels of control, a mother can feed her infant with selfless love, "genuine charity," despite the fact that the function of her behavior is to benefit her own genetic lineage (this is considered hypocrisy in the evolutionary analysis). And from maternal love, arguably, all love has evolved. Love as devotion to another is only possible insofar as its psychological workings are shielded from the adaptive basis of loving behavior; this is what evolutionary biologists call self-deception.

Still, we must admit that our motives are impure. We cannot in good conscience rescue good conscience solely by scapegoating some unintentional level of behavioral control. The selfish function of behavior so regularly informs our conscious motivations that it is likely to bleed through even when we have noble self-impressions to the contrary. As emetic #9 above suggests, the most straightforward way for a genome to achieve reproductive success is simply to produce desires and motivations that aim immediately at survival and reproduction. Accordingly, few would disagree that desires and motivations that relate simply and closely to survival or reproduction are among the strongest in human experience. However, our sociality often calls for a subtler route to reproductive success (for instance, via enhancing group stability or our own reputation [Lahti &

Weinstein 2005]), in which case those basic desires and motivations may need reining in or contravention. Even that loving mother holding an infant knows well the mixed feelings, and consequently the need for self-control, associated with giving so much of herself to one offspring. Some moralists and moral philosophies demand purity of motive or intention in order for a state of mind or a behavior to be considered good, but this situation appears to be rarely if ever actualized. That we humans are a bundle of often inconsistent motives is a lesson delivered not only by contemporary evolutionary psychologists (e.g., Kurzban 2010), but frequently in the history of wise counsel on the subject. The major religions and philosophies speak generally with one voice on this matter, thus agreeing with the last and (to some) most frustrating of the twelve points above, that we are not as good as we prefer to think. If we object to this point despite the history of introspection and now despite evolutionary prediction, we are not necessarily fighting for human dignity. In fact we may be revealing a defensiveness and unrealistically high self-image. In this the demand for purity of motive resembles the demand for free will among similarly well-meaning defenders of humanity. Most of the argument about free will stresses the contrast between it and some sense of determinism. But even if we grant a sort of agency to humans, is there any scientifically informed person today who is really prepared to defend utterly *free* will? Agency without influence? We know beyond a doubt that we have influences, both culturally acquired and inherited, and these influences sway our decisions. To demand that there is some faculty that is somehow immune to these factors is unreasonable. Any conception of the will that is worth considering will have to take this diversity of interests seriously into account. In the same way, defending a moral psychology where our motives and intentions operate in a sort of cleanroom isolated from all compromising or conflictual influence is untenable and reactionary. The spirit of the evolutionary account, and very likely literally true, is that even our noblest attitudes and actions are moved not only by worthy but also by morally neutral and even base influences. When I dissect my own emotional objection to the suggestion that posturing and manipulation are an inherent part of the human condition, I wonder how much of my objection is honest doubt as to its truth, and how much is my wish that I and *a fortiori* others not be that way. I also admit an uneasy sense, as predicted by the theory, that I don't want open discussion of this tendency to lead to the spread of an opinion that these strategies are acceptable.

This brings us to the third reason why morals can survive the evolutionary account: because our own moral rectitude is not, should not be, and never really was a condition for a functioning morality. Knowing that we are bad (to whatever extent we are) does not wreck the ideals we hold up, or prevent us from holding them up, any more than denying something makes it false. Nevertheless, the fact that tendencies we would consider good evolved hopelessly tangled up with those we would consider bad does complicate our job as moral beings. It means we are stuck assigning moral values to attitudes and actions regardless of their original evolutionary function. We may praise certain things, such as sacrifice for our social group, and despise others, such as racism, regardless of the fact that these two tendencies became adaptive in the context of precisely the same function: fostering group unity in the face of competition with other groups. We do not generally consider similarity of function to be at all a reason to assign a similar moral value, and for good reason. In the same way, the fact that cooperation with each other and a hypocritical self-representation both evolved for the same function has little relevance for how we will be constructing our ideal moral values. An equally troublesome consequence of the moral ambivalence of our history is that is that we cannot simply consult our nature to determine how we ought to live, because the average human tendency is not necessarily something we would want to encourage. Statistical tendency can be important in science but does not have to be morally important to us as individual humans. Even if psychology bears out an evolutionary prediction that the average person's motives are mixed, generosity is calculated, advice is self-serving, and modesty is false—even if this ends up being true, whoever said that goodness had to be the mean, had to be typical? That is the doctrine of a hopeful sort of humanist religion we have always had plenty of reason to doubt, long before Alexander and Trivers started talking about self-deception. The game theorists who consider nearly all of us "good guys," and get their papers into *Science* when they promise that cheats really do lose out, owe much more to that fuzzy religion than to evolutionary biology. Likewise those ecologists for social justice, by warning against Alexander, were not defending society or justice so much as their own comforting conception of human nature. Seeing ourselves as honest and caring from the core of our beings, if this is a false view, will not help us make ourselves or the world better.

If the evolutionary account is correct about the moral ambivalence of our history and nature, we can sympathize with calls to widespread empathy (e.g., Baron-Cohen 2011), while realizing, with Darwin, that such an idea has absolutely no evolutionary precedent, and so must be taught (1871, ch. 4), and also realizing, with Alexander, that bringing about such empathy will likely be arduous because of contrary influences. If we do have widespread tendencies toward manipulation or false moral pretenses, most likely we will be in a much better position to rectify social ills if we are aware of these tendencies and can take them into account. This I take to be the import of the final sentiment of the associated paper, "a society of well-meaning people who understand themselves and their history very well is a better milieu than a society of well-meaning people who do not." A bigger question is whether knowledge of the evolutionary basis of our psyche can enhance our ability to live up to whatever ideals or goals we do espouse: take "global cooperation" for instance. On this point, Richard Alexander is hopeful but realistic. In a paper he recently called his scientific swan song, he concluded by questioning whether the situation that started us competing in the first place—our genetic differences, and subsequent conflicts of interest—would ever subside enough in importance to permit peaceful coexistence as a species. As he exclaimed in the last sentence, "If only we could devise an effective way to tackle the question—and generate the solution—of how to make 'otherness' go away!" (Alexander 2012).

There are two complementary senses in which we can say that morality deals with the ideal: The first is, as Alexander said, that the moral is never actualized as it is conceived. The other spirit, the other side of the coin, is that the moral is something to which we can aspire if we so choose. By engaging in such encouraging words I fully admit to be falling right into Alexander's prediction that each individual will urge others to be more moral than oneself. Avoiding autopsychoanalysis, if I accept for the sake of argument a degree of duplicity in my moralizing, what is the next step? Should I abandon the whole project and be nihilistic? Or just keep my opinions to myself? An alternative strategy is to go with the flow in just this one area: what if I try to convince others to be better, and they try to convince me to be better, and I like a sucker fall for them rather than steel myself with cynicism. If we were to do this, society might remain a decent place to live, and might even become more so, whereas rejecting the evolutionary account or throwing up our hands in moral skepticism seem far less promising options. This is not blindness to the truth, it is merely rolling with the evolutionary punches in a case (like eating and breathing) where it is actually good for us to do so. We so often seem to be struggling to muffle or channel recalcitrant aspects of our evolutionary heritage; let's just relax in this particular area and allow ourselves to moralize and be taken in by moralizing. But of course you know there could be a devil on my shoulder when I say that.

Acknowledgments

Many thanks to the editors for their helpful suggestions, and to Andrew Richards for extensive input and discussion.

References

Alexander, R.D. 1978. Evolution, creation, and biology teaching. In: *American Biology Teacher*. pp. 91–96, 101–104, 107.

Alexander, R.D. 1979. *Darwinism and Human Affairs*. Seattle: University of Washington Press.

Alexander, R.D. 1982. Biology and the moral paradoxes. *J. Soc. Biol. Struct.* 5:389–395.

Alexander, R.D. 1987. *The Biology of Moral Systems*. Hawthorne, NY: Aldine de Gruyter.

Alexander, R.D. 1988. Evolutionary approaches to human behavior: what does the future hold? In: Betzig, L. L., Borgerhoff Mulder, M., and Turke, P. W. (eds), *Human Reproductive Behavior: A Darwinian Perspective*, Cambridge: Cambridge University Press, pp. 317–341.

Alexander, R.D. 1989. Evolution of the human psyche. In: Mellars, P. and Stringer, C. (eds.), *The Human Revolution*. Edinburgh: University of Edinburgh Press, pp. 455–513.

Alexander, R.D. 1990. How did humans evolve? Reflections on the uniquely unique species. *Univ. Michigan Mus. Zool. Spec. Publ.* 1:1–38.

Alexander, R.D. 1992. Biological considerations in the analysis of morality. In: Nitecki, M.H. and Nitecki, D.V. (eds.), *Evolutionary Ethics*, Albany, NY: SUNY Press, pp. 163–196.

Alexander, R.D. 2005. Evolutionary selection and the nature of humanity. In: Hosle, V. and Illies, C. (eds.). *Darwinism and Philosophy*. South Bend, IN: University of Notre Dame Press, pp. 424–495.

Alexander, R.D. 2009. Understanding ourselves. In: Drickamer, L. C. and Dewsbury, D. A. (eds.), *Leaders in Animal Behavior: The Second Generation*, vol. II. New York: Cambridge University Press.

Alexander, R.D. 2011. *The Mockingbird's River Song: Poems, Essays, Songs and Stories, 1946-2011*. Manchester, MI: Woodlane Farm Books.

Alexander, R.D. 2012. Darwin's challenges and the future of human society. In: Wayman, F., Williamson, P., and Bueno de Mesquita, B. (eds.), *Prediction: Breakthroughs in Science, Markets, and Politics*. Ann Arbor: University of Michigan Press.

Alexander, R.D. and Borgia, G. 1978. Group selection, altruism, and the levels of organization of life. *Annu. Rev. Ecol. System.* 9:449–474.

Baron-Cohen, S. 2011. *The Science of Evil: On Empathy and the Origins of Cruelty*. New York: Basic Books.

Batson, C.D., Thompson, E.R., Seuferling, G., Whitney, H. and Strongman, J.A. 1999. Moral hypocrisy: appearing moral to oneself without being so. *J. Pers. Soc. Psychol.* 77:525–537.

Brockmann, H. J. 2001. The evolution of alternative strategies and tactics. *Adv. Stud. Behav.* 30:1–51.

Burt, A. and Trivers, R. 2006. *Genes in Conflict: The Biology of Selfish Genetic Elements*. Cambridge: Harvard University Press.

Churchland, P.S. 2011. *Braintrust: What Neuroscience Tells Us About Morality*. Princeton, NJ: Princeton University Press.

Darwin, C. 1871. *The Descent of Man and Selection in Relation to Race*, 1874 edn. London: John Murray.

de Waal, F. 1996. *Good Natured: The Origins of Right and Wrong in Humans and Other Animals*. Cambridge: Harvard University Press.

Descartes, R. 1637. *A Discourse on Method*, 1969 edn. London: Everyman, Dent.

Ghiselin, M. 1974. *The Economy of Nature and the Evolution of Sex*. Berkeley: University of California Press.

Gross, M.R. 1996. Alternative reproductive strategies and tactics: diversity within sexes. *Trends Ecol. Evol.* 11:92–98.

Haidt, J. 2006. *The Happiness Hypothesis: Finding Modern Truth in Ancient Wisdom*. New York: Basic Books.

Hobbes, T. 1651. *Leviathan*, 1839 edn. London: John Bohn.

Hrdy, S.B. 2009. *Mothers and Others: The Evolutionary Origins of Mutual Understanding*. Cambridge, MA: Belknap Press.

Keller, L., and Ross, K.G. 1998. Selfish genes: a green beard in the red fire ant. *Nature* 394:573–575.

Kierkegaard, S. 1846. *Purity of Heart Is to Will One Thing*, 1939 edn. New York: Harper.

Kurzban, R. 2010. *Why Everyone (Else) Is a Hypocrite: Evolution and the Modular Mind*. Princeton, NJ: Princeton University Press.

Lahti, D.C. 2003. Parting with illusions in evolutionary ethics. *Biol. Phil.* 18:639–651.

Lahti, D.C., and Weinstein, B.S. 2005. The better angels of our nature: group stability and the evolution of moral tension. *Evol. Hum. Behav.* 26:47–63.

Murdoch, I. 1970. *The Sovereignty of Good: Studies in Ethics and the Philosophy of Religion*. London: Routledge and Kegan Paul.

Oakley, B. 2007. *Evil Genes: Why Rome Fell, Hitler Rose, Enron Failed, and My Sister Stole My Mother's Boyfriend*. New York: Prometheus Books.

Trivers, R. 2000. The elements of a scientific theory of self-deception. *Ann. NY Acad. Sci.* 907: 114–131.

Wittgenstein, L. 1921: *Tractatus Logico-Philosophicus*, 1974 edn. London: Routledge and Kegan Paul.

BIOLOGY AND THE MORAL PARADOXES

Alexander, R.D. Biology and the Moral Paradoxes. *Journal of Biological Structures* 5:389–395.

Considerations from biology suggest (1) that human interests can be generalized as reproductive, involving activities by individuals that tend to promote the survival of their individualized sets of genes; (2) that ethical, moral and legal questions arise out of conflicts of interest that exist because of our history of genetic differences; (3) that human behavior probably always involves egoistic tendencies and moral inconsistency; (4) that the stages of moral development described by social scientists correspond to the patterns of life effort discussed by biologists; (5) that the idealized moral systems of philosophy and religion have been developed as models that are promoted in others but not (or more than) in one's self; and (6) that what are usually seen as the closest approaches to these idealized models are the sources of our most severe problems because they involve between-group competition and strife.

Ethical, moral and legal questions arise out of conflicts of interest among human individuals and groups. Although this assertion seems to be accepted universally, those who write on ethics, morality and law rarely emphasize it (Pound, 1941; Perry, 1954; Kelsen, 1957, represent the exceptions). Evidently, no student of human behavior has undertaken the obvious challenge of explicitly identifying human interests and quantifying their conflicts. Equity theory from psychology, network and exchange theory from sociology and anthropology, and theories of interest from law, political science and economics are partial attempts. These theories, however, are all restricted to a superficial level, involving only reciprocal transfers of good or beneficent acts. They neither identify the ultimate significance of goods and beneficent acts, nor deal satisfactorily with the all-important class of interactions that frustrated equity theorists have termed 'deep and intimate' (Walster, Walster & Berscheid, 1978). In other words, these theories provide no means of defining human interests in a general or complete sense, therefore, no means of dealing generally with the intensities and directions of individual efforts (see Alexander, 1979, and references therein).

A theory of interests is a theory of lifetimes: what they are about and how their goals are achieved. A growing body of information and theory from biology now provides a reasonable and testable answer: lifetimes have been molded by natural selection to yield the greatest likelihood of survival of the individual's genetic materials. This likelihood is maximized by success in reproduction, which includes producing offspring, and assisting both descendant and non-descendant relatives. The 'deep and intimate' interactions causing difficulty to equity theorists are actually those most directly involving reproduction-those occurring between mates, potential mates and relatives. The currencies that mold the proximate mechanisms of altruism in these interactions are genetic, not a matter of returned goods or services, and this is the reason the payoffs have not been apparent to investigators outside biology. Even the investments and returns of reciprocity (exchange, equity) are ultimately comprehensible only in terms of their eventual effects on the 'deep and intimate' interactions of mates and relatives. Included are wealth, status, good will and innumerable other items.

Biologists divide lifetimes into somatic and reproductive effort: use of calories and taking of risks in (1) building the body or soma (= amassing resources) and (2) using the soma to reproduce (= redistributing resources in the interests of one's own genetic materials). Reproductive effort is in turn divisible into mating effort (on behalf of gametes), parental effort (on behalf of offspring) and extraparental nepotistic effort (on behalf of all relatives other than offspring). There are good reasons for supposing that normal lifetimes include no other kind of effort (Alexander, 1979).

I would regard the central paradoxes of moral philosophy to be those of (1) the incompatibility of egoism and utilitarianism (seeking the greatest benefits to one's self versus seeking some version of the greatest benefits to the greatest number) and (2) the associated problem of duality in human nature. These paradoxes have been developed and discussed in many forms, but always independently of the current biological view of interests and life-times. I shall argue that they remain paradoxes not because of some inherent irresolvability but because those concerned with them have not adequately discussed the costs and benefits of either egoism or altruism. Kalin (1968), for example, speaks of 'winning' and 'coming out on top', and Frankena (1973, 1981) of getting 'the best score', but neither describes the actual currency involved. Some authors speak of survival, but it is unlikely that humans or any other organism have evolved to survive (Alexander, 1979), and it is easy to show that they all do things that reduce their likelihood of survival. Essentially all authors consider pleasure or happiness as reward (benefit) and pain and suffering as punishment (cost), but none can explain in egoistic terms either the voluntary acceptance of pain or the pleasure of helping others. Because the indisputable prevalence of egoistic behavior eliminates any likelihood of a purely altruistic or utilitarian society, except as an unattained (and as yet unexplained) pursuit or ideal, the problem of duality, and of moral inconsistency as normal behavior, persists.

Biological theories of interests and lifetimes have the power to resolve these paradoxes, at least in terms of the natural history of moral systems (the 'why' of behavior in respect to morality). Thus, an organism whose interests are in its own genetic survival must first develop a soma (be a wholly or largely egoistic juvenile), then reproduce (show the 'altruism' of parenthood and nepotism) while maintaining the soma by which it continues to reproduce (thus retaining egoistic tendencies during adulthood). Direct and indirect reciprocity (Alexander, 1979) are distinctive human overlays that add to the complexity, but they create no special problems. They may be seen as indirect somatic or nepotistic efforts routed through pseudo- or temporarily-altruistic investments in the welfare of others who are expected to reciprocate with interest.

The stages of moral development in the individual, as interpreted by Kohlberg (1981) and others, are remarkably supportive of this biological view. Represented, first, is a purely somatic (selfish, 'amoral') stage. This is followed by the introduction of reciprocity through a system of rewards and punishment, usually by the parent. The individual gradually forgoes immediate rewards in favor, I would argue, of larger later ones (reciprocity). Acceptable rewards may be both increasingly later and increasingly less direct (in the senses of involving diverse currencies, and of coming from society at large rather than the person or persons directly involved in the original social act). Eventually the individual also begins to forego personal (somatic) rewards in favor of unreciprocated rewards to others (nepotism). And he becomes increasingly able to assess the profitability of social acts without outside help.

From these arguments about interests it follows that conflicts of interest arise out of the history of genetic differences. This hypothesis is strongly supported by the absence of observed conflicts among non-human individuals in clones and other cases of long-standing genetic identity, and by the general diminution of altruism with decreasing relatedness within human societies the world over. It explains human individuality, and bears upon powerful human issues, such as what Wallace called 'the impossibility, despite all the labor of God, Freud and the Devil, of one man fully understanding another, or the loneliness of existence' which he regarded as 'a pan-human theme'. It explains the unique cooperativeness of unrelated pairs pledged to lifetime monogamy, and of genetically different workers in the colonies of social wasps, bees, ants and termites. In both cases the genetically different individuals involved share interests because they reproduce through the same third parties: the offspring produced jointly by the monogamous pair and the siblings of worker insects produced by their common mother. It is significant that Kohlberg's final stage of moral development is that in which the individual has learned *for himself* how best to assess his personal costs and benefits in following (and using) whatever social rules prevail.

Viewing humans and their moral behavior in terms of natural selection provides stark and dramatic answers to some serious and very general questions: the incompleteness of justice; the persistence of conflicts of interest; the failure of idealized moral systems; and the absence of universal happiness and satisfaction. Part of the answers lie in the *relative* nature of success in evolutionary terms:

> In natural selection the likelihood of a genetic element persisting depends entirely on its rate of change in frequency in relation to its alternatives; changes in absolute numbers are irrelevant. Among the attributes of living creatures, whatever can be shown to have resulted from the action of natural selection may be expected to bear this same relationship to its alternatives. Thus, we should not be surprised to discover that the behavioral striving of individual humans during history has been explicitly formed in terms of relative success in reproductive competition, that justice is necessarily incomplete, that happiness is not easily made universal, and that ethical questions continue to plague us, and can even become more severe when everything else seems to be going well. (Alexander, 1979: 240)

I stress that our interests are not individual because of genetic differences *per se*, or current genetic differences, because such information has never been directly available to humans. Relatives are known through circumstantial evidence, and only recently have geneticists learned what the average relatedness actually is for relatives whose learned assumptions about relatedness from genealogical connections and kinds of social interactions are nevertheless usually correct. The individualized genetic constitutions of the successions of our ancestors caused natural selection to save and mold proximate mechanisms whereby appropriate efforts could be mounted by individuals in each successive generation to realize their separate and individualized interests. We *learn* who our relatives and friends are, and how to treat them; but our learning responses are themselves evolved, and often very specific and channeled.

The hypothesis that conflicts of interest derive from the history of genetic differences also generates new and sometimes startling questions: What are the benefits of the group to the selfishly reproducing individual? Why does one kind of ultrasocial group (eusocial insects) achieve its greatest numbers and unity (up to 22 million) as a single nuclear family in which one individual does all of the reproduction while the other (humans) achieves its greatest numbers and unity (now approaching one billion in China) by leveling the reproductive success and opportunities of its members (through socially-imposed monogamy, graduated income taxes, gradations of negative correlations of government support with family size, restrictions on 'free' enterprise etc)? How do these questions relate to the morality of individuals and the idealized moral systems discussed by moral philosophers?

The altruism of human nepotism and reciprocity is discriminative: Different relatives, and relatives of different needs, are distinguished. Friends are treated individually. As yet, no evidence of truly indiscriminate, species- or population-wide altruism has been reported for any organism, and there is no undisputed evidence for unlearned recognition of relatives in any species (Alexander, 1979). These facts are crucial to understanding moral paradoxes and the rise of moral systems. Indiscriminate altruism requires no special proximate mechanisms—no social learning. I would venture that without genetic individuality, and the consequent *discriminative* altruism in nepotism and reciprocity, social learning would have remained simple, and human society as we know it could not have evolved. The very concepts of ethical, moral and legal would be unknown.

To think of humans existing without conflicts of interest is to assume situations involving or mimicking group selection, in a way explicitly opposing the notion of individuals striving to maximize their separate reproductive successes. It seems to me that this is the ideal state of morality postulated by philosophers and social scientists. If so, perhaps biology gives us the reason for understanding interpretations such as that of Perry (1954: 100).

Morality is like a cultivated field in the midst of the desert. It is a partial and precarious conquest. Ground that is conquered has to be protected against the resurgence of original divisive forces. The moralized life is never immune against demoralization. At the same time that morality gains ground in one direction it may lose ground in another. Changes in the natural and historical environment and the development of man himself are perpetually introducing new factors and requiring a moral reorganization to embrace them. In the last analysis all depends on the energy, perseverance, and perpetual vigilance of the human person.

Numerous philosophers have suggested that morality, at least as expressed in the behavior of individuals, is in fact only an ideal, or a pursuit, and not something that is actually realized. This idea seems consistent with the approach from evolutionary biology that I have been describing. Thus, it is common, if not universal, to regard morality in the behavior of an individual as consisting of a kind of altruism that yields the altruist less than he gives. In a utilitarian system (defining utilitarianism as promoting the greatest good to the greatest number) morality would not always *require* that complete and indiscriminate altruism cause individual losses. This would not, for example, be the case when the interests of the group and the interests of the individuals comprising the group are the same. Such a confluence of interests would happen each time the group was threatened externally in such fashion that complete cooperation by its members would be necessary to dissipate the threat, and when failure of the group to dissipate the threat would more severely penalize any remaining individuals than would the use of all the individual's effort to (successfully) support them (this is the true, but in these times of nuclear threats forgotten, meaning of the term 'national security'). In other circumstances, as when some competitiveness has a likelihood of benefiting individuals in the group (i.e. the individuals' interests are not all completely tied up in the survival of the group or its success in dissipating some external threat), morality of an ideal sort would require the kind of genetic altruism, unlikely in evolutionary terms, in which the altruist truly gets back less than he gives. Of course, if an external threat came from another group of humans, the definition of morality as indiscriminate altruism would again be in jeopardy.

Reflecting on these circumstances, we see that if approaches to morality are expressed consistently, and to the degree usually achieved in society, because there is continual pressure to bring about a condition of morality, this pressure is likely to be applied by each individual so as to cause his neighbors, if possible, to be a little more moral than himself. To say it another way, it would be to the advantage of each individual that other individuals in his society—especially those not closely related to him—actually achieve the ideal of completely moral behavior. Any ideally moral person would incidentally 'help' every other person in the society, however slightly, to achieve the goals that evolutionists believe have driven evolution by natural selection, because he would hurt himself (a competitor) by dispensing his beneficence indiscriminately. Accordingly, one might expect that every individual in a society would gain from exerting at least a little effort toward encouraging other individuals to be a little more moral (altruistic) than they otherwise might have been. Among the many ways of furthering this aim is included the setting up of an idealized model of morality and the encouragement of everyone (else) to become like that. One way of promoting this outcome is to designate as heroes (i.e. as appropriate targets for special rewards) those who most closely approach the ideal moral condition. This line of reasoning predicts that sainthood will be awarded to individuals who spend their lives on explicitly anti-reproductive behavior. The prevalence among saints of asceticism, self-denial, isolation from relatives, devotion to the welfare of strangers, and otherwise indiscriminate tendencies to be altruistic supports this hypothesis. So does the fact that sainthood is generally awarded (long) after the death of the awardee.

So we are provided with the general hypothesis that the concept of morality, and the establishment of systems promoting ideal moralism, at least *appear* to have as their aim the support of the goals of society as a whole. For this reason, within society, each and every individual may be expected to promote in his associates tendencies to be moral. Because of continuing possibilities of differential success within groups, though, we can also understand that each and every individual may also be expected to promote a slightly greater degree of 'morality' (altruism) in his neighbor than in himself. And we can understand why the idealized morality of the philosophers is never a reality in society as a whole, and occurs only as an accident, a manipulation or in special circumstances.

The question may be raised, why anyone should be susceptible to being manipulated unduly far in the direction of morality, given that we have been subjected to such manipulations for so long? Why, in other words, should moralizing ever be effective?

I think there are at least four contributing factors. First, the degrees of morality that are actually reproductively appropriate will vary dramatically as societies move between periods of extreme danger and relative security, making it difficult to know how to behave. When will a specified degree of failure to accede to exhortations to be altruistic cost more than it yields, because of (1) failure of the group on which one depends for success, or (2) responses within the group to one's failure to be altruistic?

Second, individuals may be expected to take advantage of the dramatic shifts in most profitable degrees of altruism to deceive others about costs or dangers so as to induce in them unduly altruistic behavior. It is obvious that aspiring leaders often use such deception to promote their own leadership, as an antidote to the supposed threat and as a promoter of unity.

Third, we may expect that the individuals in a society such as we have been describing will evolve to deceive others about the degree of altruism they themselves are exhibiting: Everyone will wish to appear more altruistic than he is. There are two reasons: This appearance, if credible, is more likely to lead to direct social rewards than its alternatives. It is also more likely to encourage others to be more altruistic. If one's associates are altruistic, then he can afford to be more altruistic than if they were not. We may expect everyone to be concerned that everyone else appear altruistic so that people in general will feel comfortable with a higher degree of altruism than would otherwise be the case.

Fourth, if kin recognition is learned (Alexander, 1979), mistakes are likely in this context, and one may insinuate himself into the role of relative so as to receive inappropriate nepotism, or even to pretend to be nepotistically altruistic so as to receive the appropriate altruistic responses.

Playing upon the tendency of everyone to strive to appear more altruistic than one's self, and using the other ploys just described, may produce a considerable amount of successful social manipulation. These various factors seem to be the elements necessary to produce and maintain what we commonly call moral systems, and moral behavior in individuals. They represent the means for resolving the philosophers' paradoxes with respect to morality, and for understanding why moral systems have always fallen short of our ideals, and why we establish and maintain such ideals. If accurate, these arguments may also clarify the routes by which we can most closely approach what are seen as idealized moral systems, and perhaps most confidently avert moral disasters.

The introduction of indirect reciprocity, whereby society as a whole or some large part of it provides the reward for altruism and the punishment for selfishness, simultaneously served both society and the individuals comprising it, and provided the vehicle for socially manipulating individuals to levels and kinds of altruism detrimental to them (or their reproductive success). It is somewhat paradoxical that the tendencies and pressures in the direction of idealized moral systems should serve everyone up to a point, but then be transformed by the same forces that molded them into manipulations of the behavior of individuals that are explicitly against their interests and in the interests of those ostensibly promoting everyone's interests by promoting trends toward morality in the system.

The concept of a single just God for all people, however it is believed to have originated, implies social unity. I would regard this concept as one representation of an idealized moral system arising out of religion; and it is just as difficult to follow as those generated from moral philosophy. It is not trivial that the concept of a single God for all people differs from that of a 'tribal' God looking out for the interests of only one group or society. Adhering to this concept requires denial of practices like slavery, caste systems and other within-group discrimination. Despite its prominence and use during times when groups are threatened externally, the concept in some sense fails whenever such external threats involve (or *are*) other groups of people. This failure is, of course, denied by the invention of anti-Gods, or Devils, and the ascription of others' motivations to their control. As a US Christian picketing over the arrival of some Russians put it, 'I *could* love them if they were *my* enemies, but they are the enemies *of God!*'

The concept of God also implies *continuity* of social unity—a long-lasting, intergenerational social contract. If nepotism is our evolved function, then God (in the sense of *vex populi, vex Dei*) really can guarantee a reward 'in Heaven', or after our individual deaths—or a kind of 'everlasting life' (for our genetic materials)—as a reward for moral behavior during life. This guarantee is in the form of a renewable contract in reciprocity which occurs when those who remain after our death use our own life of 'morality' to judge our children (and other relatives who remain) as suitable risks to continue receiving (and giving, and receiving and giving, and receiving and giving…) the benefits of social reciprocity. The guarantee actually exists because, unless those in a position to honor it do so for us, the same possibility will not exist for them. The ceremonies associated with death, and the reverence given to the dead, are surely, in part, ritually related to this guarantee.

If morality tends to mimic the effects of group selection, if moralizing seeks to promote this mimicry, and if tendencies for people to be altruistic are self-reinforcing within societies, then it is not remarkable that sincere, knowledgeable and well-meaning people sometimes resent the arguments that natural selection is not powerful at group levels, and that humans, as individuals, have evolved to be interested in furthering their own reproduction. Such persons may well believe (or sense) that publicizing or stressing such arguments, even if they are correct, will diminish altruism and morality by providing an anti-moral model. The indications that humans have regarded moral models as extremely important in achieving societal goals cause such a belief to be completely understandable. Nevertheless, this attitude runs counter to the goal of diminishing human problems through improving self-knowledge.

Our truly serious problems of morality and law stem, not from the behavior of individuals, but from the behavior of groups that may show most dramatically within themselves the indiscriminate altruism that represents approaches to the idealized morality of philosophy and religion. Indeed, loyalty and patriotism are revered as the highest forms of morality and virtue *within* groups. But this same level and kind of within-group 'morality' has also created our most devastating problems—those involving intergroup conflict—that we must somehow supercede. What we seek, when we think of *world* peace and *world* law, has no precedent in the history of life, not to say that of humankind. There seems to be no evidence that humans or any other organism have achieved the species-wide indiscriminate altruism represented in the idealized moral models of philosophy and religion.

I offer only one conclusion in this brief and perhaps unsettling essay: that, in the effort to solve humanity's most profound problems, there is potentially great value in adding a perspective from modern evolutionary biology to those developing out of philosophy, the social sciences, religion, history and the humanities. This biological perspective must be added, not as an argument for determinism, but precisely to the contrary, as a possible way to greater freedom, deriving from greater knowledge of the cause-effect patterns that underlie our history and our nature. Some of my colleagues in biology, and many people outside biology, deny that humans can be understood in biological terms. Others cling to the notion that we evolved by an innocuous (and hypothetical) form of group selection and can somehow return to it. Or they argue that if this is not the case, we should deny the truth and pretend ourselves toward world peace and human justice; or that it is better to be ignorant with an idealized moral model before us than to know about an immoral history. I believe that people who think in these fashions are wrong. Worse, because of the enormity of the problems that face us, I regard approaches that deny biology, and sometimes deny reality, as potentially deadly. Essentially everyone thinks of himself as well-meaning, but from my viewpoint a society of well-meaning people who understand themselves and their history very well is a better milieu than a society of well-meaning people who do not.

Literature Cited

Alexander, R.D. 1979. *Darwinism and Human Affairs*. Seattle, WA: Univ. Washington Press.

Frankena, W.K. 1973. *Ethics*, 2nd. ed. Englewood Cliffs, NJ: Prentice Hall.

Frankena, W.K. 1980. *Thinking about Morality*. Ann Arbor, MI: Univ. Michigan Press.

Kalin, J. 1968. *Studies in Moral Philosophy*. American Philosophical Quarterly Monograph Series, N. Rescher (ed). Oxford: Basil Blackwell.

Kelsen, H. 1957. *What is justice? Justice, law, and politics in the mirror of science. Collected essays*. Berkely, CA: Univ. Calif. Press.

Kohlberg, L.L. 1981. *Essays on Moral Development. I. The Philosophy of moral development*. San Francisco, CA: Harper and Row.

Perry, R.B. 1954. *Realms of value: a critique of human civilization*. Cambridge, MA: Harvard Univ. Press.

Pound, R. 1941. My philosophy of law. *Credos of sixteen American scholars*. Boston, MA: Boston Law Book Co.

Walster, E., G.W. Walster & E. Berscheid. 1978. *Equity: Theory and Research*. Boston, MA: Allyn & Bacon, Inc.

12

Evolution and Humor

Just for fun
or the reverse.
Try falsifying this: Humor mocks reality,
social and moral reality, the kind of
reality which, in the absence of a
supernatural, necessarily has
human architects.   Humor
causes human institu-   -tions
to be modified, improved or destroyed, individuals
to be informed, sometimes mortified, also sometimes
destroyed, via toying with incongruities and the
ridiculous, exposing failures to comprehend
such boundaries,
functioning on
the author's side to elevate status.
The jocular gossip
of a good-natured,
story-telling humor session makes
everyone (?) feel better and laugh
together, forming, stabilizing,
and reinforcing coalitions,
necessarily
likely to be
at least some-
what exclusive,
everyone in-
volved

seeking

thereby to be

gainers not losers.

But contemplate falsifying this claim:

No instance of humor has ever been relished

universally: always, someone is the goat.

Alexander, *2011, p. x*

INTRODUCTION

The Adaptive Significance of Humor
Stan Braude

> So, two orthopterists walk into a bar....
> The first one didn't see it because he was busy fiddling
> with his tape recorder.
> The second was a former grad student following in his
> advisor's footsteps.

This old joke illustrates some of the core ideas in Alexander's (1986) essay, "Ostracism and indirect reciprocity: the reproductive significance of humor." If you did not find it humorous, keep that to yourself for now and keep reading. Others might have found that joke humorous on a number of levels: the recognition of the joke as a variant on a familiar class of "two guys walk into a bar" jokes, the double meaning of "walk into a bar," the slapstick image of two people bumping their heads, the allusion to Alexander being the first of the two, or to my cohort of his former students who continue to use the methods he taught us. Alexander (1986) offered an ultimate hypothesis of the adaptive value of jokes and humor. He suggested that in telling jokes we elevate our status by lowering the status of others, such as orthopterists and former graduate students. If status translates into access to resources, he argued, it ultimately affects survival and reproduction.

In his 1986 essay, Alexander outlined an exhaustive hypothetical framework for understanding the adaptive value of humor. However, this essay has been cited almost exclusively for his discussion of indirect reciprocity (Dugatkin, 1998, Hamilton and Taborsky, 2005, Killingback and Doebeli, 2002). The discussion of humor has been mostly ignored, even by Gervais and Sloan-Wilson (2005) in their recent *Quarterly Review* article on the evolution of humor. I intend to begin rectifying this oversight right here and now.

OSTRACIZING THE OUTGROUP OR BONDING WITH THE INGROUP?

What could be less funny than explaining a joke? Perhaps explaining all jokes? In a parallel fashion to his habit of exhaustively laying out alternate hypotheses, Alexander attempted to categorize all the different types of jokes. This was a key step in testing his hypothesis that jokes are essentially ostracizing because that model would have to fit all sorts of jokes. The model is built on the assumptions that jokes involve breaking rules in order to trick a victim, that being tricked lowers the status of the victim, and that jockeying for status is a zero-sum game. While some jokes do this explicitly and target one or few victims, others lower the status of a whole class of individuals, and by contrast might raise the status of the joker and his audience. Slapstick humor is an example where the status of a single individual is lowered and ethnic jokes or those about the opposite sex demean or ostracize larger groups.

Leacock (1938) first distinguished ostracizing humor from affiliative humor and it has since been demonstrated experimentally that humor can enhance group cohesion (Banning and Nelson, 1987; Greatbach and Clark, 2003; Jung, 2003; Storey, 2003; Vinton, 1989). Alexander recognized Leacock's distinction but pointed out that affiliative behavior can only spread by selection if affiliation enhances survival and reproduction relative to members of other groups. Thus he offers the distinction between jokes that are directly or indirectly ostracizing. At first, it seems to be quite a stretch to imagine puns, clever turns of phrase, or humorous observations on the absurdity of modern life, as ostracizing anyone, even indirectly. But there is a much simpler and more compelling reason why all humor is inherently ostracizing.

All humor involves recognizing a scenario and then being surprised by a twist or unexpected ending: the punch line (Howe, 2002; Gamble, 2001; Kuhn, 1962; Lefcourt, 2000; Ramachandran, 1998; Shulz, 1976). The ingroup in the audience are those who get the joke. They are clearly distinguished from the outgroup who do not get the joke, either because they do not understand the allusions or the social context of the joke, or because they do not have sufficient mastery of the language (Flamson and Barrett, 2008). Interestingly, demonstration that one can express and comprehend humor in a foreign language is among the skills necessary for qualifying as having the highest level of proficiency in that language (Alderson and Huhta, 2005).

Thus all humor is inherently ostracizing because the audience is made up of two types of people: those who get the joke, those who don't get it, and those who just can't count. There could not be an ingroup without an outgroup: people who do not see the double meaning of "walk into a bar," do not know that Alexander is an orthopterist, might not even know what an orthopterist is, or how former students behave in the presence of a famous, dominant, professor.

A number of specific predictions follow from the ostracism hypothesis. Jokes that overtly target an outgroup should elicit a hostile response (or at least a stiff "that's not funny") from members of that group or their allies. On the other hand, jokes that use special knowledge to highlight an outgroup, should elicit a very different response by the victims (i.e., false laughter to hide the fact that you are a member of the outgroup). This perspective on the ostracism that is inherent in humor leads to the prediction that jokes that involve more restricted insider information and more subtle tricks will be judged as funnier. On the other hand, we would expect a joke that is so obvious that anyone can recognize the scenario and the twist would be dismissed as juvenile. Of course testing these predictions would require finding a way to distinguish those who don't get the joke from those who don't think it's funny because we expect subjects will be reticent to admit when they don't get a joke. Butcher and Whissell, (1984) as well as Levy and Fenley (1979), have used laughter as outcome data to get around some of the problems of self-reporting.

Darwin suggested that laughter arose before smiling, but Van Hoof (1971), and Lockard (1977) argued that the smile arose from the bared-teeth grin, while the laugh develops from the open mouth display. Alexander (1986) discusses laughter and smiling as stages in his hypothetical scenario of the evolution of humor and notes that although a baby's smile is generally an attractive signal, an open mouthed laugh is not particularly attractive. He suggests that laughing in the presence of others shows that you are not fooled by the trick that is inherent in any joke and that you are an ally of the trickster. Similarly, the ingroup/outgroup hypothesis suggests that we laugh to indicate that we are part of the ingroup that gets the joke. In contrast, Gervais and Sloan-Wilson (2005) as well as Weisfeld (1993) suggest that laughing only occurs in the company of others because the function of laughter is to reinforce the joker with positive feedback, to encourage him or her to continue entertaining us. But this proximate hypothesis offers no explanation for why we would want to encourage the joker. Furthermore, if the target of laughter as communication is the joker, and the function of laughter is to reinforce the joker, as Gervais and Sloan-Wilson suggest, we would not expect an increase in individual laughter in larger groups and might even predict that individuals would be able to decrease the intensity of their laughter when the contributions of other audience members can be pooled in signaling the joker. However, Butcher and Whissell, (1984) as well as Levy and Fenley (1979) found that comic video clips evoke greater laughter per person as audience size increases. If the ingroup/outgroup hypothesis is correct, the next time you hear the orthopterist joke I suggest that you laugh loud enough that everyone in the room can tell that you get all the allusions. On the other hand, Alexander might suggest that you respond with a groan. This not only communicates that you get the joke, but also that you think it is just too simplistic, too obvious, or too tortured to be funny. And in the eyes of your audience, you would raise your own status by lowering mine.

WHY DO WE ENJOY HUMOR?

Whether the ostracism is about being the butt of the joke or about being in the group that doesn't get the joke, we are left wondering why we enjoy humor and actively seek it out. Alexander (1986) dismisses the problem of enjoying humor as proximate feedback, like the pain of bumping your head into a bar. This begs the question of what benefit anyone gains by seeking out humor. Alexander (1989) offers the suggestion that, like art, music, literature, and theater, attending to humor can help us develop or enhance our social-scenario building abilities that are critical in a system of indirect reciprocity. Attending to humor may also help us develop our sense of humor, which apparently makes us much more attractive (Bresler et al., 2005; Bressler and Balshine, 2006; Howe, 2002; Li et al., 2009; McGee and Shevlin, 2008; Miller, 2000). Not surprisingly, "sense of humor" is a highly rated trait on internet dating sites.

No doubt reading novels and watching theater can foster social scenario building skills. However, the joy we experience with humor, in both social and nonsocial settings, points to a more basic, and highly adaptive, attraction and reward system. This system rewards us for attending to subtle abstract patterns and surprises all around us. Alexander's (2008) general theory of the arts suggested that the common elements of "rhythm, rhyme, melody, repetition, inspiration, and novelty" might be "proximate devices that cause attentiveness" (page 19 of Alexander's supplementary footnotes). Surprisingly he does not dismiss the importance of these proximate cues but looks for the common theme. Although Alexander suggests that attention to these cues helps us learn to better anticipate the future, an alternative is that they hone our ability to detect patterns and surprises in the world around us. This is consistent with Levitin's (2007) work on the neurobiology of music.

Menon and Levitin (2005) and Blood and Zatore (2001) have shown that our enjoyment of music comes from internal rewards in the mesolymbic system of the brain when we recognize patterns and when our expectations are met as a musical piece continues. Levitin (2007) further suggests that sophisticated music requires surprising twists in the melody or chord progression. But the surprise must still fit an unanticipated pattern (Levitin and Menon, 2005; Koelsch, et al., 2008). Recognizing the new pattern and how it fits is rewarded, but if we cannot recognize how the surprise fits in, the music seems discordant and unpleasant (Blood et al., 1999; Fritz et al., 2009). At a young age and when we are musically unsophisticated, we enjoy simple melodies and rhythms. As we gain musical experience we are able to recognize more subtle patterns and surprises, which provide the dopamine reward and other positive feedback.

If humor and music share these common features of identifying an abstract pattern and noticing a subtle surprise, perhaps our enjoyment of art and poetry are also manifestations of a general reward system that hones our ever improving skills at detecting subtle anomalies in the world around us. This general skill at detecting a subtle discontinuity within an abstract pattern would improve with practice and would be highly adaptive. It would make us better at tracking prey, detecting predators and identifying deception in social situations. This hypothesis might lead to the prediction that the most successful fishermen, infantry soldiers, poker players, or police detectives would be people who spend their leisure time listening to music, reading poetry, or viewing sophisticated art and comedy. Although I have not been able to uncover data with which to test these predictions, Conan Doyle's fictional detective, Sherlock Holmes, frequently spent his afternoons at the symphony, and Robert Parker's Spencer and Hawk often quote classic poetry. Nonetheless, the commonality in the underlying adaptive value of music, humor, art, and poetry is supported by a number of brain imaging studies that suggest a common underlying proximate mechanism.

Music, humor, and visual art all elicit marked activity of the limbic system associated with emotion and in the dopamine reward centers in the prefrontal cortex (Blood et al., 1999; Cella-Conde, 2004; Menon and Levitin, 2005; Mitterschiffthaler et al., 2007; Mobbs, et al., 2002; Watson et al., 2007; Wild et al., 2003). The prefrontal cortex is generally associated with learning, problem solving, and puzzle solving, but music and humor both activate more specific subregions within the prefrontal cortex: the nucleus accumbens as well as the dorsal and ventral tegumental areas. The subregions of the prefrontal cortex involved in appreciation of visual art have not been as precisely mapped and there is no imaging data on our response to poetry in general. However, Thierry (2008) has noted Shakespeare used surprising word order to attract the attention of the audience or the reader and Petterson (2004) noted that Ogden Nash's poetry uses the devise of surprise and incongruity in familiar scenarios.

A reward system for practicing and improving the ability to detect abstract patterns and subtle surprising incongruities should be so fundamental that we would expect to see evidence of it outside of humans. Like young children, Irene Pepperberg's parrot, Alex, clearly enjoyed the game of identifying elements of a set that do not match. Once he mastered those games they were no longer inherently rewarding despite the extrinsic food rewards he was offered. Alex demanded more sophisticated games where the surprise was more and more difficult to detect (Pepperberg, 1999). And even if Alex did not exhibit a sense of humor, McGee (1979) suggests that apes do. He notes that gorillas and chimps that sign will laugh after giving incongruent answers to questions and chimps will give the sign for funny after throwing feces at people.

CONCLUSION

Alexander's goal in writing many of the essays that have been collected in this volume was to generate an exhaustive list of alternate hypotheses and predictions in the best Darwinian tradition and to stimulate the scientific discussion. Hence he concluded the 1986 essay on humor, "I end this speculative essay on the note that whether or not the particular arguments presented here are correct in any significant way, they may support the notion that it is appropriate to examine topics like humor in the context of evolved biological (reproductive) function."

Although Gervais and Sloan-Wilson's (2005) snub of Alexander (1986) might have been a petty attempt at ostracism, it more likely results from the fact that Alexander's essay was published in *Ethology and Sociobiology*, a journal that is not indexed in PubMed. Of the 71 articles citing Alexander 1986 (uncovered by Google Scholar), only a handful took up Alexander's challenge (Nesse and Lloyd, 1992; Palmer, 1993;

Polimeni and Reiss, 2006; Weisfeld, 1993) with Palmer offering the only attempt to test of Alexander's ostracism hypothesis in an amusing behavioral analysis of amateur hockey players in Newfoundland. Nonetheless, Polimeni and Reiss demonstrated that they were better google scholars than Gervais and Sloan-Wilson, when they credited Alexander as "the first to methodically analyze humor and laughter within an evolutionary context" (page 351). Perhaps the republication of this essay, and the others in this volume, will help fulfill Alexander's goal of provoking us to get to work and test his hypotheses. And let's hope that the next time two orthopterists walk into a bar, they not only get a cold drink, but they also get the bartender's jokes.

References

Alderson, J.C. and Huhta, A. 2005. The development of a suite of computer-based diagnostic tests based on the Common European Framework. *Lang. Test.* 22:301–320.

Alexander, R.D. 1986. Ostracism and indirect reciprocity: the reproductive significance of humor. *Ethol. Sociobiol.* 7:253–270.

Alexander, R.D. 1989. Evolution of the human psyche. In: P. Mellars and C. Stringer (eds.), *The Human Revolution* Chicago: University of Chicago Press, pp. 455–513.

Alexander, R.D. 2008. Unpublished supplemental footnotes to: Evolution and human society. *Hum. Behav. Evol. Soc. Newsl.* Summer Issue.

Banning, M., and Nelson, D. 1987. The effects of activity-elicited humor and group structure on group cohesion and affective responses. *Am. J. Occup. Ther.* 41:510–514.

Blood, A., and Zatore, R., 2001. Intensely pleasurable responses to music correlate with activity in brain regions implicated in reward and emotion. *Proc. Nat. Acad. Sci. USA* 98:11818–11823.

Blood, A., Zatore, R., Bermudez, P., and Evans, A. 1999. Emotional responses to pleasant and unpleasant music correlate with activity in paralimbic brain regions. *Nat. Neurosci.* 2:382–387.

Bressler, E. R., and Balshine, S. (2006). The influence of humor on desirability. *Evol. Hum. Behav.* 27:29–39.

Bressler, E.R., Martin, R. and Balshine, S. 2005. Production and appreciation of humor as sexually selected traits. *Evol. Hum. Behav.* 27:121–130.

Butcher, J. and Whissell, C. 1984. Laughter as a function of audience size, sex of the audience, and segments of the short film "Duck Soup." *Percep. Motor Skills.* 59:949–950.

Cela-Conde, C., Marty, G., Maestu, F., Ortiz, T., Munar, E., Fernandez, A., Roca, M., Rossello, J., and Quesney, F. 2004. Activation of the prefrontal cortex in the human visual aesthetic perception. *Proc. Nat. Acad. Sci. USA* 101:6321–6325.

Dugatkin, L. 1998. Game theory and cooperation. In: L. Dugatkin and H. Reeve (eds.), *Game Theory and Animal Behavior*. New York: Oxford University Press.

Flamson, T., and Barrett, H.C. 2008. The encryption theory of humor: A knowledge based mechanism of honest signaling. *J. Evol. Psychol.* 6:261–281.

Fritz, T., Jentschke, S., Gosselin, N., Sammler, D., Peretz, I., Turner, R., Friederici, A., and Koelsch, S. 2009. Universal recognition of three basic emotions in music. *Curr. Biol.* 19:1–4.

Gamble, J. 2001. Humor in apes. *Humor* 14:163–179.

Gervais, M., and Sloan-Wilson. D. 2005. The evolution and functions of laughter and humor: a synthetic approach. *Q. Rev. Biol.* 80:395–430.

Greatbach, D., and Clark, T. 2003. Displaying group cohesiveness: humour and laughter in the public lectures of management gurus. *Hum. Relations* 56:1515–1544.

Hamilton, I., and Taborsky, M. 2005. Contingent movement and cooperation evolve under generalized reciprocity. *Proc. R. Soc. Lond. B* 272:2259–2267.

Howe, N. 2002. The origin of humor. *Med. Hypoth.* 59(3):252–254.

Jung, W. 2003. The inner eye theory of laughter: mindreader signals cooperator value. *Evol. Psychol.* 1:214–253.

Killingback, T., and Doebeli, M. 2002. The Continuous Prisoner's Dilemma and the evolution of cooperation through reciprocal altruism with variable investment. *Am. Nat.* 160:421–438.

Koelsch, S. Fritz, T., Schlaug, G. 2008. Amygdala activity can be modulated by unexpected chord functions during music listening. *NeuroReport* 19:1815–1819.

Kuhn, T. 1962. *The Structure of Scientific Revolutions.* Chicago: University of Chicago Press.

Leacock, S. 1938. *Humor and Humanity. An Introduction to the Study of Humor.* New York: Holt.

Lefcourt, R. 2000. *Humor: The Psychology of Living Buoyantly.* New York: Plenum.

Levitin, D. 2007. *This Is Your Brain on Music: The Science of a Human Obsession.* New York: Penguin.

Levitin, D., and Menon, V. 2005. The neural locus of temporal structure and expectancies in music: Evidence from functional neuroimaging at 3 Tesla. *Music Percep.* 22:563–575.

Levy, S. and Fenley, W. 1979. Audience size and the likelihood and intensity of response during a humorous movie. *Bull. Psychonom. Soc.* 13:409–412.

Li, N., Griskevicius, V., Durante, K., Jonason, P., Pasisz, D., and Aumer, K. 2009. An evolutionary perspective on humor: sexual selection or interest indication? *Pers. Soc. Psychol. Bull.* 35:923–936.

Lockard, J.S., Fahrcnbrnch, C.E., Smith, J.L., Morgan, C.J. 1977. Smiling and laughter: different phyletic origins? *Bull. Psychonom. Soc.* 10:183–186.

McGee, E., and Shevlin, M. 2008. Effect of humor on interpersonal attraction and mate selection. *J. Psychol.* 143:67–77.

Menon, V., and Levitin, D. 2005. The rewards of music listening: response and physiological connectivity of the mesolimbic system. *NeuroImage* 28:175–184.

Miller, G. 2000. *The Mating Mind: How Sexual Choice Shaped the Evolution of Human Nature.* New York: Anchor Books.

Mitterschiffthaler, M., Fu, C. H. Y., Dalton, J., Andrew, C., and Williams, S. 2007. A functional MRI study of happy and sad affective states induced by classical music. *Hum. Brain Map.* 28:1150–1162.

Mobbs, D., Greicius, M., Abdel-Azim, E., Menon, V., Reiss, A. 2003. Humor modulates the mesolimbic reward centers. *Neuron.* 40:1041–1048.

Nesse, R. and Lloyd, A. 1992. The evolution of psychodynamic mechanisms. In: J. Barkow, L. Cosmides, and J. Tooby (eds.), *The Adapted Mind.* New York: Oxford University Press.

Palmer, C. 1993. Anger, aggression, and humor in Newfoundland floor hockey: an evolutionary analysis. *Agress. Behav.* 19:167–173.

Pepperberg, I. 2002. *The Alex Studies.* Cambridge: Harvard University Press.

Petterson, B. 2004. Exploring the common ground: sensus communis, humor and the interpretation of comic poetry. *J. Lit. Seman.* 33:155–167.

Polimeni, J., and Reiss, J. 2006. The first joke: exploring the evolutionary origins of humor. *Evol. Psychol.* 4:347–366.

Ramachandran, V.S. 1998. The neurology and evolution of humor, laughter, and smiling: the false alarm theory. *Med. Hypoth.* 51:351–354.

Shulz, T. 1976. A cross-cultural study of the structure of humour. In: A. Chapman and H. Foot (eds.), *It's a Funny Thing, Humour.* New York: Pergamon Press.

Storey, R. 2003. Humor and sexual selection. *Hum Nature* 14:319–336.

Thierry, G., Martin, C., Gonzalez-Diaz, V., Rezaie, R., Roberts, N., and Davis, P. 2008. Event-related potential characterization of the Shakespearean functional shift in narrative sentence structure. *Neuroimage* 40:923–931.

Van Hoof, J. 1971. A comparative approach to the phylogeny of laughter and smiling. In: R. Hinde (ed.), *Non-verbal Communication.* Cambridge: Cambridge University Press, pp 209–243.

Vinton, K. 1989. Humor in the work place: is it more telling than jokes? *Small Group Behav.* 20:151–166.

Watson, K., Matthews, B., and Allman, J. 2007. Brain activation during sight gags and language-dependent humor. *Cerebral Cortex* 17:314–324.

Weisfeld, G. 1993. The adaptive value of humor and laughter. *Ethol. Sociobiol.* 14:141–169.

Wild, B., Rodden, F., Grodd, W., and Ruch, W. 2003. Neural correlates of laughter and humor. *Brain* 126:2121–2138.

OSTRACISM AND INDIRECT RECIPROCITY

Alexander, R.D. Ostracism and Indirect Reciprocity: The Reproductive Significance of Humor. 1986. *Ethology and Sociobiology* 7: 253–270.

Humor is hypothesized to be a social activity that alters the status of the humorist positively and that of the object or victim negatively. Of the two traditionally distingushed classes of humor, "ostracizing" humor singles out a victim, with others present or absent either incidental affiliates of the humorist (and one another) or unaffected. "Affiliative" humor, on the other hand, is focused on creating or maintaining group cohesiveness, with the identity of the victim more or less incidental.

INTRODUCTION

Cynic

> A misanthrope; spec., one who believes that human conduct is motivated wholly by self-interest.
>
> —WEBSTER'S COLLEGIATE DICTIONARY 1947 EDITION

> A blackguard whose faulty vision sees things as they are, not as they ought to be.
>
> —AMBROSE BIERCE'S DEVIL'S DICTIONARY 1911 EDITION

In developing a topic for my contribution to this issue, I selected humor because it is one of several attributes commonly seen as either antithetical or irrelevant to natural selection or reproductive success (music, art, aesthetics, humor...), and committed myself explicitly to discussing it in terms of reproductive significance. I did this for two reasons: first, I thought it would be useful to extend my efforts toward topics likely to be most difficult to relate to evolution by natural selection; and second, in my course on human behavior and evolution, I had already developed a fairly detailed hypothesis about humor that related it to ostracism. Indeed, as will become clear, I unexpectedly found the analysis converging on conclusions I had reached earlier with respect to morality (Alexander 1982, 1985, in press).

For the purposes of my discussion, I define ostracism from my 1947 edition of Webster's Collegiate Dictionary as "A method of temporary banishment by popular vote.... Exclusion by general consent from common privileges, favor, etc. [read: resources of reproduction]...as, social *ostracism.*"

Ostracism is a topic of almost unbelievably broad significance. I see it as varying from such extremes as shunning, excommunication, and designation of "outlaw" to the most subtle forms of status shifting through implied or real, partial or complete exclusion from temporary or even momentary and casual groupings of social interactants. I see ostracism as an instrument for the manipulation of conflicts and confluences of interest through adjusting access to resources. Conflicts and confluences of interest, I believe, underlie everything that is social about humans. What sets us apart from other organisms more than anything else seems to be (1) the astonishing complexity of our conflicts and confluences of interest (deriving from the fact that we continue our lives together as large groups of long-lived adults and children mixed as relatives of varying degrees—also the reason for the prominence and elaborateness of incest avoidance—and as accomplished social cooperators, reciprocators, and competitors with countless ways of helping and hurting one another); and (2) the extraordinary array of proximate mechanisms we have evolved for assessing and dealing with our conflicts and confluences of interest. Underlying it all, I believe (cf. Alexander 1979), is the fact that humans achieved, apparently a very long time ago, a peculiar situation in which the greatest threats (and aids) to individuals and groups come from other humans rather than other species. The nature and complexity of the human psyche, I believe further, with its aspects designated as conscious, preconscious, subconscious, and nonconscious—as conscience, intelligence, self-awareness, foresight, and all the rest relate powerfully to the problems of dealing appropriately with conflicts and confluences of interest within the human social scene. Thus I see consciousness, self-awareness, foresight, and conscience as "overrides" of more ancient and more immediate indicators of costs and benefits (such as pain and pleasure). Humans use consciousness, self-awareness, foresight, and conscience to estimate long-term costs and benefits and to make decisions about rejecting short-term pleasures or accepting short-term pains. The special condition favoring such attributes, I hypothesize, is the ability of competing and cooperating humans to adjust continually the relationships between short and long-term costs and benefits so that intelligence, foresight, and deliberate planning have been the best available tools for realizing one's own interests (for fuller discussion, see Alexander 1979, and in press).

I have used the term "indirect reciprocity" in my title, and this also deserves some explanation (see Alexander 1977, 1979, 1985, in press; also see Trivers 1971, under the term "generalized reciprocity"—but not generalized reciprocity as used by Sahlins 1965).

Direct reciprocity occurs when an individual (or a group) is beneficent toward another and is repaid for his temporary altruism (or social investment) by a parallel act of beneficence, not necessarily involving the same currency, but typically resulting in gains for both interactants (Trivers 1971, defined this condition formally and referred to it as "reciprocal altruism").

Indirect reciprocity occurs when interested people observe direct reciprocity between others and use the observations to determine who will be their own future associates and how they will interact subsequently with the observed parties. Indirect reciprocity occurs whenever rewards or punishments come from individuals or groups other than those directly involved in a social interaction involving investment or exploitation. It includes public and private opinions, and status. Indirect reciprocity is the foundation of moral, ethical, and legal systems. Its existence and pervasiveness in human social life, I believe, are the most important factors to consider in an analysis of the nature and complexity of the human psyche. I think they account for human interest in theater in all of its guises, from soap operas to Shakespeare, poetry to sociology, neighborhood parties to the Olympic games. Indirect reciprocity is the reason that very few things are more relevant to our individual social success than the ability to see ourselves as others see us and respond appropriately (which means, I think, to cause them to see us as we wish them to, and not otherwise).

In this article I consider the hypothesis that humor is a principle according to which the evolved abilities and tendencies of people to see themselves as others see them, to use ostracism to their own advantage, are manipulated so as to induce status shifts—both subtle and not so subtle. My general hypothesis is that humor has developed as a form of ostracism and that, historically, at least, ostracism has tended to affect the reproduction of the ostracized individual (or group) deleteriously, especially in relation to the reproduction of the ostracizers, by restricting access to significant resources.

To my knowledge there is no well-developed previous theory of the function of humor, in the sense of evolved or "ultimate" function, even though the literature on humor is filled with hints in the direction I take this article. [Thus, Robinson (1977) says that "studies...describe the function of humor to solidify the in-group, to attain gratification at the expense of another group..." and "There is a pecking order to joke-telling. The joketeller is the dominant one; the joke is his weapon; his laughter is a sign of victory. The audience is submissive; their laughter is the sign of their acceptance of defeat."] Absence of explicit theories of function seems to result partly because previous authors have either attributed "function" solely or in part to the satisfaction of some proximate system or mechanism, or because they have avoided the question of ultimate (usually given as "survival") function, sometimes giving the reason that the question is not experimentally testable (for reviews, see McGhee 1979; Schmidt and Williams 1971).

To say that a particular behavior or tendency exists or is carried out to satisfy pleasure, relieve frustration, or even to help one deal with an immediate situation in the sense of adjusting one's frame of mind (e.g., gallows humor, as exemplified by Freud's description of the man who, on his way to the gallows on a Monday, steps out into the sunshine and remarks, "Well, the week is beginning nicely.") begs the question of the reason for the existence of the effect (pleasure, relief, comfort) or the recognition of a mental "problem" (frustration, wrong frame of mind). Pleasure and pain presumably exist because, respectively, they cause us to repeat beneficial actions and avoid repeating deleterious ones (Dawkins 1976; Alexander 1979). Similarly, frustration exists, I would speculate, because of the importance of solving problems that may be difficult. The question we have to deal with eventually, and on which I concentrate here, is precisely what such immediate mechanisms are programmed to accomplish—that is, how are actions determined to be "beneficial" and "deleterious"?

Typically we suffer pain when we incur an injury that, prior to medical technology, was reparable provided certain actions were taken and others avoided (as in protecting an injured part). We typically do not suffer pain when injuries irreparable prior to medical technology occur (e.g., object thrust into the brain, damage to the spinal cord). I assume that mental pain and pleasure analogues serve similar functions in the social scene—e.g., that the only way, in the end, to deal with frustration and distress is to solve the problem that is causing it. In other words, I see frustration and distress as mechanisms serving some function, not as either incidental or pathological conditions to be relieved per se, without connection to other difficulties. Thus, I do not use the word "function" as Flugel (1954) used it when he said that "it seems clear that one important function of the humorous attitude at all levels is to relieve us from the burden of reality..." (p. 713). I will argue, in the end, that the kind of humor or the aspects of humor that in Flugel's sense make us feel good (e.g., humor among patients in a cancer ward) stem from group-unifying aspects of humor that originally gave us pleasure because they cause groups to be more effective competitive units in intergroup competition (hence, as in the cancer ward, seem to reduce some other hostile force by providing additional motivation to deflect it).

I do not wish to underestimate the complexity of relationships between proximate and ultimate aspects of humor and its correlates, such as changes in facial expression and laughter. Zajonc (1985) gives reasons for believing from physiological effects that "laughing must be healthy..." and Roger Masters (personal communication) has similarly reminded me of the saying that "laughter is the best medicine," adding: "Laughter, as a predictable consequence of humor, has a physiological effect on the organism [see also Cousins 1979]. We know that agonic emotional states inhibit learning...and...fear is generally associated with both subordinate status and ineffectual coping with environmental novelty. It follows that hedonic states of emotion have, in the absence of life-threatening situations, a likelihood of improving reproductive success. If so, laughter could...be positively reinforcing because it is associated with the effective coping behavior."

The question one has to consider is why laughter should be healthy or have beneficial effects on our physiology or behavior? If, as Zajonc and Masters suggest, the reason is purely physiological—say, through effects on blood flow to or from the brain—then one has to ask why we do not laugh all the time or modify the involved physiological conditions so that we experience the beneficial effects without having to laugh. The alternative is that beneficial effects of laughter or humor, measured as physiological, are only beneficial in social situations that evoke laughter. That is, laughter is caused by social situations that in turn cause the laughing individual correctly to have a kind of confidence that allows concentration on things like learning or coping with novelty or whatever social effects accrue from hedonic feelings. In other words, whatever the original physiological causes or effects of laughter, its social effects must have led to feedbacks that altered the physiological effects and associated behaviors appropriately (moreover, once humor and laughter had come to have beneficial physiological effects, new vistas would be opened for the "entertainer" who could raise his own status by creating both the social and the physiological effects in situations in which they would otherwise not occur). This view lays great emphasis on the social effects of humor and laughter, and I am convinced that this emphasis is proper. It does not alter significantly the search for ultimate function, which is important because it is also the ultimate shaper of the trait.

HUMOR AND STATUS
General Hypotheses

Note: I have tried to make the following set of interrelated hypotheses internally consistent, meaning that each subhypothesis, if true, should support the main hypothesis and, if false, deny it. I have also tried to make this set of hypotheses exhaustive—that is, I have tried to include every situation I can think of that involves humor. In addition I have attempted to use Darwin's (1859) two methods of (1) describing phenomena that, if observed, would deny my hypothesis; and (2) analyzing observed phenomena that seem difficult to explain by my hypothesis. The reader will see that I have not accomplished all of these goals, but perhaps my efforts will help others who analyze humor to achieve them later.

Status Effects

I use status according to the definitions "position of affairs" or "state or condition of a person" (Webster's Collegiate Dictionary, 1947 ed.). My hypotheses assume that the following are desirable (sought-after) effects on status, meaning as well that I assume that these effects typically influence reproduction favorably by improving access to resources:

1. Elevating of one's own status in relation to
   a. Part of the group (audience)
   b. All of the group (audience)
   c. One or more third parties not present
2. Lowering of someone else's (the "victim's") status in relation to one's self
   a. The victim is the only individual present.
   b. The victim is not the only one present.
   c. The victim is not present.
3. Reinforcing (maintaining) a presumably favorable status relationship with
   a. Part of the group (audience)
   b. All of the group (audience)
   c. One or more third parties not present

The basic hypotheses, then, from which all those that follow are derived, are as follows:

> I. *Jokes involve tricks.* Trick is defined as "An artifice or strategem; crafty procedure or practice; a cheating device" (Webster's Collegiate Dictionary, 1947 ed.). In my opinion, dictionary definitions of cheating are unsatisfactory, so I define cheating here as breaking rules or manipulating them in an unacceptable fashion; I define rules as established procedures or contracts.
>
> II. *Tricks are devices for lowering status of those on whom they are played and raising the status of those who play them.* Telling jokes, and laughing at them, are ways of adjusting status in one's own favor.
>
> III. *Based on the hypothesis of ostracism or status-shifting, humor seems to develop as two related forms:*

> A. Jokes that explicitly exclude or lower the status of a party or parties (by representing a trick played successfully upon the demoted party) and thereby also indirectly or seemingly incidentally (implicitly) bond together those who are party to the joke (trick) or share it (e.g., ethnic, racist, or sexist jokes).
> B. Jokes that implicitly exclude or lower the status of some individual or recognizable group by explicitly reinforcing fellowship or cohesiveness or unity in the group (among the individuals) sharing the joke. These include the kinds of jokes that influenced Stephen Leacock to write that "Humor may be defined as the kindly contemplation of the incongruities of life, and the artistic expression thereof." (See also the dictionary definition as "that quality which appeals to a sense of the ludicrous or absurdly incongruous.")

Leacock (1938) wrote in a way that anticipates (but obviously does not correspond precisely to) the dichotomy in humor proposed here:

> One is tempted to think that perhaps the original source [of humor] parted into two streams. In one direction flowed, clear and undefiled, the humor of human kindliness. In the other, the polluted waters of mockery and sarcasm, the "humor" that turned to the cruel sports of rough ages, the infliction of pain as a perverted source of pleasure, and even the rough horseplay, the practical jokes, and the impish malice of the schoolboy. Here belongs "sarcasm"—that scrapes the flesh of human feelings with a hoe—the sardonic laugh...the sneer of the scoffer, and the snarl of the literary critic as opposed to the kindly tolerance of the humorist.

Similarly, O'Connell (1960) stated that, following Freud, "humor" and "wit" separate into approximately the two general kinds of humor I have postulated here, humor being associated with empathy and wit with hostility. Robinson (1977) wrote that "Studies...describe the function of humor to solidify the in-group, to attain gratification at the expense of another group, and, more recently, as a way to create a new image and as an agent of social change. The role of the fool in society has been described as a means of enforcing group norms of propriety."

Robinson also noted, tellingly, that in the medical community "the critical issue is that it is never justifiable to make fun of or laugh at the patient or his symptoms. Laughing with someone rarely does harm" (Robinson 1977, p. 79).

Freud (see Brill 1938), Eysenck (1947), and Flugel (1954) looked for three levels or kinds of humor, variously termed conative (wit), affective (humor), and cognitive (comic). I cannot distinguish the second and third (see also below).

> *IV. Humor is associated with smiling and laughing.* Smiling (visual) and laughing (auditory and visual) are ways of communicating pleasure (truthfully or deceptively). The pleasure of smiling and laughter is a social phenomenon.

Laughing probably occurred first as a result of physical events like tickling, which also has social significance, and which occurs in chimpanzees (at least) as well as in humans (Darwin 1899; Yerkes and Learned 1925; Goodall 1968). Goodall (1968, p. 258) stated that "'Laughing' (a series of staccato panting grunts) frequently accompanied bouts of wrestling and tickling." Goodall also noted that, in "greeting behavior...as two individuals approach each other they may utter soft or loud panting sounds...particularly the subordinate as it bows, crouches, or bobs. Sometimes both the dominant and the subordinate individuals may "grin." [See McGhee (1979) for numerous examples from primates.] Today laughter among humans probably occurs most frequently during social communication without physical contact.

It is worth stressing that what we are required to explain is not only why what we call humor causes pleasure but (especially) why special mechanisms exist for the communication of the pleasure that derives from humor (eventually, the same problem must be taken up with respect to grief and crying). In the case of tickling, it seems a reasonable hypothesis that vocal and other physical responses originally functioned (i.e., had as their evolved significance) to keep the tickler tickling. Two curious aspects are that (1) we cannot successfully tickle ourselves [Flugel (1954) remarked on this fact] and (2) our tendency to be ticklish renders us vulnerable to a kind of cruelty in the form of unwanted tickling. I see this vulnerability as paralleling the vulnerability of humans who have evolved to appreciate and use humor that does not involve physical events like tickling to having this appreciation and sensitivity as well turned against them. Vulnerability to excessive tickling, or excessive responsiveness to tickling, sometimes takes the form of being "goosey," meaning to be so intolerant of tickling—or even the threat of tickling—that a word, gesture, body movement, or simply a stare can be a form of torture and can cause a susceptible individual to do extraordinary things like leap out of a window or injure himself in a frantic effort to escape, even when the tormentor is some distance away.

Evolutionary Origins of Smiling and Laughter

Smiling and laughter have been postulated either to represent a "continuum of graded intensities" (Andrew 1963; Hinde 1974) or to have different phyletic origins (Van Hoof 1967, 1971; Lockard et al. 1977). Following the first alternative, Darwin (1899) seemed to postulate that laughing preceded smiling, and Hayworth (1928) agreed. Van Hoof (1971) and Lockard et al. (1977) suggested that smiling evolved from the "silent bared-teeth submissive grimace...of primates, and laughing.. ., from the relaxed open-mouth display...of play." Both displays are known in several primates (*Macaca, Cercopithecus, Pan, Mandrillus,* and *Theropithecus*). Although I am inclined to agree with Darwin and Hayworth, the two alternatives may not affect significantly the arguments presented here.

Ontogeny of the Sense of Humor

Presumably, the evolution of smiling and laughing in infants, as with many other aspects of infant social behavior, followed the evolution of functions in smiling among adults. In this hypothesis the infant would be enhancing its attractiveness to those responsible for its future by mimicking social responses that in adults signify good will and comaraderie. The alternative hypothesis would be that smiling originated as a part of the attractiveness of infants and acquired its social significance among adults (acknowledgement of subordinance?) later, and perhaps as a result of its significance in the parent-offspring interaction. In a sense both hypotheses may be correct in this case. Thus, smiling may have evolved out of a grimace associated with being tickled, and tickling may have evolved out of physical interactions between parent and offspring. As smiling acquired more profound implications in the complex social world of adults, there may have been significant feedback enhancing smiling and laughing in infants.

McGhee (1971) says "Grotjahn (1957) argued that the child first discovers comic situations when he begins to master and enjoy body movements. When he begins to feel superior to other children in this respect, he is likely to see their mistakes or weakness as funny." McGhee (1971) reported that "While speech mistakes and other bumbling errors (e.g., slipping on a banana peel) are funny in their own right to the healthy child, they may become the source of cruel and derisive laughter in the child who feels unloved or unsure of himself." Leuba (1941) noted that if a playful attack (e.g., of tickling) becomes too serious, laughter in younger children turns into expressions of fear. Wolff et al. (1934) similarly stated that "A child, for instance, who has made a remark which because of its naive cleverness evokes laughter may burst into tears if the mirth is too openly expressed." Jones (1926) stated that the same stimuli might arouse laughter in 16–36-month-old children on one occasion but crying in another. Justin (1932) found that only incongruity "was found to increase in its effectiveness in producing laughter as a function of age" (among surprise or deflated expectation, superiority and degradation, incongruity and contrast, social smile as a stimulus, relief from strain and play situations; these did not change during ages 3, 4, 5, 6). This last finding suggests that the directly integrating aspect of humor, aside from the infant's smile, appears later in development than the directly ostracizing aspect. The other findings indicate that children have more difficulty than adults separating the two kinds or effects of humor.

Robinson (1977) says of "mature humor" that it "signifies . . . emotional maturity and . . . is based upon deeper life experiences and kindly, tolerant acceptance of oneself and therefore of others."

These remarks seem to me to integrate interestingly with Flugel's 1954 note that McDougall (1923) "draws attention to the aesthetically interesting fact that smiling is beautiful, whereas laughter is ugly. Both smiling and laughter appear in human infants at an early age, and all observers seem to agree that developmentally the smile precedes the laugh." If smiling associates with the "mature" humor of integration and laughter with the explicitly ostracizing function of humor, then the hypothesis is supported that smiling by infants without physical contact may indeed have evolved after the integrative function of humor was established. A test would be whether or not infants of nonhuman primates smile without physical contact. The alternative is that laughter evolved after smiling, and the ostracizing function of humor after the integrating function. This argument is supported by Laing's (1939) finding that the unusual "arouses laughter earlier than the discomfiture of others, and that both precede anything which might be called '*wit*,' which in turn is, in its early stages, visual rather than verbal" (Flugel 1954, p. 712). Laing supposedly found that the developmental order could be described as "absurdity," "slapstick," "satire," or "whimsey. " But none of these words—except possibly the first—seems to describe the integrative aspect of humor, or that associated with smiles as opposed to laughter.

Sex Differences in Humor

The only sex differences I have found so far seem to support the general hypothesis of ostracism. Thus, O'Connell (1960) found that men appreciate "hostile wit" more than women do, while women prefer "nonsense humor." Jones (1926) found that girls smile more while boys laugh more. Laing (1939) found that "girls more often deprecated 'unfeeling laughter.'" It has also been argued that males tell jokes more often than females, and that they tell more sexual jokes (for the latter, see Flugel 1954). These meager findings are consistent with the prevailing opinion that men compete more intensely than women and that they tend to do so more frequently in coalitions (cf. Alexander 1979; Symons 1979).

Hypothesized Stages in the Evolution of Humor

> **Stage 1**. It becomes useful to scratch or groom oneself to remove parasites or for other reasons.
>
> **Stage 2**. In social organisms—such as between parents and offspring, mates, or siblings (probably in all extensively parental species)—it becomes useful (because of kin selection) to groom relatives or mates (initially for the same reasons as above).

Stage 3. Grooming and similar activities acquire social significance beyond removal of parasites and other original functions. They represent or suggest a willingness to invest in the groomed individual. Thereby they also signify a kind of exclusivity; the implication that there is a greater or an exclusive willingness to invest in the groomed individual rather than in other individuals. This public willingness already implies ostracism by defining a group with a certain relationship that includes the groomer and the groomee—the tickler and the ticklee. If all individuals were equally willing to invest in all other individuals this implication would not arise. If it were not important for a relationship to be exclusive, I am saying, all grooming would not take the forms it does. The suggestion of exclusivity may have had its initial significance for the groomee only, but in group-living species with complex shifts of interests, where other individuals could observe grooming, it would very quickly acquire significance for observers as well as participants. I suggest that the intimacy and exclusiveness of tickling and grooming interactions are why they usually cause uneasiness in some observers (necessarily being excluded), especially when they become intense or are long continued. [elsewhere—Alexander (in press)—I argue that because of indirect reciprocity public aspects of nepotistic and reciprocal interactions are crucial in understanding their overall significance.] Radcliffe-Brown's (1965) discussion of the "joking relationship," which appears to have ritualized significance between particular kinds of relatives in certain societies, is relevant here.

Stage 4. Special responses to grooming begin to evolve (This effect probably started earlier, but the aspects of most concern to us here would likely become complex and significant as Stage 3, above, developed.) These aspects could include appreciative stances, movements, or vocalizations. As such responses evolve, there will be tendencies to try to elicit them. Tickling and its associated laughter, squirming, and focusing by the tickler on the ticklish spots make up one example. At this stage both grooming (and its relatives such as tickling) and responses to grooming may begin to acquire social significance beyond either (a) the willingness of the groomer to invest and the groomee to accept the commitments and (b) the observation by others of the mutual commitment. That is, the grooming (tickling, horseplay, necking, petting, whatever) may become a game (or a deception) in which the principal significance for one or both interactants is not to develop a deep or long-lasting commitment to the other but to attract the attention of observers who may be better partners in such investments. Similarly, such interactions (as in preadults) may be primarily practice or learning experiences useful in later repetitions.

It has been suggested that tickling typically involves stimulation of body areas that would be vulnerable in combat. This suggestion is consistent with the notions that (1) tickling is play (and play is practice), (2) tickling and laughter are parts of interactions involving trusted associates, and (3) tickling and laughter reassure (e.g., Hayworth, 1928). In effect, "affiliative" humor may be derived from play, or take its form from play. (I do not think this possible interpretation changes my arguments, but it requires more development than I can give it here.)

> **Stage 5**. Laughter and expressions of pleasure become liberated from the context of physical grooming, tickling, etc.—and also from such contexts as courtship—and begin to be expressed in other social situations. This step can only be taken as social reciprocity becomes important as the binding cement of sociality. That step, in turn, depends upon the organism living in social groups composed at least of a complex mix of relatives of varying degree and different and fluctuating reproductive potentials (hence, not uniformly sharing interests or requiring assistance) and probably as well including sets of nonrelatives. [See Alexander (in prep.) for a fuller justification.] As earlier arguments have emphasized, there are always special reasons for such group-living, and because group-living entails automatic expenses that must be compensated before it can evolve, these take the form of one or another kind of hostile force (Alexander 1974, 1979). In humans, probably from the earliest times, these hostile forces are likely to have included other human groups. This fifth stage may be said to represent the appearance of humor in its modern form.

Is Humor Ever Nonostracizing?

As related above, humor is seen here as a status-altering or ostracizing form of activity. One is tempted to ask whether, as in Leacock's view, a sixth stage has been achieved in which some humor has acquired a social significance that rises loftily above the two categories here described (III A,B), both of which seem to come off as a bit grimy and contemptible. Flugel (1954) raises essentially the same question but does not resolve it (see also Freud, in Brill 1938; and Eysenck 1947). In the arguments presented here the question becomes whether or not "affiliative" actions (humor) have as their ultimate function the competitive success of the thereby unified cooperative group as compared to other groups (or individuals). In evolutionary terms I believe the argument can be sustained that cooperative group-living can only evolve by increasing the reproductive success of the group members as compared to those living alone or in other groups. The reasons are that (1) reproductive success is relative and (2) cooperative group-living cannot be maintained and elaborated unless all of the participants somehow improve their reproductive success. If all group members accomplish this, then the relevant comparison can only be between members of different groups (see also Alexander 1979, in press). This means that group cooperativeness always implies at least indirect intergroup competitiveness; humans, at least, have obviously not left intergroup interactions in this realm but placed them at center stage by making them direct, elaborate, and continual.

Interestingly enough, especially considering another humorist's high opinion of humorists (Leacock, quoted earlier), Ambrose Bierce (1911) defines one word in his Devil's Dictionary in a way that is neither cynical nor humorous. The word is "humorist," which he defines as "A plague that would have softened down the hoar austerity of Pharoah's heart and persuaded him to dismiss Israel with his best wishes, cat-quick." As a contrast, especially for the purpose of this essay, the word just before "humorist," which is "humanity," he defines as "The human race, collectively, exclusive of the anthropoid poets." (Bierce undoubtedly saw himself as a humorist, as did Leacock. Even cynical humorists, evidently, do not typically enjoy jokes on themselves.)

It is most paradoxical, at first, to imagine that which seems to bring us considerable social reward and pure pleasure as unpraiseworthy. But we already know that the paradox exists. All of us have at one time or another been filled with mirth at an episode or story that we would not share completely and that we knew mortified or denigrated some other, and would do more severely (*too* severely?) if it were known to all. What I am hypothesizing here is that in some sense this is true of every aspect of humor, and that not only human social structure in general but the human psyche as well has evolved in a milieu in which subtle and complex forms of ostracism were inevitably a principal ingredient. Elsewhere (Alexander, in press), I argue similarly that morality and justice are concepts founded on the idea, not of equality for all, but of ostracism and exclusivity; they represent either gestures or convictions with respect to some kind of equality within a group, but explicitly not beyond it and in fact for the purpose of excluding some others. The problem in our modern, dangerous world seems to be in diminishing the one tendency while retaining and expanding the other; Darwin (1871) recognized and described this problem in almost the same terms as modern writers (e.g., Singer 1981).

I emphasize that the argument that humor is invariably either direct or indirect ostracism (as opposed to being solely ostracizing in one situation and solely integrating in the other) stands or fails on the assumptions that (1) cooperative groups form as defenses against hostile forces and (2) human groups have for a very long time had as their raison d'etre the existence of other competitive and hostile groups (for references and a development of the argument, see Alexander 1979; Strate 1982).

This is not to suggest that what would typically be seen as a cynical view means that there is no escape—no way to alter any inelegant or pain-causing activities or attitudes of humanity. But I do mean to imply that recognizing such, when and if they exist, has a certain likelihood of assisting in efforts toward social harmony on grander and more nearly universal scales, perhaps through explicit and deliberate promotion of effects of humor that integrate and diminution of effects that ostracize.

> The joking relationship is a ... relation between two persons in which
> one is by custom permitted, and, in some instances required, to tease or

make fun of the other, who in turn is required to take no offense.

RADCLIFFE-BROWN, *1965, P. 90*

Greenland Eskimos...resolve their quarrels by duels of laughter.
The one who gets the most laughs from the audience wins. The other,
humiliated, often goes into exile.

ROBINSON, *1977, P. 103*

Sudden glory is the passion which maketh those grimaces called
laughter.

THOMAS HOBBES *1651, P: 57*

Kinds of Jokes

What follows is an effort to see if different kinds of jokes in different situations
seem to suppor t or deny ideas expres sed earlier. It is not a particularly good effort
at falsification, but it may point the way to better ones.

A. Puns and shaggy dog stories are regarded as the "worst" or "lowest"
   forms of humor because the trick is on the listener.

   1. Puns are "worst" when told to a single person, who is obligatorily the
      object of ridicule. This does not mean that the greatest status shift
      will take place when a pun is told to a single individual but that it is
      difficult for either individual to gain when a pun is told to a single
      individual. Jokes told to single individuals typically are the kind that
      lower the status of a third party not present.

   2. Puns are "best" (i.e., most liked by listeners) when told to several
      listening parties. The reason is that the others can laugh at the one to
      whom the joke seems directed (which means as well that the joke can
      serve the "integrating" or "unifying" function). This hypothesis seems
      to assume that jokes involving tricks on listeners will be directed at
      one member when the audience includes several individuals, and
      usually not to the one with the highest prestige or the one least vul-
      nerable to a downward status shift. The except ion to the second part
      of the exclusion is when the joke-teller has very high status and is
      explicitly trying to lower the status of the individual cur rent ly hold-
      ing the highest status in the group.

B. The "best" jokes are those that seem to elevate the status of the listener
   (e.g., in relation to some third party, by putting that party down). For
   example, James Herriott, in a series of humorous books about his life as
   a veterinarian in Scotland (e.g., Herriott 1972), has the ability to cause
   the episodes of his life to appear as a series of jokes upon himself, in

which "tricks" on the other participants typically seem known only to the readers of Herriott's books (i.e., the "trick" is that the others are being "observed" in their idiosyncrasies by both Herriott and his readers). The effect is to give an impression of raising the status of the readers, and I speculate that this is one cause for the enormous success of Herriott's books. On the other hand, Ben K. Green, in a series of books about his life as a veterinarian in West Texas (e.g., Green 1971) typically portrays himself as the clever winner of every set-to, with the other parties getting their just due and knowing it. Even if the individual stories are inherently funny and well-told, the overall effect on the reader is quite different, an uneasy feeling appears, and the reader tends not to like Ben K. Green nearly as well as he does James Herriott, who has generated truly memorable feelings of affection among his readers. (Perhaps, as one reader suggests, the reason for Green's failure is that he is not as effective as, say, Mark Twain at his cynical "against the world" best at causing the reader to identify with him.)

C. Telling jokes on others is a way of:

   1. elevating one's own status;

   2. lowering the status of the butt of the joke;

   3. elevating the status of the listener by:

   a. allowing him to be in the right situation to laugh;

   b. lowering the status of the object of ridicule;

   4. increasing camaraderie or unity by identifying the butt of the joke as a member of an adversary group or a common object of ridicule.

D. Telling jokes on oneself is a way of:

   1. trying to elevate the status of the listener so as to set up a better relationship between one's self and the listener;

   2. trying to show that one's own status is so high that it can be lowered without changing the basic nature of the relationships;

   3. trying to channel an already existing joke on one's self so that its effect on one's status is less than would otherwise be the case.

E. Laughing at jokes on oneself is a way of:

   1. counteracting the effect of the trick, showing that it causes pleasure, therefore could not be detrimental—i.e., turning the trick back on the joke-teller (Rodney Dangerfield does not lower his status by describing the endless ways in which he purportedly has failed to get respect; that he does not, while pretending to, is, of course, his best joke);

   2. accepting the implicit status shift in the interests of a better future relationship with the joke-teller or listeners (or both).

F. Laughing at jokes in the presence of others is a way of saving status in the situation through:

1. showing (claiming) that you recognize the trick and would never be so naive as to be taken in the same fashion;

2. telling the joke-teller (and other listeners) that you and he (they) are on the same side and see others (the objects of ridicule) the same way;

3. elevating (or assuring) the joke-teller's status in relation to you or some others (A joke can be given as a gift to a friend—i.e., as an item to be used by the recipient to his own advantage. In my experience such transfers are frequently followed by the receiver of the "gift" relating to the donor how effectively it worked.)

G. Jokes on sacred topics are a way of showing that:

I. one is extremely sophisticated (has high status) and can afford to be unconcerned about the sacredness of the topic;

2. the sacred topic is not so important (to anyone).

H. Failing to laugh at someone else's jokes is a way of:

1. putting down the story-teller;

2. shewing one's own sophistication, dominance, or independence.

I. Laughing "too" hard suggests a very strong ("too" strong) effort to realize the above functions, hence implies a feeling of low status and special need to elevate one's own status.

J. The ultimate putdown is to see humor in a situation, or in a joking effort, when it is not seen by the person being put down. That person has two choices:

1. he can laugh and pretend he "got the joke"; or

2. he can sniff and pretend that he understood the joke but did not regard it as funny.

(*Note*: The above situations do not exclude circumstances in which a purveyor of humor elevates the status of an object of his humor in relation to others in a group, but it does seem to preclude raising of the status of an object of his humor in relation to the humorist's status *unless* the status of the others is so lowered as to result in a net elevation of the humorist's status. In cases that may seem to the observer to have nothing to do with status shifts it is usually revealing to visualize a reversal of the identities of the teller of the joke and its object.)

Obviously if the situation is such that the status of the listener is not being threatened in respect to the joke-teller and many others in the group, but instead the general nature of the joke promises to elevate the listener's status—in relation to say, some third party or class of people—then the listener may accept the tiny loss of status from admitting that he did not "get" the joke and reap the rewards of being able to laugh along with everyone else at the object of ridicule.

DISCUSSION

None of the above is an argument that laughter cannot be truly pleasurable, that humor is never sincerely funny, or that any of the above functions or outcomes are in the conscious minds or motivations of people who tell or respond to jokes. To contrast or oppose such proximate results of humor to its possible reproductive significance is to muddle the relationship between functions and the mechanisms which bring them about, It also overlooks the potential significance of self-deception in a world in which deliberateness in deception of others is the worst of all social transgressions, and sincerity (even what the psychologist Donald T. Campbell calls "sincere hypocrisy") is viewed as a noble virtue. The question is whether or not the above outline of hypotheses tends to clarify cases of "sincerely funny" humor as well as the kinds that some of us believe we already see as pernicious or as serving those who actively perpetuate them and sympathetically respond to them.

This raises another issue, emphasized by the fact that people chuckle to themselves when alone, and also seek out cartoons and other humor to read and enjoy in solitary, even if they do not intend to (and do not) mention these experiences to others. Depending on the precise circumstances and consequences, this fact may not be negative to the hypotheses discussed here. The reason is that social success may be powerfully enhanced by adjusting one's outcomes as a result of social scenario-building. There is no reason to believe that humor is exempt, and the degree to which it actually does contribute to status shifts should reflect the degree to which it has become involved in the scenario-building. In other words, if understanding and responding to jokes is an important kind of social behavior, then a little practice may be useful. I would hypothesize, then, that when laughter and humor are expressed by solitary individuals (and even when they seem to the individual to be completely internal and personal), they are secondary derivatives of social situations, representing scenario-building that will have its effect in later social situations, including the indirect and anticipated communication of the writer or performer for an absent reader, viewer, or interactant

I end this speculative essay on the note that whether or not the particular arguments presented here are correct in any significant way, they may support the notion that it is appropriate to examine topics like humor in the context of evolved biological (reproductive) function, and that humor has an aspect of ostracism, and of coalition formation and maintenance, and is a topic worth careful scrutiny as we intensify our concern with the problem of peaceful coexistence in an ever-more dangerous world of technological sophistication and balance-of-power races.

> God works wonders when he can
> Here lies a lawyer, an honest man
> Gracious! The two of them buried in the same grave!

> —MODIFIED FROM AYE (1931)

Acknowledgements

For criticisms and assistance I am grateful to the other participants in this project and to the members of the Human Behavior and Evolution Group at the University of Michigan. I particularly thank Bobbi S. Low, Laura Betzig, Paul Turke, Beverly Strassmann, and Roger Masters.

References

Alexander, R.D. The evolution of social behavior. *Annual Review of Ecology and Systematics* 5: 325–383, 1974.

—— Natural selection and the analysis of human sociality. In *Changing Scenes in the Natural Sciences, 1776-1976*, C.E. Goulden (Ed.). Philadelphia: Bicentennial Symposium Monograph, Philadelphia Academy of Natural Science, Special Publication (12), 1977, pp. 282–337.

—— *Darwinism and Human Affairs*. Seattle: University of Washington Press, 1979.

—— Biology and the moral paradoxes. In *Law, Biology, and Culture: The Evolution of Law*, M. Gruter and P. Bohannon (Eds.). Santa Barbara, CA: Ross-Erikson, 1982, pp. 101–110. (Also *Journal of Social and Biological Structures* 5: 389–395, 1982.)

—— A biological interpretation of moral systems. *Zygon* 20: 3–20, 1985.

—— *The Biology of Moral Systems*. Hawthorne, New York: Aldine Press.

Andrew, R.J. Evolution of facial expression. *Science* 142: 1034–1041, 1963.

Aye, J. *Humour among the Lawyers*. London: The Universal Press, 1931.

Bierce, A. *The Devil's Dictionary*. New York: Thomas Y. Crowell, 1911.

Brill, A.A. *The Basic Writings of Sigmund Freud*. New York: The Modern Library, 1938.

Cousins, N. *Anatomy o f an Illness*, New York: Norton, 1979.

Darwin, C.R. 1859. *On the Origin of Species" A facsimile of the first edition with an introduction by Ernest Mayr, published in 1967*. Cambridge, MA: Harvard University Press.

Darwin, C.R. *The Descent of Man and Selection in Relation to Sex*. New York: Appleton, 1871, 2 vols.

—— *The Expression of the Emotions in Man and Animals*. New York: Appleton, 1899.

Dawkins, R. *The Selfish Gene*. New York: Oxford University Press, 1976.

Eysenck, H.J. *Dimensions of Personality*. London: Kegan Paul, 1947.

Flugel, J.C. Humor and laughter. In *Handbook of Social Psychology*. Cambridge, MA: Addison-Wesley, 1954, Vol. 2, pp. 709–734.

Goodall, J. The behavior of free-living chimpanzees in the Gombe Stream Reserve. *Animal Behavior Monographs* 1: 165–311, 1968.

Green, B.K. *The Village Horse Doctor West of the Pecos*. New York: Knopf, 1971.

Grotjahn, M. *Beyond Laughter*. New York: Macmillan, 1957.

Hayworth, D. The social origin and function of laughter. *Psychological Review* 35: 367–384, 1928.

Herriott, J. *All Creatures Great and Small*. New York: Bantam, 1972.

Hinde, R.A. *The Biological Bases of Human Social Behavior*. New York: McGraw-Hill, 1974.

Hobbes, T. *Leviathan; or the Matter, Form and Power of a Commonwealth*. Indianapolis: Bobbs-Merrill, 1958 (original publication 1651).

Humphrey, N.K. The social function of intellect. In *Growing Points in Ethology*, P.P.G. Bateson and R.A. Hinde (Eds.). Cambridge: Cambridge University Press, 1976, pp. 303–321.

Jones, M.C. The development of early behavior patterns in young children. *Pedagogical Seminary* 33: 537–585, 1926.

Justin, F. A genetic study of laughter provoking stimuli. *Child Development* 3:114–136, 1932.

Laing, A. The sense of humour in childhood and adolescence. *British Journal of Educational Psychology* 9: 201, 1939 (abst).

Leacock, S. *Humor and Humanity. An Introduction to the Study of Humor*. New York: Henry Holt, 1938.

Leuba, C. Tickling and laughter. *Journal Genetic Psych* 58: 201–209, 1941.

Lockard, J.S., Fahrcnbrnch, C.E., Smith, J.L., Morgan, C.J. Smiling and laughter: Different phyletic origins? *Bulletin of the Psychonomic Society* 10: 183–186, 1977.

McDougall, W. *An Outline of Psychology*. London: Methuen, 1923.

McGhee, P.E. Development of the humor response: A review of the literature. *Psychological Bulletin* 76: 328–348, 1971.

——*Humor, Its Origin and Development*. San Francisco: Freeman, 1979.

O'Connell, W.E. The adaptive functions of wit and humor. *Journal of Abnormal Social Psychology* 61: 263–270, 1960.

Radcliffe-Brown, A.R. *Structure and Function in Primitive Society*. New York: Free Press, 1965.

Robinson, V.M. *Humor and the Health Professions*. Thorofare, NJ: Slack, 1977.

Sahlins, M.D. On the sociology of primitive exchange. In *The Relevance of Models for Social Anthropology*, M. Banton (Ed.). London: Tavistock, 1965, pp. 139–236.

Schmidt, H.E., Williams, D.I. The evolution of theories of humour. *Journal of Behavioral Science* 1: 95–106, 1971.

Singer, P. *The Expanding Circle. Ethics and Sociobiology*. New York: Farrar, Strauss, and Giroux, 1981.

Strate, J.M. *An Evolutionary View of Political Culture*. Ph.D. Dissertation, University of Michigan, Ann Arbor, 1982.

Symons, D. *The Evolution of Human Sexuality*. New York: Oxford University Press, 1979.

Trivers, R.L. The evolution of reciprocal altruism. *Quarterly Review of Biology* 46: 35–57, 1971.

Van Hooff, J.A.R.A.M. The facial displays of the catarrhine monkeys and apes. In *Primate Ethology*, D. Morris (Ed.). London: Widenfeld and Nicolson, 1967, pp. 7–67.

—— Aspects of the Social Behavior and Communication in Human and Higher Primates. Rotterdam: Bronder-offset, 1971.

—— A comparative approach to the phylogeny of laughter and smiling. In *Non-Verbal Communication*, R.A. Hinde (Ed.). Cambridge: Cambridge University Press, 1972.

Wolff, H.A., Smith, C.E., Murray, H.A. A study of the responses to race-disparagement jokes. *Journal of Abnormal and Social Psychology* 28: 341–366, 1934.

Yerkes, R.M., Learned, B.W. *Chimpanzee Intelligence and its Vocal expressions*. Baltimore: Williams and Wilkins, 1925.

Zajonc, R.B. Emotion and facial efference: A theory reclaimed. *Science* 228:115–221, 1985.

Ecological Constraints and Human Cooperation

People! People!

People, people, everywhere,
and not a break in sight,
they copulate and propagate,
congest with all their might.
Across the planet many starve
but too few grasp the lesson;
to our festering egos fewer humans
is an abominable suggestion.
We're everything that matters here,
all else worldly is merely tools
for us to tweak and twist at will
while we multiply like fools.
It's quantity, more quantity,
not quality we strive for.
More macadam! More concrete!
That's what we're really alive for.
But those submerged in dark despair
can take this consolation:
the trend can go only so far
in a single

     o

     o

     o

     o

generation...

          —ALEXANDER, *2011, P. 39*

INTRODUCTION

Darwin's Question: How Can Sterility Evolve?
Laura Betzig

DARWIN'S PROBLEM: NEUTER BEES

More than two years before his book *On the Origin of Species* came out, Darwin sent a letter to his bulldog, Thomas Henry Huxley, about sterile castes. "Bees offer in one respect by far my greatest theoretical difficulty" he admitted. Then, just short of two months before *The Origin* went to press, he sent another letter to his Lord High Chancellor in Natural Science, Charles Lyell. "I fairly struck my colours before the case of neuter-insects," he confessed (Burkhardt et al. 1985-, Letters 2017 and 2496).

Even before he opened his first notebook on the transmutation of species, Darwin had considered a solution to that problem. Two years into his trip aboard HMS *Beagle,* after a March, 1834 stop off the South American coast, he wrote: "The bee could not live by itself. And in the neuter, we see an individual produced which is not fitted for the reproduction of its kind—that highest point at which the organization of all animals, especially the lower ones, tends—therefore such neuters are born as much for the good of the community, as the leaf-bud is for the tree" (Darwin 1839:262).

A quarter of a century later, Darwin devoted a whole section of his book *On the Origin of Species* to the "one special difficulty" that had once seemed fatal to his whole theory. "I allude to the neuters or sterile females in insect-communities: for these neuters often differ widely in instinct and in structure from both the males and fertile females, and yet, from being sterile, they cannot propagate their kind." His simple solution to that most serious problem, the one he'd come up with in the Falkland Islands, was that sterility might benefit relatives. "This difficulty, though appearing insuperable, is lessened, or, as I believe, disappears, when it is remembered that selection may be applied to the family, as well as to the individual, and may thus gain the desired end." Breeders with sterile family members flourish (Darwin 1859:236–38).

But as R. D. Alexander was among the first to point out, sterile castes are not restricted to social insects. Members of another arthropod genus, *Synalpheus*—including the snapping shrimps, *S. regalis, S. filidigitus,* and *S. chacei*—have been considered eusocial. So have members of the vertebrate genera, *Heterocephalis* and

Cryptomys—including the naked mole-rat, *H. glaber*, and the Damaraland mole-rat, *C. damarensis* (Alexander et al. 1991; Sherman et al. 1991; Sherman, this volume: compare Jarvis 1981; Duffy 1992; Duffy et al. 2000; Bennett et al. 2005).

And as Alexander and others have argued, sterile workers are not always the close relatives of breeding queens and kings. In most cases, Darwin's "magnificent hypothesis" holds: colony members are family members. But costs and benefits also matter. Eusocial animals can be haplodiploid, diploid or clonal; many have different mothers or fathers; some are outbred, and others are unrelated altogether—like termites in fused groups, or cofoundress paper wasps, or slave ants (Hölldobler and Wilson 1990; Queller et al. 2000; Korb and Schneider 2007). Wherever the benefit-to-cost ratio is high enough, individuals should be selected to sacrifice breeding opportunities in order to help others—regardless of relatedness—reproduce (Alexander et al. 1991:4 and Sherman this volume: compare Fisher 1930; Williams and Williams 1957; Hamilton 1964).

Eusociality, or "true" sociality, always includes a reproductive division of labor (Batra 1966; Wilson 1971). But sterility can be a matter of degree. At one end of the sociality continuum, workers are obligately, or permanently, sterile. And at the other end, helpers are facultatively, or temporarily, sterile, or otherwise reproductively suppressed (Sherman et al. 1995; Crespi 2005).

H. sapiens societies span that continuum—from ancient civilizations, with their hundreds of thousands of eunuch workers; to feudal societies, with their hundreds or thousands of celibate helpers; to the more egalitarian societies we lived in as hunter-gatherers, and live in now. After writing, or history, began, roughly 5,000 years ago in the Near East, literate societies were often eusocial. Emperors from the Atlantic to the Pacific raised hundreds of children by thousands of women, who were protected and provided for by eunuchs, or obligately sterile worker castes. And after around 2,000 years ago in Western Europe, aristocrats raised dozens of children by hundreds of women, who were protected and provisioned by celibates, or facultatively sterile helpers-at-the-nest. But for more than 100,000 years before history began, most people were mothers or fathers. And for the last few hundred years, more of us have been again.

Monarchy

In the beginning, in the first histories, Saul became Israel's first king. He was succeeded by his lyre player, David, who was reassured by his God: "I anointed you king over Israel, and I delivered you out of the hand of Saul; and I gave you your master's house, and your master's wives into your bosom" (2 Samuel 12:7). David surrounded himself with ladies of honor and virgin companions, and fathered a daughter and 19 named sons. He was succeeded by his son Solomon, who kept 300 concubines and 700 wives; and Solomon was succeeded by his son, Rehoboam, who fathered 60 daughters and 28 sons.

They were protected and defended by a sterile caste. Samuel warned the people of Israel before he anointed Saul: "He will take the best of your fields and vineyards and olive orchards and give them to his servants. He will take the tenth of your grain and of your vineyards and give it to his מִסָרִים," or *sarisim,* or officers, or eunuchs. And later, when he made the announcement that Solomon would be his successor, "David assembled at Jerusalem all the officials of Israel, the officials of the tribes, the officers of the divisions that served the king, the commanders of thousands, the commanders of hundreds, the stewards of all the property and cattle of the king and his sons, together with the מִסָרִים," or *sarisim,* or officials, or eunuchs. The Hebrew words סָרִים, or *saris,* and יסרב-סר, or *rab-saris,* can be found 45 times in the Hebrew Bible, variously rendered as officer, official, eunuch and Rabsaris, or chief eunuch, by the translators who worked for King James. Some of those eunuchs belonged to Egyptian pharaohs; others belonged to Assyrian, Babylonian, Persian and Hebrew kings (1 Samuel 8:14-15; 1 Chronicles 28:1; Betzig 2005, 2009a).

In China, there are *huan guan,* or eunuchs, on Shang Dynasty oracle bones; and there are eunuchs in Zhou Dynasty *Shi jing* odes. Qin Shihuangdi, the First August Emperor of Qin, established a eunuch agency, the *Zhongchangshi;* and Ming Dynasty emperors employed 100,000 eunuchs in 12 Directorates, 4 Offices, and 8 Bureaus, who were appointed as Grand Commandants over the court at Nanjing, or installed as Grand Defenders over provincial armies. In India, *varshadharas* (or rain holders), *shandhas* (or effeminates), *tritiya prakritis* (or third genders) and *klibas* (or impotents) worked under Maurya and Gupta Dynasty emperors, as spies and harem guards. In Greece, the first ευνουχοι or bedkeepers, worked in private houses; but they ran the empire in Rome. Eunuchs worked under every emperor after Caesar: Augustus, the first emperor, was attended in public by a pair of eunuchs; and after Constantine moved the capital to Constantinople, eunuchs worked as the emperors' ambassadors, and commanded the imperial armies—a *spartharius* headed security, a *sacellarius* kept the purse, a *castrensis sacri palatii* worked as palace steward, a *comes sacrae vestis* attended the wardrobe, a *comes domorum* managed the imperial estates, and a *praepositus sacri cubiculi* administered the emperor's sacred bedchamber.

Others looked after emperors' women and children. The Caesars were provided with women by their senators and praetorian guardsmen, and had access to thousands of *ancillae,* or female slaves, whose children—*vernae* (or homebred slaves) and *liberti* (or freed slaves)—filled the *Familia Caesaris,* or civil service. Greeks had access to *hetairai* (or courtesans) and *pallakai* (or concubines), *pornai* (or prostitutes), and *douloi* (or slaves)—besides their *gynaikes,* or wives. Sanskrit texts from the *Arthashastra,* written for the Maurya emperors, to the *Kamasutra,* probably finished under the Gupta emperors, mention *anthapuras,* or harems, for thousands of women, with maternity wards and dormitories for princes and princesses. And Chinese emperors collected women from the provinces—from the 10,000 Qin Shihuangdi herded together in 270 palaces taken from the feudal rulers, to the 100,000 assembled at Yangzhou by Yangdi. Kublai Khan may have kept another 100,000 in his summer palace at Xanadu, or Shangdu: A Y chromosome linkage found in 16 contemporary Asian populations, and in 1/200 of all late-20th-century men, is arguably borne by male line descendants of his grandfather, Genghis ("Golden Lineage") Khan (Zerjal et al. 2003; Betzig 2010, 2012, 2013).

Theocracy

Roughly a generation after Jesus of Nazareth was hung on a cross, Paul of Tarsus sent a letter to his friends in the Corinthian church. "Are you free from a wife?" he asked. And if so, "To the unmarried and the widows I say that it is well for them to remain single as I do"—though it was better to marry than to burn (1 Corinthians 7:8, 27). Roughly another generation later, gospel writers had Jesus tell his followers to leave their families behind, in order to follow him. "The sons of this age marry and are given in marriage; but those who are accounted worthy to attain to that age and to the resurrection from the dead neither marry nor are given in marriage, for they cannot die anymore, because they are equal to angels and are sons of God" (Luke 20:34–36).

Helpers-at-the-nest, they filled the medieval Church. In the first few hundred years after Jesus was born, in the first Roman emperor's reign, a few *monachi,* or "lonely ones," left their families and went off to live alone in the deserts. Some of those monks were driven into the wilderness by famine; but many were the *oblates,* or "offerings," of rich parents. By the 6th century, St. Benedict had forsaken his father's house to found monasteries in the Italian forests, where he took in other men's younger sons, and asked them—in his *regula monachorum,* or rules for monks—to chastise their bodies and avoid all sins of the flesh.

So the Republic of St. Peter was carried on by monogamously married, but promiscuous, first born sons. One day in 1148 or '49, Christine, the last heiress of the house of Ardres, married an heir to the count of Guines, Baldwin II. She gave him 10 sons and daughters, and her husband added a few bastards. "Because of the intemperate tumult of his loins he was of ungovernable lust from his first adolescent impulses to his old age; very young women, especially virgins, aroused him" (Lambert of Ardres, *History,* 127). There were beautiful women from St Omer; there were noble girls from Louches. Some of the 23 or more children they gave birth to married heirs or heiresses themselves, or became knights; other spurious issue ended up in the church.

Gregory, the Tours bishop whose *Ten Books of History* tell the story of the first Frankish kings, remembered Clovis' father as a seducer: "His private life was one long debauch" (Gregory of Tours, *History,* 2.12). Charlemagne, who descended from Clovis, had 4 known queens and 5 known concubines: Sons of his wives inherited his empire, or raised insurrections; sons of his mistresses became abbots or archbishops. And the Holy Roman Emperors who descended from Otto the Great, who descended from Charlemagne, were the fathers of swarms of bastards. They kept *gynaecea,* or women's workshops, all over the countryside, on hundreds of estates; and they kept ladies of pleasure in palaces from Paris, to Aachen, to Magdeburg. Every day, 1,000 pigs and sheep, 1,000 measures of grain,

10 wagonloads of beer, 10 wagonloads of wine, and miscellaneous chickens, eggs, fish and produce were shipped out to Otto the Great's itinerant court. His descendants would have consumed at least as much (Betzig 1995, 2009b).

Democracy

On a June day in 1525, Martin Luther, an Augustinian monk, married Katharina von Bora, one of 9 nuns hustled out of their Cistercian convent in a Wittenberg fish cart. The second son of an upwardly mobile city council member, Luther insisted that celibacy was an unnatural state: "The devil must have ordered it," he wrote. In an *Open Letter To the Christian Nobility of the German Nation Concerning the Reform of the Christian Estate,* he argued that not one in a hundred monks looked for a living in church. The rest had been put there by their propertied fathers. Not every one of a nobleman's children could become a landowner (Luther, 1520:14).

Then on a cold day in January of 1649, seven years after the royal standard was raised at Nottingham, the House of Commons ordered the execution of King Charles I as a "tyrant, traitor, murderer and public enemy to the good people" of England (Rushworth, *Historical Collections,* 7.1418). One-hundred-forty years later, the Bastille was stormed; and on another cold January day of 1793, "for conspiracy against the public liberty and general safety," the French National Convention had King Louis XVI guillotined (Robespierre 1794). Across the Atlantic, on a warm July day in 1776, signers of the *Declaration of Independence* had agreed, by then, that it was the right of the people to alter, or to abolish, bad government.

For the more than 100,000 years before writing began, we were fairly egalitarian. Hoarders and freeloaders, bullies and braggarts were looked down on: "We refuse one who boasts," is how a Kalahari Bushman put it (Lee 1979: 246). Most foragers lived hand to mouth; they stored very little food, and owned only as much as they could carry. Any kind of leadership was temporary. And most people were monogamous, most of the time (Betzig 1986, 1997, 2012: compare Irons 1979; Flinn and Low 1986; Low 2000).

Then after 1492, we were more egalitarian again.

DARWIN'S SOLUTION: A FIXED ABODE

Four years before he boarded the *Beagle* as a biologist, Darwin took up divinity at Cambridge. He had the life of a country parson in mind. "I find I steadily have a distant prospect of a very quiet parsonage, & I can see it even through a grove of Palms," he wrote home to his sister Caroline, from Botofogo Bay. "To a person fit to take the office, the life of a Clergyman is a type of all that is respectable & happy: & if he is a Naturalist & has the 'Diamond Beetle' ave Maria; I do not know what to say," he wrote later, from Lima, to his cousin, William Fox. In the autobiography he put together toward the end of his life, Darwin remembers that, if the phrenologists were to be trusted, he was destined to become a clergyman: "I had the bump of Reverence developed enough for ten Priests." But he never intended to live without a wife. He worried about the loss of time and money, and the added anxiety and responsibility, of having a family; as a father, he might never go up in a balloon, or learn French. But, "My God, it is intolerable to think of spending ones whole life, like a neuter bee, working, working, & nothing after all," he scribbled in a memorandum to himself (Darwin 1838, 1882:57; Burkhardt et al. 1985-, Letters 166 and 282).

Sterility, as Darwin noted in passing, seemed to be an artifact of civilized life. Most of the "savages" he met on his *Beagle* trip, or had read about since, were married, and they generally married young. As he put it, a little unkindly, in his *Descent of Man,* "the greatest intemperance with savages is no reproach." Chastity belonged to a "very early period in the moral history of civilized man," he thought; and celibacy, "ranked from a remote period as a virtue," but to him "a senseless practice," had come after that (Darwin 1871:96).

As Darwin knew, and R. D. Alexander agreed, cooperation is mitigated by kinship. In social insects, and in other sterile castes, family members help other family members reproduce. But kinship isn't always enough.

As Darwin guessed, and R. D. Alexander was among many to point out, eusociality is often an effect of ecological constraints. In December of 1832, a year after he set sail out of Davenport, the *Beagle's* naturalist made a few notes about the "savages" that inhabited the southernmost part of the Americas at Tierra del Fuego, in groups with no government or chief. Three and a half decades after his ship finally docked at Falmouth, toward the end of his 1871 book on *The Descent of Man,* Darwin suggested that civilized societies required a fixed abode. "Nomadic habits, whether over wide plains, or through the dense forests of the tropics, or along the shores of the sea, have in every case been highly detrimental," he wrote (Darwin 1839:234; 1871:167). Civilizations, and their sterile castes, tend to exist in saturated habitats (Alexander 1974, 1979; Alexander et al. 1991; Sherman, this volume: compare Emlen 1982; Vehrencamp 1983; Wilson and Hölldobler 2005; Hölldobler and Wilson 2008).

Many eusocial animals live and breed in their food. Aphids feed in the galls of trees or shrubs; Australian thrips eat in the galls of Acacia leaves; Australian ambrosia beetles bore galleries into Eucalpytus trees; termites devour decaying logs; snapping shrimp scavenge in the currents of coral reef sponges; mole-rats chew into enormous tubers under arid East African landscapes (Aoki 1977; Crespi 1992; Kent and Simpson 1992; Faulkes et al. 1997; Thorne 1997; Duffy et al. 2000).

Other eusocial animals living in rich, patchy habitats learn to grow their own food. Ants of a variety of species herd other insects, and eat "honeydew" from the ends of their alimentary canals; other ant species sow flowering plant gardens, and harvest fruit pulp and nectar; honeybees forage for plant nectar, but live on the honey they store in their hives; a variety of beetles, termites and ants raise fungus; and *H. sapiens* practices agriculture (Davidson 1988, Davidson et al. 2003; Mueller et al. 1998, 2005; Seeley 1995: compare Brock et al. 2011).

Our eusociality started in the alluvial area between the Tigris and Euphrates, on the black soil of the Nile, along Indus and Ganges tributaries, and around the

Yellow River, or Huanghe, where people were hemmed in by the Syrian or Saharan Deserts, and the Himalayas or Zagros Mountains (Betzig 1994, 1996, 2012). As Robert Carneiro once famously summed up, civilizations began along rivers rich in life, set off by mountains and deserts, where "escape in every direction" was blocked (Carneiro 1970:375: compare Turke 1988; Strassmann and Clarke 1993; Summers 2005; Crespi 2008; and review in Hrdy 2009). Or, in the words of China's first historian, Sima Qian, who was castrated by the Han emperor, Wudi: "The old territory of Qin is well protected by mountains and girdled by the Yellow River, a state fenced in on four sides. From the time of Duke Mu to that of the First Emperor, Qin had over twenty rulers, and at all times they were leaders among the feudal lords. Surely this was not because generation after generation they were worthy men, but because of the strategic position they occupied" (Sima Qian, *Shi ji*, 6).

Our eusociality ended after Christopher Columbus discovered a New World (Betzig 2009b, 2010). Just over 500 years ago, in August of 1492, with the backing of Ferdinand and Isabella of Spain, he started the first of four trips across the Atlantic in three ships, sailing as far as the Caribbean and the American mainlands. Four years later, John Cabot got a commission from Henry VII of England to seek out and discover whatever countries or islands he could find, and made it as far as Newfoundland. Almost a century later, Sir Walter Raleigh got a charter from Elizabeth I to set up a colony on Roanoke Island, off the Virginia coast. Then in May of 1607, on a commission from James I, John Smith established a permanent settlement on the Chesapeake Bay. There were close to 4 million people in the United States, mostly of British descent, by the July 4th, 1776, when 56 members of the Continental Congress signed the *Declaration of Independence* (Jefferson 1776). They agreed that: "We hold these truths to be self-evident, that all men are created equal, that they are endowed by their Creator with certain unalienable Rights, that among these are Life, Liberty and the pursuit of Happiness.—That to secure these rights, Governments are instituted among Men, deriving their just powers from the consent of the governed,—That whenever any Form of Government becomes destructive of these ends, it is the Right of the People to alter or to abolish it."

Acknowledgment

Dick Alexander introduced me to evolutionary theory, posed the question of reproductive opportunity leveling in human societies, and for many years made it possible for me to work in the Insect Division of the Ruthven Museums. The rest, as they say, is history.

References

Alexander, R.D. 1974. The evolution of social behavior. *Ann. Rev. Ecol. Syst.* 5:325–83.

Alexander, R.D. 1979. *Darwinism and Human Affairs*. Seattle: University of Washington Press.

Alexander, R.D. 2011. *The Mockingbird's River Song: Poems, Essays, Songs and Stories, 1946-2011*. Manchester, MI: Woodlane Farm Books.

Alexander, R.D., Noonan, K. M., Crespi, B. J. 1991. The evolution of eusociality. In: P.W. Sherman et al. (eds.) *The Biology of the Naked Mole-Rat*. Princeton, NJ: Princeton University Press, pp. 3–44.

Aoki, S. 1977. *Colophina clematis* (Homoptera, Pemphigidae), an aphid species with soldiers. *Japan. J. Entomol.* 45:276–82.

Batra, S. 1966. Nests and social behavior of halictine bees of India. *Ind. J. Entomol.* 28:373–93.

Bennett, N.C., Faulkes, G.C., and Jarvis, J. 2005. *African Mole-Rats: Ecology and Eusociality*. Cambridge: Cambridge University Press.

Betzig, L.L. 1986. *Despotism and Differential Reproduction: A Darwinian View of History*. New York: Aldine-de Gruyter.

Betzig, L.L. 1994. The point of politics. *Analyse & Kritik* 16:20–37.

Betzig, L.L. 1995. Medieval monogamy. *J. Fam. Hist.* 20:181–215.

Betzig, L.L. 1996. Monarchy. In: D. Levin and M. Ember (eds.), *Encyclopedia of Cultural Anthropology*. New York: Henry Holt, pp. 803–805.

Betzig, L.L. 1997. People are animals. In: L. Betzig (ed.), *Human Nature: A Critical Reader*. New York: Oxford University Press, pp. 1–13.

Betzig, L.L. 2005. Politics as sex: the Old Testament case. *Evol. Psych.* 3:326–46.

Betzig, L.L. 2009a. Sex and politics in insects, crustaceans, birds, mammals, the Ancient Near East and the Bible. *Scand. J. Old Test.* 23:208–32.

Betzig, L.L. 2009b. But what is government itself but the greatest of all reflections on human nature? *Pol. Life Sci.* 28:102–105.

Betzig, L.L. 2010. The end of the republic. In: P. Kappeler and J. Silk (eds.), *Mind the Gap: Primate Behavior and Human Universals*. Berlin: Springer Verlag, pp. 153–168.

Betzig, L.L. 2012. Means, variances and ranges in reproductive success: comparative evidence. *Evol. Hum. Behav.* 33:309–317.

Betzig, L.L. 2013. Eusociality in history. *Hum. Nat.*, in press.

Brock D., Douglas T., Queller D., and Strassmann J. 2011. Primitive agriculture in a social amoeba. *Nature* 469:393–96.

Burkhardt, F. and S. Smith eds. 1985-. *The Correspondence of Charles Darwin*. Cambridge: Cambridge University Press.

Carneiro, R.L. 1970. A theory of the origin of the state. *Science* 169:733–38.

Crespi, B.J. 1992. Eusociality in Australian gall thrips. *Nature* 359:724–26.

Crespi, B.J. 2008. Social conflict resolution, life history theory, and the reconstruction of skew. In: C. Jones and R. Hager (eds.), *Reproductive Skew in Vertebrates*. Cambridge: CambridgeUniversity Press, pp. 480–507.

Crespi, B.J. 2005. Social sophistry: logos and mythos in the forms of cooperation. *Ann. Zool. Fennici* 42:569–71.

Darwin, C.R. 1838. *Notes on Marriage*. CambridgeUniversityDarwin Archive.

Darwin, C.R. 1839. *Narrative of the Voyages of His Majesty's Ships Adventure and Beagle: Journal and Remarks*. London: Henry Colburn.

Darwin, C.R. 1859. *On the Origin of Species*. London: John Murray.

Darwin, C.R. 1871. *The Descent of Man and Selection in Relation to Sex*. London: John Murray.

Darwin, C.R. 1882. *Autobiography*, edited by N. Barlow. London: Collins Press, 1958.

Davidson, D.W. 1988. Ecological studies of neotropical ant gardens. *Ecology* 69:1138–52.

Davidson, D.W., Cook, S.C., Snelling, R. R., and Chua, T. H. 2003. Explaining the abundance of ants in lowland tropical rainforest canopies. *Science* 300:969–72.

Duffy, J.E. 1992. Eusociality in a coral-reef shrimp. *Nature*, 381:512–14.

Duffy, J.E., Morrison, C. L., and Ríos, R. 2000. Multiple origins of eusociality among sponge-dwelling shrimp (*Synalpheus*). *Evolution* 54:503–16.

Emlen, S.T. 1982. The evolution of helping, I, II. *Am. Nat.* 119:29–53.

Faulkes, C.G., N.C. Bennett, M.W. Bruford, H.P. O'Brien, G. H. Aguilar, and J. U. Jarvis. 1997. Ecological constraints drive social evolution in the African mole-rats. *Proc. Biol. Sci.*, 264:1619–27.

Fisher, R.A. 1930. *The Genetical Theory of Natural Selection*. London: Clarendon.

Flinn, M. V. and Low, B. S. 1986. Resource distribution, social competition, and mating patterns in human societies. In: R. Wrangham and D. Rubenstein (eds.), *Ecological Aspects of Social Evolution*. Princeton, NJ: Princeton University Pressn.

Gregory of Tours. *History of the Franks*, translated by L. Thorpe. Harmondsworth, England: Penguin, 1974.

Hamilton, W.D. 1964. The genetical evolution of social behavior. *J. Theoret. Biol.* 7:1–52.

Hölldobler, B., and Wilson, E.O. 1990. *The Ants*. Cambridge: Harvard University Press.

Hölldobler, B., and Wilson, E.O. 2008. *Superorganism*. Cambridge: Harvard University Press.

Hrdy, S.B. 2009. *Mothers and Others*. Cambridge: HarvardUniversity Press.

Irons, W.G. 1979. Cultural and biological success. In: N. Chagnon and W. Irons (eds.), *Evolutionary Biology and Human Social Behavior: An Anthropological Perspective*. North Scituate, MA: Duxbury, pp. 257–272.

Jarvis, J.U.M. 1981. Eusociality in a mammal: cooperative breeding in naked mole-rat colonies. *Science* 212:571–73.

Jefferson, T. 1776. Declaration of independence. In: J. Boyd and G. Gawalt (eds.), *The Declaration of Independence: The Evolution of the Text*. Washington, DC: Library of Congress, 1999.

Kent, D., and Simpson, J.A. 1992. Eusociality in the beetle *Australoplatypus incompertus*. *Naturwissenschaften*. 79:86–87.

Korb, J. and Schneider, K. 2007. Does kin structure explain the occurrence of workers in a lower termite? *Evol. Ecol.* 27:817–28.

Lambert of Ardres. *History of the Counts of Guines and Lords of Ardres*, translated by L. Shopkow. Philadelphia: University of Pennsylvania Press, 2000.

Lee, R.B. 1979. *The !Kung San*. Cambridge: Cambridge University Press.

Low, B.S. 2000. *Why Sex Matters*. Princeton, NJ: Princeton University Press.

Luther, M. 1520. *Open Letter to the Christian Nobility of the Germany Nation*, translated by C. M. Jacobs and James Atkinson. Philadelphia: Fortress Press, 1520/1966.

Mueller U., Rehner S., and Schultz, T. 1998. The evolution of agriculture in ants. *Science* 281:2034–2038.

Mueller, U., Gerardo, N., Aanen, D., Six, D., and Schultz, T. 2005. The evolution of agriculture in insects. *Ann. Rev. Ecol. Syst.* 36:563–595.

Queller, D., Zacchi, F., Cervo, R., Turillazzi, S., Henshaw, M., Santorelli, L., and Strassmann, J. 2000. Unrelated helpers in a social insect. *Science* 405:784–787.

Robespierre, M. 1794. Speech. In: J. Hardman (ed), *The French Revolution Sourcebook*. London: Benjamin Arnold, 1999, pp. 406–408.

Rushworth, J. *Historical Collections*. London: Stationery Office, 1701.

Seeley, T.D. 1995. *The Wisdom of the Hive*. Cambridge: Harvard University Press,.

Sherman, P.W., Jarvis, J.U.M., and Alexander, R.D. 1991. Preface. In: P. Sherman et al. (eds), *The Biology of the Naked Mole-Rat*. Princeton, NJ: Princeton University Press, pp. vii–xii.

Sherman, P.W., Lacey, E.A., Reeve, H.K., and Keller, L. 1995. The eusociality continuum. *Behav. Ecol.* 6:102–108.

Sima Qian. *Shi ji*, translated by B. Watson. New York: Columbia University Press, 1993.

Strassmann, B., and Clarke, A. 1998. Ecological constraints on marriage in rural Ireland. *Evol. Hum. Beh.* 19:33–55.

Summers, K. 2005. The evolutionary ecology of despotism. *Evol. Hum. Behav.* 26:106–35.

Thorne, B. 1997. Evolution of eusociality in termites. *Ann. Rev. Ecol. Sys.* 28:27–54.

Turke, P. 1988. Helpers at the nest: Childcare networks on Ifaluk. In: L. Betzig, M. Borgerhoff Mulder, and P. Turke (eds.), *Human Reproductive Behavior: A Darwinian Perspective*. London: Cambridge University Press.

Vehrencamp, S.L. 1983. A model for the evolution of despotic versus egalitarian societies. *An. Behav.* 31:667–82.

Williams, G.C. and Williams, D. 1957. Pleiotropy, natural selection, and the evolution of Senescence. *Evolution* 11:398–411.

Wilson, E.O. 1971. *The Insect Societies*. Cambridge: Harvard University Press.

Wilson, E.O., and Hölldobler, B. 2005. Eusociality: Origin and consequences. *Proc. Nat. Acad. Sci. USA* 102:7411–14.

Zerjal, T., Xue, Y., Bertorelle, G., Wells, R.S., Bao, W., Zhu, S., Qamar, R., Ayub, Q., Mohyuddin, A., Fu, S., Li, P., Yuldasheva, N., Ruzibakiev, R., Xu, J., Shu, Q., Du, R., Yang, H, Hurles, M.E., Robinson, E., Gerelsaikhan, T., Dashnyam, B., Mehdi, S.Q., and Tyler-Smith, C. 2003. The genetic legacy of the Mongols. *Am. J. Hum. Evol.* 72:717–21.

THE EVOLUTION OF EUSOCIALITY

Excerpt from Alexander, R.D., Noonan, K.M., and Crespi, B.J. 1991. The Evolution of Eusociality. In: P. Sherman, J. Jarvis, and R.D. Alexander (eds), *The Biology of the Naked Mole-Rat*. Princeton, NJ: Princeton University Press, pp. 3–44.

Eusociality is a remarkable topic in evolutionary biology. The term, introduced by Michener (1969), refers to species that live in colonies of overlapping generations in which one or a few individuals produce all the offspring and the rest serve as functionally sterile helpers (workers, soldiers) in rearing juveniles and protecting the colony. The wasps, bees, ants, and termites known to live this way had previously been called the "social" insects.

The recent discovery of eusociality in aphids (Aoki 1977, 1979, 1982) and naked mole-rats (Jarvis 1981, this volume) has provided biologists with new impetus to understand more fully the origins and selective background of this phenomenon, which has already played a central role in the analyses of sociality in all animals (Hamilton 1964) and, indeed, of evolution itself (Darwin 1859). These two new instances both broaden the search for correlates of eusociality in the widely different groups in which it has evolved independently and stimulate comparative study of related species of insects and vertebrates with homologous behaviors verging on eusociality (Eickwort 1981; J. L. Brown 1987; Lacey and Sherman 1991, chap. 10).

An unusual and complicated form of sociality has thus evolved independently in four different groups, and in one, the Hymenoptera, has persisted from perhaps a dozen independent origins (F. M. Carpenter 1953; Evans 1958; Michener 1958; Wilson 1971). Explaining this phenomenon requires attention to a number of different questions. Darwin (1859) answered the basic one, How can natural selection produce forms that would give up the opportunity to reproduce, instead using their lives to contribute to the success of the offspring of another individual?

Darwin's Question: How Can Sterility Evolve?

Darwin used the origin of sterile castes as a potential falsifying proposition for his theory of evolution by natural selection. He referred to "the neuters or sterile castes in insect-communities ... [which] from being sterile... cannot propagate their kind" as "the one special difficulty, which at first appeared to me insuperable, and actually fatal to my whole theory" (1859, p. 236). To solve the problem of how the sterile castes could evolve, he generated the magnificent hypothesis, which still stands, that if sterility (or any trait of a sterile form) can be carried without being expressed, then if those who express it contribute enough to the reproduction of others who carry the trait but do not express it, the trait itself can be "advanced by natural selection" (p. 236). In other words, if functionally sterile individuals help relatives produce offspring and thereby cause enough copies of the helping tendency to be created, then the tendency (ability, potential) can spread. Darwin was particularly concerned with how the sterile castes could evolve their own sets of attributes; his statements indicate that when he spoke of selection at the level of the "family" and "community" in eusocial insects, he was referring to the spread and preservation of traits that exist among the members of groups of related individuals. Thus, in the same context, he noted that "A breed of cattle always yielding oxen [castrates] with extraordinarily long horns could be slowly formed by carefully watching which individual bulls and cows, when matched, produced oxen with the longest horns; and yet no one ox could ever have propagated its kind" (p. 238). Similarly, he remarked that tasty vegetables could be produced by saving seeds from relatives of the vegetables already tasted or eaten and therefore unable to produce seeds. He also noted that cattle with "the flesh and fat...well marbled together" could be bred although "the animal has been slaughtered" if "the breeder goes...to the same family" (p. 238).

Darwin's hypothesis could scarcely be improved on today, even though, not knowing about genes, he had to rely on the concept of trait survival, and he had no way of being quantitative. His various remarks taken together are quite close to what modern investigators such as Hamilton (1964) and D. S. Wilson (1980) mean when they refer, respectively, to "inclusive-fitness maximizing" and "trait-group selection." Darwin's "family" method of selection to preserve traits is one of those long advocated by agricultural scientists (e.g., Lush 1947). His remarks cited here demonstrate the error of assertions either that Darwin invoked (a simplistic and unsupportable kind of) group selection to explain eusociality or that he did not discuss selection above or below the level of the individual. Darwin also showed in these statements that he understood how organisms can carry the potential (which we now know to be genetic) for varying their phenotypes between profoundly different states, depending on environmental circumstances.

Fisher (1930, p. 177) began the quantification of Darwin's idea of reproduction via collateral relatives (although he gave no evidence of being aware of Darwin's discussion when he did so) by developing a hypothesis to explain how bright coloration that attracted (and taught) predators could evolve in distasteful or poisonous caterpillars. He noted that if bright coloration were to spread among distasteful or poisonous caterpillars traveling in sibling groups, then a caterpillar with a new allele making it slightly more noticeable and thus more likely to give its life being tested could thereby teach a predator to avoid the entire sibling group. But, that caterpillar would have to save more than two full siblings, since each would have only a 50% chance of carrying the same allele for brighter color. (Using phylogenetic inference, Sillén-Tullberg [1988] argued that distastefulness and bright coloration often preceded gregariousness in lepidopterous larvae, but this argument does not negate the possibility of continued exaggeration of these traits among gregarious forms.) Fisher (1930, p. 181) also remarked that tendencies of humans to risk their lives in heroic acts are most likely to have spread and become exaggerated because of the beneficial effects on copies of the genes responsible located in the collection of the hero's relatives.

Haldane (1932) carried the arguments about reproduction via collateral relatives further and also related them to the eusocial insects. (Haldane is reported to have commented [Maynard Smith 1975; pers. comm.] that we should expect individuals in species like our own to have evolved to give their lives only for more than two brothers or more than eight cousins, since brothers have a one in two chance of carrying alleles for such bravery and cousins a one in eight chance. This comment is said by Maynard Smith to have been made sometime in the early 1950s in a pub with only Maynard Smith and Helen Spurway Haldane present [see also Haldane 1955]. The close resemblance of this reported statement to Hamilton's [1964] statement has aroused some attention [see also Hamilton 1976]. In any case, the original idea of reproduction via collateral relatives was Darwin's, its initial quantification was by Fisher, and, as discussed later, Hamilton [1964] first developed it extensively.) Williams and Williams (1957) discussed the evolution of eusocial insects, approximately in Darwin's terms, but they were unaware of Fisher's discussions (G. C. Williams, pers. comm.) and added no new arguments.

Hamilton (1964) not only developed the ideas of Darwin, Fisher, and Haldane extensively, but he also showed that maximization of what he called *inclusive fitness* (a process some others have called *kin selection*, following Maynard Smith 1964) really applies to all social species. The general principle, familiar now to nearly all biologists, is that one can reproduce not only by creating and assisting descendants but also by assisting nondescendant or collateral relatives, and, other things being equal, it pays more to help closer relatives than to help more distant ones.

Literature Cited

Aoki, S. 1977. *Colophinia clematis* (Homoptera, Pemphigidae) an aphid species with "soldiers." *Kontyû* 45:276–282.

Aoki, S. 1979. Further observations on Astegopteryx styracicola (Homoptera: Pemphigidae), an aphid species with soldiers biting man. *Kontyû* 47:99–104.

Aoki, S. 1982. Soldiers and altruistic dispersal in aphids. In *Biology of Social Insects*. M.D. Breed, C.D. Michener, and H.E. Evans, eds., pp. 154–158. Boulder, Colo.: Westview Press.

Brown, J.L. 1987. *Helping and Communal Breeding in Birds*. Princeton, N.J.: Princeton University Press.

Carpenter, F.M. 1953. The geological history and evolution of insects. *Am. Sci.* 41:256–270.

Darwin, C.R. 1859 (1967). *On the Origin of Species: a Fascimile of the First Edition and with an Introduction by Ernst Mayr*. Cambridge, Mass.: Harvard University Press.

Eickwort, G.C. 1981. Presocial insects. In *Social Insects*. Vol. 2. H.R. Hermann, ed., pp. 199–280. New York: Academic Press.

Evans, H.E. 1958. The evolution of social life in wasps. *Proc. 10th Int. Congr. Entomol.* 2:449–457.

Fisher, R.A. 1930. *The Genetical Theory of Natural Selection*. 2d. ed., 1958. New York: Dover.

Haldane, J.B.S. 1932. *The Causes of Evolution*. London: Longman, Green. Reprinted 1966. Ithaca, N.Y.: Cornell University Press.

Haldane, J.B.S. 1955. Population Genetics. *New Biol.* 18:34–51.

Hamilton, W.D. 1964. The genetical evolution of social behavior. I, II. *J. Theor. Biol.* 7:1–52.

Hamilton, W.D. 1976. Haldane and altruism. *New Sci.* 71:40.

Jarvis, J.U.M. 1981. Eusociality in a mammal: Cooperative breeding in naked mole-rat colonies. *Science (Wash., D.C.)* 212:571–573.

Lacey, E.A. and Sherman, P.W. 1991. Social organization of naked mole-rat colonies: evidence for division of labor. In *The Biology of the Naked Mole-Rat*. P.W. Sherman, J.U.M. Jarvis, R.D. Alexander, eds., pp. 274–357. Princeton, N.J.: Princeton University Press.

Lush, J.L. 1947. Family merit and individual merit as bases for selection. *Am. Nat.* 81:241–261.

Maynard Smith, J. 1964. Group selection and kin selection. *Nature (Lond.)* 201:1145–1147.

Maynard Smith, J. 1975. Survival through suicide. *New Sci.* 67:496–497.

Michener, C.D. 1958. The evolution of social behavior in bees. *Proc. 10th Int. Congr. Entomol.* (Montreal, 1956) 2:441–447.

Michener, C.D. 1969. Comparative social behavior of bees. *Annu. Rev. Entomol.* 14:299–342.

Sillén-Tullberg, B. 1988. Evolution of gregariousness in aposematic butterfly larvae: a phylogenetic analysis. *Evolution* 42:293–305.

Williams, G.C. and D.C. Williams. 1957. Natural selection of individually harmful social adaptations among sibs with special reference to social insects. *Evolution* 11:32–39.

Wilson, D.S. 1980. *The Natural Selection of Populations and Communities*. Menlo Park, Calif.: Benjamin/Cummings.

Wilson, E.O. 1971. *The Insect Societies*. Cambridge, Mass.: Belknap Press of Harvard University Press.

Evolution and Religion

God, Most Recently

To every reason for cooperating uninhibitedly,
praise the power of the Lord, our useful
and admirable metaphorical Spirit Father Figure.
Yet be conservative about passing the ammunition,
brother: genocide, though not our stated aim,
has too long been a claim to fame.
With and despite God's existence and help,
unfortunately, we remain personally
and collectively engaged in changing
the measurements of the lifetimes
and comforts of our fellow beings in the species
that alone among the known apes of history
have gained the capability of traveling and living globally
but so far have been unable to cooperate globally
about anything at all—excepting, perhaps,
the worthiness, indeed, the seemingly shared necessity
of strong patriotism, patriotism that
continues to evolve and use its skills
and unwavering determination to destroy
every opposed and opposing side.
Asking for God's help as our collective guiding
spirit of cooperativeness, we may convince ourselves
that we persist only toward honorable outcomes.
Sadly, it appears that we may not for some time

find ways to desist from continuing efforts
to adjust the lifetimes of our fellow humans
so frequently in the wrong direction.

—ALEXANDER, *2011, P. 295*

INTRODUCTION

The Concept of God as a Metaphor for Social Unity:
Richard Alexander's Hypothesis
William Irons

Richard Alexander's hypothesis about the core meaning of the concept of God is unique among evolutionary theories of religion and performs a valuable service to the evolutionary study of religion by focusing attention on something that most other theories of religion have ignored. This is the question of the central values which religion celebrates and reinforces.

There are numerous contemporary evolutionary theories of religion, but few have focused primarily on this question, the question of how religion defines the highest good. In a basic way, Alexander's theory is closer to Durkheim's (1915) theory of religion that was put forward in the early twentieth century. Durkheim defined religion as "a unified system of beliefs and practices related to sacred things...that unite into one single moral community all those who adhere to them" (page 62). Here the attention is clearly focused on what is valued most, and it is the unity of the local community of close kin and neighbors which is the primary vehicle for the maximization of an individual's inclusive fitness.

In my opinion, Alexander's concept-of-God hypothesis needs to be expanded to take into account how communities and societies with different social structures modify the concept of God (Wright 2009), and more importantly it needs a complimentary theory of how religions conceive of the greatest evil. Most, if not all, religions define not only the highest good, but also the greatest evil. Most contemporary theories of religion focus on one of two questions: (1) How do religious beliefs arise? (2) What are the social consequences of religion?

The first set of theories tends to explain religion by appeal to psychological mechanisms which cause a belief in something unseen: spirits, gods, or other invisible agents. A good example of such a theory is the theory that religion arises from a hyperactive agency detection device, a HADD (Dennett 2006). This has often been explained by exploring a hypothetical but probable situation in which evolving human beings had to decide whether they are encountering something with agency. While sleeping in a rock shelter, a member of an early human population might hear a noise outside the shelter that could be wind blowing branches against the entrance to the shelter, or could be a large predator entering the shelter.

Errors in evaluating this question had very different consequences. Falsely assuming it was wind when in fact it was a predator, could lead to death for oneself and one's family. Falsely assuming it was a predator would only lead to a restless night's sleep. The asymmetry of consequence favored the evolution of a psychological bias toward ascribing agency to phenomena that were not completely understood and might be hostile agents like predators. This bias, which is labeled a HADD, led to the cultural evolution of beliefs in elaborate unseen agent: spirits, gods, and so forth.

The second set of theories tends to see the consequence of religious beliefs as enhancing social cooperation within groups. Such a benefit is usually seen as beneficial to the survival of the social group and of the individual members of the group. Logically such theories can be seen as an extension of Alexander's earlier theory of morality in which he argued that intergroup competition in human evolution caused natural selection to favor traits that enhanced the cohesiveness of human social groups (Alexander 1987). He pointed to morality based on indirect reciprocity as the primary mechanism created by selection for larger and better-united groups that would be more successful in inter-group competition. Religion can be seen as a way of enhancing the signaling involved in indirect reciprocity.

Thus theories of this second variety tend to see religion as adaptive in ancestral environments and perhaps still in contemporary ones. Human beings have evolved psychological mechanisms that cause them to invent and maintain beliefs and rituals that unify social groups and encourage cooperation within them. The systems of religious beliefs, rituals, and sacred stories—the religious traditions— are complexes of interrelated memes resting on top of the evolved propensities to absorb and practice the religion of one's community of kith and kin. These meme complexes also evolve as individuals find variations in them more satisfactory in meeting the need to celebrate their most basic core value. Like other cultural institutions, they are continually changing as the members of various communities negotiate new understandings of what means the most to them in ever-changing circumstances of life. Thus Alexander's concept of God hypothesis nicely complements those theories of religion which see religion as a device for forming and maintaining cohesive social groups (Bulbulia 2004a, 2004b; Irons 2001, 2008; Sosis, 2000, 2005; Sosis and Bressler 2003; and Wilson 2002, 2005).

Most importantly, it calls attention to the core values contained in religious traditions—something which those studying religion from an evolutionary point of view need to keep in mind. Being aware of the centrality of the concept of God to the most basic values of human communities should make us appreciate why religious beliefs are so persistent when faced by scientific challenges. This should lead us to see the foolishness of head-on aggressive attacks on religion and suggest instead that scientists troubled by the role of religion in discouraging acceptance of scientific findings should seek ways to minimize the conflict between science and literal, traditional religious beliefs.

Alexander's concept of God hypothesis should assume a central place in the further development of the evolutionary study of religion. It focuses attention on a central feature of religion which most other evolutionary theories of religion ignore.

References

Alexander, Richard D. 1987. *The Biology of Moral Systems*. New York: Aldine De Gruyter.

Alexander, R.D. 2011. *The Mockingbird's River Song: Poems, Essays, Songs and Stories, 1946-2011*. Manchester, MI: Woodlane Farm Books.

Bulbulia, Joseph. 2004a. Religious costs as adaptations that signal altruistic intention. *Evol. Cog.* 10:19–38.

Bulbulia, Joseph. 2004b. The cognitive and evolutionary psychology of religion. *Biol. Phil.* 18:655–686.

Dennett, Daniel C. 2006. *Breaking the Spell: Religion as a Natural Phenomenon*. New York: Viking.

Durkheim, Emile. 1915. *Elementary Forms of the Religious Life*. New York: Free Press.

Irons, William. 2001. Religion as a hard-to-fake sign of commitment. In: R.M. Nesse (ed.), *Evolution and the Capacity for Commitment*. New York: Russell Sage Foundation, pp. 292–309.

Irons, William. 2008. Why people believe (what some other people see as) crazy ideas, In: J. Bulbulia, R. Sosis, E. Harris, R. Genet, C. Genet, and K. Wyman (eds.), *The Evolution of Religion: Studies, Theories, and Critiques*. Santa Margarita, CA: Collins Foundation Press.

Sosis, Richard. 2000. Religion and intragroup cooperation: preliminary results of a comparative analysis of utopian communities. *Cross-Cult. Res.* 34(1):70–87.

Sosis, Richard. 2005. Does religion promote trust? The role of signaling, reputation, and punishment. *Interdisc. J. Res. Relig.* 1:1–30.

Sosis, Richard, and E. Bressler. 2003. Cooperation and commune longevity: A test of the costly signaling theory of religion. *Cross-Cult. Res.* 37:211–239.

Wilson, David Sloan. 2002. *Darwin's Cathedral: Evolution, Religion, and the Nature of Society*. Chicago: University of Chicago Press.

Wilson, David Sloan. 2005. Testing major evolutionary hypotheses about religion with a random sample. *Hum. Nat.* 16:382–409.

Wright, Robert. 2009. *The Evolution of God*. New York: Little Brown.

RELIGION, EVOLUTION, AND THE QUEST FOR GLOBAL HARMONY

Richard D. Alexander
(Essay original to this volume)

Introduction

This essay is an effort to bring together aspects of human existence that have proceeded more or less separately, and even antithetically. They are (1) religion, in its principal components, and comprising the most widespread, divergent, and tenaciously authoritative defenses of morality; (2) organic evolution, as the science of all life; and (3) by far the most important and difficult, the effort (or at least, a hope or desire!) to work toward world-wide social harmony. It seems to me that the relationships of these and several other problems need to be considered together if humans in general are to moderate their hyper-competitiveness and hyper-patriotism, their theatrical attraction to violence, murder, and destruction, and the world's continuing scourge of deadly conflicts. On the one hand is the universal and familiar coordination of personal and collective musings, beliefs, and efforts by which humans have for centuries sought to understand themselves and their associates; and on the other hand are the results and consequences of the more recently recognized and analyzed process of organic evolution. To every indication the evolutionary process has been responsible for the nature of all life, including the scope and diversity of human sociality and the consequences of the myriads of never-ending split-second and unexpected environmental changes that continually modify our performances by relentlessly racing across our human lifetimes.

The several questions I am setting out to discuss in this essay are, I think, more difficult and more important than any others I have ever even imagined undertaking. I am essentially certain that I will be unable to bring solutions to my questions together in ways that will create satisfying syntheses. But the questions are intertwined with one another, and they might well be the most important set of questions anyone can try to put together on almost any topic. I have started parts of this project repeatedly, and I continue to regard my efforts as inadequate. For example, I have tried to identify what might be termed the concept of God. But, at least until now, my efforts (my fantasies!?) have failed to convince me that a universal concept of God—whether judged natural or supernatural—can ever be applied globally, faced as we are with continuing limited or local maps of disagreement that cannot serve comfortably in scatterings of differently organized parts of the world, let alone any that can change the world as a whole. I have also tried to think about pathways that could lead the more than seven billion people of the world to reduce the continuing development of ever more horrendous and, almost certainly, someday, irrevocably catastrophic weapons, and instead spread congeniality and cooperation that could direct us away from the bitterness of serious competitions. Of course, my thoughts about those pathways have always been less than adequate. The size of the human population and its potential and willingness to engage in wars and genocides seem continually to leave only hopeless discouragement. But perhaps we can at least mull over the seeming inescapability of our fate.

So I am trying again. Perhaps there will be readers who can transform some of the problems that have stymied me, and enable us all to realize what wonderful times the world could experience if the lives of people in general could be changed in directions of cooperativeness that over-ride their opposite competitiveness, wars, senseless murders, and bullying. I apologize for my inadequacies, but surely we should not be reluctant to open the doors for whatever potentially satisfying future the people of the world may someday contemplate to build for us.

THE SCIENCE OF EVOLUTION AND ITS RECENTNESS

The three introductory topics, and questions about them, have had their beginnings buried in such different backgrounds, and across so many millions of years that, despite an almost endless literature, we can scarcely imagine their specific origins. Within the last century and a half, however, evolutionary biology has splashed new knowledge of human social life on the scene, initially with Charles Darwin (e.g., 1859, 1871, 1872), and subsequently with Sir Ronald A. Fisher (especially 1930, 1958), in particular discovering how the evolutionary process works (i.e., what adaptation means, and how it takes place). Approximately a century after Darwin, several important and broadly credible additional steps were presented (e.g., Williams and Williams 1957; Hamilton 1963–1971; Williams 1957, 1966; Trivers 1971). And of course there has been a long-continuing explosion of an incredibly prolific literature, some of which is referenced both directly and indirectly at the end of this essay.

Unfortunately, evolution has failed to become widely accepted as the central aspect of human understanding. Nevertheless, despite its neglect—or reluctance—by humanity in general, organic evolution indisputably continues as the universal process underlying and shaping the existence, nature, and patterning of all forms, constituents, and divisions of life, including religious and other human social endeavors, many of which remain unexplained. Across most or all of history, humans appear to have been largely oblivious to the workings of organic evolution. Many have consistently shunned approaches to evolution, partly because they have gained few opportunities to comprehend and use evolution to contemplate the adaptive processes influencing and shaping the lives of the enormous human population. Perhaps it is also partly because many humans, especially those who are strongly religious, simply believe that we have been getting along just fine without assistance from a great deal of scientific knowledge (e.g., excerpts and references from Alexander 1967–2011).

Charles Darwin's 1859 Challenge

> If it could be demonstrated that any complex organ existed, which could not possibly have been formed by numerous, successive, slight modifications, my theory would absolutely break down.
>
> —ON THE ORIGIN OF SPECIES BY MEANS OF NATURAL SELECTION: P. *189*

Despite many efforts to deny Darwin's magnificent challenge, or declare it wrong, it has not been dismissed or falsified across the past 152 years of its existence. In a no-holds-barred challenge to the entire world of humanity, and with hundreds of thousands of non-human species living in every direction, all of them presumably available for endless testing, Darwin placed his entire theory in complete jeopardy. He announced an incredible bet that, if it held, would be telling every honest and open-minded person how to consider and identify the truth about organic evolution—about the background of all living creatures—including humans. After more than a century and a half without a misstep, Darwin's daring challenge has been demonstrated unequivocally. In the process he laid out for us the basic nature of organisms, from genes to fertilized eggs, genomes to individuals, and including populations predicting and describing the whole of humanity that lies before us.

Although Darwin did not generate the permanent name for what we now call genes, he carefully and brilliantly posited precisely the requirement of *"numerous, successive, slight modifications,"* a remarkably special set of conditions. In 1859, Darwin could not have gotten much closer to the later-arriving concept of genes and mutations. Ironically, a packaged copy of Gregor Mendel's paper on the patterning of the genetics of garden peas was found on Darwin's desk after his death, still unopened. Mendel referred to "factors" that behaved appropriately to Darwin's challenge, and were also references to what would later be labeled as "genes." But Darwin evidently never learned about all of that, and as a result the concept of the gene did not acquire its permanent label until around the turn of the twentieth century.

CONCEPTS OF GOD AND THE ANTIQUITY OF RELIGION
Natural or Supernatural: Does It Make a Difference?

In this essay I suggest a possible origin, and usefulness, of the concept of God that (1) is based on natural causes, with or without requirements matching supernatural causes; (2) is reasonably consistent with historical and current usages, interpretations, and beliefs about the concept of God (there is no reason to expect complete correspondence at this point); (3) is entirely in agreement with what we know about the sometimes distressing outcomes of natural selection via organic evolution; (4) is consistent with all that we know about how the human species has evolved; and (5) proposes a concept of God as a universal entity, or spirit, that—we can hope—is capable eventually of serving the entire human population, without unduly restrictive ceremonies, narrowly ritualized authority, or onerous opposition to widely acceptable and reasonably fair and improvable tenets underlying established laws and social behavior.

I suggest further that, with mindfulness, the science of evolution and the concept of God may be linked to make useful contributions to acceptable versions of the formidable question: What is the meaning of life? As well, part of the key to reducing hostility and divisiveness among religions, and between religious and non-religious people, may derive from active attempts to generate tolerance of a greater acceptance and diversity in concepts of God. I have tried to make my thoughts on these and other topics include efforts to reflect some of the similarities and differences between religion and science, and between religion and secularism, in ways that relate to the structure of this essay.

Although everything I will say here has been touched upon by countless thinkers across a long history, not surprisingly it is difficult to locate arguments that begin from the relatively recent process of evolutionary selection, with the consequence that few combinations of arguments arrive at syntheses similar to those presented here (see references, including the appended volumes). After all, from long before knowledge of organic evolution, the world has been filled with people who know religion, think religion, and do so alongside most of the rest of the human population, imbedded in many aspects of life, in practices of honesty, cooperativeness, congeniality, all the rest of positive sociality and comradeship; and, of course, many less comfortable features of humanity. But formal science is not only far more recent than is religion; much of its working is also extremely slow, compared to the flow of rapid changes in human behavior, while, perhaps ironically, religious morality tends to cling to constancy. The fairly recent disciplines of biology and evolution are often misinterpreted by their connections to the long-term histories of religions, as with, to a lesser extent, the social sciences. The significance of natural selection is often ignored or judged negatively for that reason.

* * *

There are at least two alternatives that may explain the nature of religion and God. One possible alternative is that religious ceremonies and beliefs tend to be accurate and factual, deriving from the pre-existence of a supernatural God, everlasting life, and other special features, including the value of adhering to moral rules imposed by authority figures accepting supernatural causes.

A second possible alternative is that religious ceremonies and beliefs have been generated gradually and cumulatively, by straightforward expansions of the extraordinary imagination and foresight of the human collective, or its leaders, in the absence of anything that can legitimately be regarded as necessarily supernatural, and therefore understandable as more or less metaphorical, and useful and effective as such.

Either alternative may have been generated by people, not necessarily formally religious, who had begun to live in organized groups for social reasons, because of loneliness or uneasiness about proximities of neighboring aggressive or resource-competitive groups and rising beliefs in the powers and values of willingly single or combined authorities.

It is entirely possible that today's various claims of religious significance have generated, survived, and persisted—indeed, thrived—because they enabled people to succeed, especially in closely-knit groups created by adding and enhancing changes religious in nature.

It is not difficult to find ourselves paying close attention to unusually capable group members willing to discuss serious questions and as a result becoming leaders. Modern human groups have little trouble identifying outstanding leaders in virtually any situation involving groups needing assistance and guidance. Such leaders are most likely older men such as fathers or grandfathers, well known for their special abilities (and without accounting for the curious asymmetry of access between the sexes!). When such leaders die, it is not surprising that group members may continue thinking about the advice they had received prior to the deceased person's loss. Nor is it surprising that the reputations of deceased leaders tend to grow. I suspect that virtually everyone pays special attention to the thoughts that may be passed around almost indefinitely by group members. Offspring, grandchildren, or other close relatives of a deceased leader may take up examples—both old and new—from a deceased leader, and either become a new leader themselves or continue to elevate a deceased relative. Examples may begin to derive from relatives of a deceased leader, or sometimes from almost any members of the group. It seems likely that the more elevated the reputation of a deceased leader, the more likely it becomes that, especially in close-knit or small groups, the deceased leader continues to lead, by virtue of tightly-knit group actions that elevate the deceased leader even further. How difficult should it be, then, for alert or apprehensive group members to begin to honor their deceased leader as a Spirit or a God? Nor is it surprising that supernaturalism can be accepted in such situations, strengthening the wills of group members to maximize the capabilities of their group and determining to build even further on the image of the original leader. I am sympathetic, and in no way surprised, that any and all versions of God have the potential to be generated—and to be exceedingly useful—as elevated and effective leaders, whether regarded as single or multiple, or metaphorical or supernatural, at least in the confident minds of the seekers of peacefulness and successful group cohesion.

* * *

I reiterate that, despite its neglect by humans in general, and some negative social (and other) results deriving from novel environmental effects and evolved consequences of gene effects, organic evolution indisputably continues as the universal process underlying all aspects of life, including religious and all other human social endeavors. It is scarcely plausible to assume that the increasingly well studied and well understood human organism has little or nothing to do with the straightforward facts of organic evolution; single organisms beginning, as they do, from combinings of tiny and complex particles, reasonably well under-stood through the decades of development of the human individual's lifetime, and including myriad expressions of social behavior across the everyday interactions of human throngs, to finally comprehend the inexorable deterioration of the later lives of individuals.

GROUP-LIVING: ITS REASONS AND CONSEQUENCES

If we wish to understand ourselves, across all of our history, we must eventu-ally consider the reasons why human populations live in diverse special groups. There is much to understand about different forms of group-living, and there have been many efforts to describe and explain their variations (e.g., Flannery 1972; Alexander 1979; Ember and Ember 1990; Diamond 1997; Boehm 2003, 2012, and many others; see also the appended list of relevant books on these topics). It will profit us to learn much about different behaviors of group members under dif-ferent conditions of life. It will also be necessary to review the known and likely histories of population variations, the availability and accesses of particular forms of environments, and how group-living has changed across the centuries. These ancient, extraordinary, and frequently changing topics have become virtually complete journeys—indeed, complete sets of diverse journeys.

It seems almost customary to describe and explain the long period of human distribution and social organization by starting with the earliest and least well understood of historical questions, then working toward the present situation. It may sometimes be more effective, however, to begin by examining the currently existing diverse, abundant, and accessible human patterns of social life, then work backward toward whatever can be gleaned from the necessarily less obvious, less well understood stages from early patterns of human social life. I will attempt to discuss a few aspects of early group-living by early humans, but I cannot provide a detailed review of the history of human groupings across millions of years.

The kinds of group-living across the span of human history surely have been extremely variable and distinctive, with the behavior of current groupings dramat-ically different from the estimates of early humans. Early humans almost certainly lived in small, separated, perhaps scattered groups, and would have been required to be extremely cautious about predators from other species. When humans man-aged to reduce their problems with serious large predators, as they quite obviously have done, their groups expanded and became more mobile, and also more con-cerned with restricting barriers, such as rivers, lakes, mountains, and other physi-cal obstructions. Eventually, such barriers came to be crossable, allowing increased movements and expansion of groups. Separated groups, especially those coming from distant or different places, can be expected to view the continuing flow of human migrants as alien or dangerous whenever they looked or acted significantly different. Later on, barriers would have become less significant, sizes of groups would have increased, and competition between groups, in at least some cases, would have been more likely to treat other groups as serious adversaries or enemies. Because humans have been able to reduce threats from non-human enemies, allowing humans to attend more closely the presence of strange humans, human groups have more or less covered the earth, and have both fought and mixed in such ways as to live and hybridize with humans from different locations, habitats, tribes, and other groups. Various visible or other differences among the members of other tribes or groups created the current populations of mixed groups and diverse nations as people moving into new locations began to live together, producing many varieties of hybridized humans from different areas and backgrounds. Such trends will surely continue to diminish tendencies for humans to reject or shun individuals from other parts of the world.

The results from these rather simple suggestions have ultimately produced what we are seeing today: crowds of diverse people, with many groups still tending to separate themselves from other groups. Such groups have continued to overlap extensively in regions across much of the world. It would seem that humans would have done exactly what we might presume from our everyday observations of appearances and performances of the people around us. It is also clear that many or most of the different human forms, sometimes considered to be distinct species, have been able to hybridize with groups initially from other locations. Finding small differences in the remains of ancient anthropoids, however, is not necessarily evidence of separate or distinct species. Nor are groups of people who suppose that others who look or act differently, or come from distant or quite different places, likely to represent distinct species. Our modern world demonstrates unequivocally pitched battles, deadly fights, extended warring periods, and striking differences between unusual people from different locations; but this is not convincing evidence or demonstration of evolved and completed speciation. Humans from all over the world have come together and hybridized, producing astonishing (and eventually comforting!) mergings of different peoples that have mingled from the ends of the earth, and now comprise the general world population of ordinary human beings. As a result we may be moving toward behaviors favoring global social harmony, while nevertheless living with constant anxiety about the current and future development of increasingly devastating weapons of war.

Today, we modern humans think mostly, I suspect, that we are simply spending our lives in quiet, peaceful, and enjoyable social groups. It is indeed possible to converse almost continually in many comfortable places, with congenial, friendly people enjoying the pleasantries that surround them, and only infrequently engaging in tense or serious conflicts. We tend to accept such situations almost as if they represent the entire spectrum of human existence. Under such circumstances, for example, everything about wars, genocides, or murderous attacks seem far away, sometimes even absent everywhere, especially, perhaps, in our everyday thoughts. All too often we don't feel a necessity to dwell upon war, or other kinds of upcoming or ongoing conflict, although sometimes across long periods large numbers of human beings just like ourselves are being slaughtered by dozens, hundreds, thousands, or more, somewhere else. We think of ourselves as different because of our separateness from such situations, and we mostly imagine ourselves to never again be involved in serious negative confrontations. Many of us do indeed pass through our lifetimes without being required to participate in military conflicts, or without having to seek frantically to avoid the pain and deaths of innocent individuals threatened by opponents, or others threatening to use the terrible devices we humans have constructed and applied. But we continue to invent, modify, extend, and accept new weapons in efforts to be ever more prepared to kill small or large numbers of people, in either wars or genocides—or perhaps as attacks or defenses against individuals and groups regarded as villainous and evil, and small groups that continue using the same weapons created and utilized in both defense and offense, within our own societies. We cannot rest as if what appear to be new conflicts will continue according to well-known strategies and changes in weaponry.

Although humans in general are scarcely aware of it, they are somewhat like the thousands of other species that can change dramatically between different patterns of development, quickly and definitely, in ways that fit different life patterns of peaceful harmony in different environments. Unfortunately, humans have also evolved to make incredible changes, but all too often as part of attacking and destroying others of their own species and their own phenotypes. Our largest problem may be to eliminate that second, militant version of the human phenotype.

* * *

Let us understand, once and for all, that the ethical progress of society depends, not on imitating the cosmic process, still less in running away from it, but in combating it.

—THOMAS H. HUXLEY, *1894*

* * *

Different kinds of group-living have special significance in their relationship with other human groups, or with individuals that do not live in groups. We might find ourselves, in turn, in (1) small, two-generation families, (2) large families with multiple generations and large numbers of family members, (3) mixed, outbreeding groups—including unrelated or less closely related families that include members of other families—and eventually (4) mixtures of large numbers of individuals among distant relatives and non-relatives. As already noted, human groups appear at first to be peaceful cooperative members of tightly knit groups that rarely exhibit the negative attributes listed earlier. But when members of such groups are faced with hostile forces from other groups living relatively close by, intergroup strife obviously can break out and result in devastating conflicts. If natural selection tends to increase reproductive success, then we should expect to find ways to either avoid group conflicts or else identify (not necessarily consciously) other ways to use group conflicts to increase reproduction. This is not because we can only function to increase reproductive success; obviously, we can consciously reduce our own rates of reproduction. But the slow-acting, more or less unconscious changes that maximize reproductive success, even without extreme competitiveness and deliberate killing within our own species, may not be sufficiently obvious to us (e.g., Chagnon 1968; Kelly 2000, 2005; Boehm 2012; Pinker 2011; and many others).

All organisms may be expected to do whatever gives them the greatest possible advantage over other organisms that might compete with them, perhaps by displacing or banishing them, or by seeking to destroy them. In other words, any behaviors that disturb other organisms in ways that are sufficiently severe may identify the winner—and the loser—thereby influencing the results of the conflict. In these respects, humans exceed what non-human organisms can do because humans have evolved extreme abilities for considering the future, calculating, and pondering. They have evolved such abilities as consciousness, foresight, scenario-building, planning, and long-term memory. Such traits have enabled humans to increase continually their abilities to improve and expand almost every potentially advantageous capability of importance to other humans—some traits that we all admire and regard as our most distinctive and effective features, and others that we rarely if ever are able to consider such features consciously. Across time, individuals, and later small and large groups of humans, have been able to depend on their increasing capabilities in finding better ways to acquire food, shelter, and clothing, and to identify, improve, and learn how to create and modify weapons and other ways to win. The human ability continually to expand understanding and skills to overcome other organisms, and sometimes members of their own species, has enabled humans to evolve the ability to change in directions that create serious competitions against other members of their own species. As long as these competitive races continue, and concentrate on group-living, situations involving wars and their relatives may continue to become increasingly unsettled, along with either hostile neighboring groups or, in today's world, even distant groups. Such differences, no matter how trivial, have often become the machineries that have triggered, stimulated, and conducted the worst of our wars.

It is also obvious that humans, still living in groups, have expanded across virtually the entire earth. During these expansions, various barriers have separated families, clans, and other groups, often in ways that reduced or ostracized different groups. When long periods of time separated groups, people in different groups could differentiate, not merely in appearance, but also in their arrays of weapons and competitive strategies. These different groups were sometimes composed of people recognizable because of their historical backgrounds. Such differences could come about because small migrating groups had diverged as results of being isolated across lengthy periods of time.

Why do we apply such extraordinary varieties of different terms to modes of deliberate killings? Groups did not always combine, and, as noted earlier, were sometimes more likely to treat the members of different groups almost as if they represented distinct species. Examples come from such as the Europeans migrating to North America and Australia, where the immigrants believed that they had to attack and try to destroy the American Indians and Australian Aboriginals. Ember and Ember (1990) have estimated that 30 per cent of human males in early groups have been killed by other males. In these and in other instances, earlier and more distinctive or unusual inhabitants may have forced into extinction.

From early beginnings, humans have subsequently generated a virtually worldwide variety of large and small, stable and unstable, most often overlapping populations. It is this "sea" of people that has by now exceeded more than seven billion human individuals in the world. Such abundance of varyingly contiguous, overlapping, and separated populations is extremely complex in its overall structure. Nevertheless, virtually every kind of human population now can be identified and arranged from within the current world.

Modern people who function in groups have developed social control via numerous organizations: governments, religions, clubs, military and administrative organizations, unions, courts, teachers, farmers, businessmen, team game players, scientists, artists, musicians, families, academies, clans, nations, kindreds, protest groups, political societies, schools, selfish herds, and alliances of any of these and many others. The members of such groups continually change and divide, overlapping multiply, typically in many different arrangements. Such overlapping groups give opportunity for connections and cooperation of numerous groups to respond successfully to one another, and to the collection of groups as a unit. One result is that very large groups can be formed, as when modern nations compete and often combine to engage in warfare as a single unit. Such expansions can result in extensions of positive sociality and the potential for ultimate global harmony.

Note regarding positive sociality: The somewhat indefinite but frequently used concept of "altruism" implies that a gift-giver is prepared to contribute to others without reciprocation. But the shakiness and early termination of one-way flows of benefits make little sense, except to "free riders." Sense takes hold when gifts begin to be reciprocated—when altruism is converted to reciprocity. Reciprocating positive interactions tend to become the rule in successful social life. Perhaps we can simplify and clarify by considering somewhat different terminologies. Thus, *net-cost altruism* straightforwardly offers little or no likelihood of return benefits, as its name acknowledges; it is the definite version of a solid concept, but not a useful vehicle in the organization of sociality, and not likely to represent expanding and cooperative sociality. *Social investment,* as a replacement for the ambiguous "altruism" (used frequently in quotes because of its vagueness), is obviously prepared, and is expected to pass benefits to needy or willing reciprocators. Net cost altruism turns into social investment when either reciprocators or incidental observers of social investment decide to give or return benefits to social investors. Reciprocating partners, providing *return beneficence* to social investors, create return flows of benefits to apparent or obvious social investors. Social investments, in responses with return beneficence, can support extensive, or even vast, networks of continuing and ever-expanding positive sociality that may have the wherewithal to generate and expand pockets of cooperativeness and unification as prospects building toward global harmony. *Mutualism* is the term applied when individuals gain from uniting (or cooperating), or when members of different species do so, because in each of such cases the combined benefits of the involved individuals can contribute, without risk, to partners that produce reliably rewarding, useful evolved benefits.

MYSTERIES OF THE MIND, AND THE MUTING OF CONSCIOUSNESS

When people do spend most or all of their time peacefully, we are likely to ask two questions: First, why is it that wars and other serious conflicts seem always to exist, somewhere along the horizon, reminding us that even our most congenial friends may either succeed in war or disappear prematurely as a result of participation in war? Second, how and why do people often, or virtually all of the time, forget or ignore entirely the sometimes horrendous events that loom almost permanently somewhere along those personal horizons?

For a moment we can pause and remember that we humans have evolved in the same ways that other organisms have evolved. However, unlike the millions of individuals in species that maintain themselves despite diseases and predation, along with high rates of accidental or incidental deaths across the lifetimes of individuals, large numbers of humans are all too likely to die from deliberate attacks at the hands of members of their own species. Why is it that we are so adept at forgetting or ignoring the trials, terrors, and truths that affect our fellow humans, and ourselves?

Surely, not everyone is prepared to argue that natural selection can reduce sensitivity or recall specific knowledge of previous unpleasant or deadly activities affecting ourselves or others. Most likely, we think that we simply forget. Or such activities may be removed from our consciousness because they are unpleasant, or because we become convinced that those events were not nearly as traumatic as it seemed when they were happening. But consciousness muted by natural selection is not likely to be a simple error, and it can be a much more significant change than is easy to comprehend.

Not long ago Dr. Billy E. Frye, former Provost of the University of Michigan and Chancellor of Emory University, wrote to me as follows (I have paraphrased slightly in some places):

> *A recent note from an elderly friend reminded me that many people, even well educated people whom I respect very much and who have no difficulty*

accepting the idea of human evolution (vs. miraculous creation), still cling to some sort of metaphysical idea about the mystery of humans, or the mystery of life. On several occasions this question has come out in a context that makes it clear that at some level people resent, or regard as futile and undesirable, attempts to explain human nature in evolutionary terms. They seem to hold on to the notion that there is something spiritual about us that cannot and should not be "explained away" by evolutionary considerations. In at least some of these instances, perhaps all, it is clear that people like and are wedded to the idea of grand mysteries that will never be resolved—to the notion that there are things about us that we will never understand.

Two questions occur to me: First, why do people like the mystery of the unknown? In this case they clearly do not regard it simply as an exciting challenge. Is it because we can cling to the feeling that something is unchanging and will always be as it is now? If that is so, it strikes me that this is religion in its most basic essence. It seems to me that mystery in this sense may be a way of preserving as untouchable some sentiments about a spiritual realm without ever naming it as such, without becoming literal about religious belief in the way that, say, biblical fundamentalists do. I have the impression that some of us want our universe ultimately to be infinite and unchanging, and in a way the notion of the grand, unsolvable (perhaps spiritual?) mystery puts a finite boundary around the universe of our imaginations and our intellectual probings.

Second, is it possible that part of the resistance to acceptance of evolution by many social scientists and humanists originates in this kind of desire for mystery—for unsolvable, unapproachable mystery? In support of this notion I will only observe that at least some of those who have resisted the recent emergence of evolutionary explanations have exhibited overt resentment, and even hostility and anger, toward both these ideas and the people who expound them, and sometimes have openly scoffed at the very idea of trying to explain the "inexplicable mystery" of humans. It is in this sense that I ask, why do we cling to mystery with so much passion? (Dr. Billy E. Frye, personal communication)

My first response to Professor Frye's relevant and provocative comments is that everyone may, in fact, have some realization, expectation, desire, and capacity for mystery. Perhaps that is part of the feeling expressed by a minister who declared that he was not interested in a concept but the real thing—the real God. Perhaps he meant, at least in part, "a persisting and unchanging mystery," perhaps the concept of God that has been accepted and formed mentally as matching the understanding of the minister's own life, and seeming to hover continually over all of humanity. In my view it is part of the argument for acceptance of diverse concepts of God that our feelings about the mysteries surrounding religion—and our uses of such concepts as God, Eternity, Heaven, Hell, and Purgatory—deserve the attention of anyone interested in the topics of this essay.

For me the next question is whether there is a relationship between the mysteries of religion and its various contexts, and the everyday activities, thoughts, and wonderings associated with the life activities of humans, in particular their kin interactions and other kinds of social affiliations. I think there is such a relationship, and I also think it is not entirely a consciously accepted or analyzed phenomenon—hence, perhaps, in that sense it may remain a mystery that can potentially serve the combined interests of all people via a universal or global version of God—or via prominent and influential individuals or groups that have incorporated such universal or global mysteries under the rubric of a supernatural God.

We are strange organisms, both immensely social and intensely confrontational within our own species—indeed, within our own social groups. It is as if humans have evolved to maximize two phenotypes, in the way that thousands of organisms have been able to emerge as either of two (or more!) different phenotypes, sufficiently distinctive that the different forms often have been regarded as different species. Humans, too, have evolved so as to be knotted up in the differences and significance of dualities—not merely war and peace, but such as conscious and unconscious, honest and dishonest, other-deception and self-deception, and with kin-group and friend-helping of our "in" or "we" group, as contrasted with the competitive and adversarial "they" parts of other groups—we-they or amity-enmity axes. Biologists have been able to explain the dualities and otherwise different phenotypes in terms of environmental differences, but they have not yet clearly and completely explained how humans can change themselves more or less suddenly from gentle, helpful, congenial people into extreme military warriors.

Part of the reason for thinking about mysteries derives from the fact that the kin circle or any social group is itself likely to be somewhat mysterious, perhaps because it is a social unit composed of numerous genetically or socially connected individuals. Trying to cause any such social group to function as a unit can involve many kinds of agreements and disagreements, influenced by overlapping social or kin relations, and reflecting as well the hierarchically arranged organizations and other elaborations of religious demands.

Part of the mystery of such social groups exists because, other things being equal, humans are evolved to treat their kin or their associates according to the extent of their social history or their overlap in genes identical by immediate descent; but the people involved may or may not be entirely aware of the relationship. Considering kin, we can know the order by which familiar relatives are in fact related to us; and we can be conscious that the more generational links that descend between us, the more distant is the genetic relationship. In this sense, and in the sense that a kin group consists of genetic relatives, the mystery is not so deep. But, to repeat, a good deal of what makes a kin or other social group function surely is not consciously grasped by all of its members.

When one considers these topics, it becomes obvious that we humans are always thriving on mystery—in part at least mysteries within ourselves that we typically do not wish to reveal completely. In some sense those mysteries relate back to that "nasty hoard" in the "secret cellar" that Stanley Elkin (1993) mentioned when commenting on the incompleteness of all autobiographies, and what we don't want to reveal (cf. Alexander 2010). And we are back to the fact that we do not always know what to do or why we (or religious leaders) do certain things. We don't know consciously how much or what to reveal to others, not even to close kin, and not even if we sense unconsciously what maximizes the quantification that was never known to us before the recent rise of the sciences of biology and evolution—and that quantification is a consistency in the behavior of genes, termed the maximizing of inclusive fitness. There are many situations, indeed, in which the biological principles of organic evolution are the only way to explain how organisms, including the human organism, can be understood.

People may want the concept of God to be true, at least partly so, because God is a concept that has come to represent the holder of secrets about ourselves— actually, is typically considered the keeper of the secrets—the source of confidentiality, the reasons for almost every aspect of accepted knowledge, and the means of making everything right for us when the time comes. I have used "secrets" here in a very broad way, but the word "mystery" is probably better. Take that concept away, and we may believe we have somehow diminished the quality of existence. Kin groups can have similar effects (leave aside, for now, the rest of our lifetime associates). Examples can be a slight or great mystery, a "storehouse" of knowledge that can be used either against or for nearly anyone in the group, depending on circumstances and the interactions of members of the kin group. Perhaps it is merely that we are acutely aware of our continuing inability to enter, absorb, and fully understand the workings of the minds of our individual and congenial companions, not to mention hordes of acquaintances, passersby, and complete strangers. After all, human individuals function constantly, necessarily mysteriously, and in domineering and moral elaborations within the continuing performances of their own uniquely extensive, and always to some extent, private desires and intellects; their extensive, complicated, and long-lasting memories surely function unlike the members of all other species.

I think there is a way to understand why we don't want ourselves to be entirely non-mysterious, even for the people to whom we are closest, and certainly with respect to everyone else. Our hesitance can be partly as if, should we reveal ourselves to anyone (or everyone) completely, we would also be "completely" vulnerable—for example, to reputational derogation, whether justifiable or not (we are good at manufacturing the unjustifiable kind!); and scorn and the opportunity for all of our associates and potential enemies to threaten our ruin by telling our enemies about everything negative or demeaning about us that they can muster, along routes that affect us in the worst possible ways. Perhaps some people are convinced that science will "out" all of their personal privacies, and all of the special (good and bad) things they think about themselves. It becomes a global proposition to protect ourselves from that kind of competitiveness and one-upmanship, although protection is also available because of revealed knowledge of demeaning actions susceptible to open disclosures.

I think people see as threatening—direly threatening—any possibility of complete unraveling of our individual complexities, such as many people may think will happen if scientists keep probing the nature of humans. My view, however, is that the probing can never be completed, and one reason is that the probing—the continuing exploration within the human mind—will inevitably extend, with hope, and beneficially, the array of cloistered, sequestered, changing mental strategies and subterfuges. This is a pattern that returns us to the realization that we have evolved to compete—literally to outcompete others attempting to outcompete ourselves.

It may be slightly disheartening that individuals of any species have evolved to control or outdo other individuals, despite group cooperation, and in humans, despite denials. *After all, the intentions and consequences deriving from the evolutionary selection that continually influences us all as individuals are, in some way, whether clever or clumsy, evolved to outcompete any and every other associated individual that does not provide return beneficence satisfying the silent and potentially negative or critical thoughts of affiliative but also secretive individuals.* This is a basic relationship of all individual humans, whether or not it is recognized or accepted widely, and despite the parallelism in the congeniality of profitable group-living. It is a relationship that may explain the universality of both human mysteries in general, and the consequences of utilization of such mysteries by those who may seek reasons for promoting group behaviors that can develop hierarchically domineering and extravagant ceremonies. Included are the almost universal establishments of lavish and ostentatious houses of worship, generated and operated by individuals behaving as morally superior leaders representing themselves as God's close aides and collaborators—and with little doubt considering themselves to be special individuals interpreting messages transmitted directly and frequently from a supernatural God.

I do not wish that the aspects of religion just described, or religion in general, should wane or disappear. Instead I wish that church officials and members alike can begin to regard themselves and their particular religions and activities as potentially students of social and universal success, as results of science as well as religions, and including concepts of a universal God that serves a universality of social success, and does not fragment the world of religion into prideful factions, sometimes bitterly disagreeing and war-like.

In any case, it seems clear that humans demonstrate multiple features that are collectively relevant to the questions of religion: mystery, human consciousness and its changes, and human competition and cooperation. Natural selection has caused human individuals not only to work together in appropriate circumstances, but as well to compete with other individuals, and to compete group-against-group. Competition among humans has surely caused natural selection to continually expand and magnify human mental activity. Unlike most other species, and in some ways like all other species, humans have become capable of unique, complex, and continual mental activities. Some of such mental understanding and activities of humans are likely to be withheld from other humans, withholdings that maintain and influence much of the information in the mental activity of humans. This combination of hidden human traits will always create and maintain mystery. As long as the brains of humans continue to expand their views, and their understanding of themselves and their associates and competitors, no one will escape the presence and continuance of mystery—the mystery of personally functioning human minds, working both singly and collectively in their own local environments.

It therefore seems entirely possible that mysteries which puzzle us, and confront us continually, are results of (1) the competitiveness of everyday ordinary individual humans, and (2) congenial groups of cooperative humans and congenial groups of competitive humans. Even more fascinating is the probability that the current mysteries of humanity were absorbed and became—or expanded into—the mysteries that have generated and continued to elaborate the basis of religion and the concept of God.

It should not be surprising that people, as individuals or in groups, have been ready to accept special individuals with skills to claim knowing of attractive and unworldly mysteries, presumably from their own experiences and backgrounds; special individuals that use their special qualities to connect with people that in some circumstances wish to control while making others into subordinate beings, thereby generating organized systems of domination, explicitly glorifying differently special individuals or groups that can be made unusually useful. Can we deny that virtually the entire program of religion could have originated from the value of its organization of the various groups that have spread across the planet? How might we discover whether such effects have served the elevated and impressive ceremonies that have persisted and spread since, explicitly because they were attracted to the mysteries that have been carried always within the souls and consciences of humans? Is it not possible that these arguments, these not merely particular but as well universal human mysteries, may be constantly reflecting a mixed combination of secular and formally religious backgrounds that together could become parallel services across the entire population of the world?

* * *

> Humans collect within their brains all manners of learning that they hold on to indefinitely, without necessarily thinking about them—items of knowledge that people can bring back to their consciousness, but do not always do so. Examples are foresight, memory, cognition, reflecting, itemizing, organizing, storing, combining, interpreting, scenario-building, and many additional intellectual capabilities that can remain functional across impressive portions of the durations of our active lives.

* * *

WARS, GENOCIDES, AND MURDERS

The Most Important Failures of Humanity, and What Can Be Done about Them

As readers look through this essay, I would wish to be able to ask a series of questions about the topic of this section. Specifically, I would like readers to (1) ask themselves to estimate the numbers of population deaths in various different situations known to involve wars or genocides (for examples, see below); and (2) continue to ask other people the same question. See for yourselves the accuracy, or absence of accuracy, from those situations, and consider reflecting on the topic of muted consciousness (see also, Pinker 2011).

* * *

Humans are collectively, uniquely, and by far the most ferocious, hyper-competitive, hyper-patriotic, war-mongering, genocidal, and murderous species on earth. Unlike all other species, we humans have become (or maintained ourselves as) our own worst enemies. We have made it so by dramatically diminishing the significance of other-species enemies and increasing the numbers, crowding, competitiveness, and conflicts of members of the human species. As an example of one such consequence, one study of deaths from wars and genocides alone, during the twentieth century, has yielded a maximum figure of 150–160 million such deaths, averaging approximately 4,000 deaths per day—counting every day of every year across those 100 successive years (Scaruffi 2006; see also Goldhagen 2009; Pinker 2011). During the United States' Civil War, a war with far fewer citizens, war deaths were calculated at 618,222 (Encyclopedia Americana).

* * *

In the Ann Arbor News of March 29, 2012, The United Nations has estimated that more than 9,000 individuals have been killed in recent continuing disagreements between the Syrian dictatorship and its citizens. The UN has since increased the death count in Syria's civil war to at least 60,000 people.

* * *

On the morning that I wrote this paragraph, 74 people were reported to have died in Egypt during a relatively brief turmoil involving primarily a single controversial soccer game.

* * *

The city of Detroit, Michigan, reported on ABC television that, during the past year, 344 people were murdered—only 21 fewer than one person every day across the entire year.

* * *

Kill a man, and you are an assassin. Kill millions of men, and you are a conquerer. Kill everyone, and you are a god.
—BEILBY PORTEUS *(1731–1809) Bishop of London*

* * *

A hydrogen bomb is an example of mankind's enormous capacity for friendly cooperation. Its construction requires an intricate network of human teams, all working with single-minded devotion toward a common goal. Let us pause and savor the glow of self-congratulation we deserve for belonging to such an intelligent and sociable species.
—ROBERT S. BIGELOW, 1969: *The Dawn Warriors*

* * *

Movies, operas, fictional and nonfictional stories, books, and other accounts and documents demonstrate the hyper-competitiveness and hyper-patriotism that reflect the duality of geniality and ferocity, of loving and hating, which seem to be maintained, and in certain ways are not only considered acceptable, but are relished, at least sometimes, in the minds of most humans. Over and above wars, all across the United States, the morning news lists small and large numbers of recent deaths by murders of single individuals, families, or other small numbers of people, accomplished with definite purpose, despite the strange fact that most publicly revealed felony murderers have to realize, prior to their decision to kill, that they are predictably doomed to lifelong incarceration—or else lose their awareness of their oncoming mistake(s).

* * *

How did this situation come about? What are the continuing reasons for it? How can it continue to be prevalent? Why do so many people tune in every day to trace ongoing, potential, and fictional(!) wars, and to hear about the details of morning and evening single or group murders, and attempts at murder? Why do humans multiply and sustain heinous and despicable behaviors, emphasizing and detailing endless varieties of films, fictional programs, and other performances in which multiple murders are made to appear as horrific and chilling as possible, using words like slaughter, assassination, lynching, decapitation, throat cutting, strangling, suffocation, drowning, bludgeoning, poisoning, disembowelment, serial killings, and many other incredibly diverse and sickening ways to vilify, nauseate, terrify, and destroy members of our own species? Why do real and imaginary happenings appeal so completely as to make even fictitious horror publications, movies, and shows simultaneously seem both devastating and thrilling? On what basis can we imagine movement toward a global harmony when so much of our interest is linked so frequently to such almost constantly negative, yet thrilling experiences?

Consider the great number of old-time western films, or "cowboy movies," in which nearly every actor is likely to carry ostentatiously at least two guns, and maybe knives as well, and in which small and large numbers of seemingly unfortunate actors and actresses are taught to pretend that they are being slaughtered in the most horrendous fashions. Why are virtually all such performances so persistent, and so incredibly and abundantly popular? Is it simply because people have acquired deadly weapons, and have learned with practice, and in the absence of sufficient legal control? Or is it that they believe they can slaughter those who are apparently or potentially competitive, unfriendly, or vulnerable? What has caused the success that seems to represent repulsive, disgusting, and appalling aspects of human nature? Why do we apply such extraordinary varieties of different terms to modes of deliberate killings? Why do such attitudes continue to persist, and to be regarded as honorable or laudable? How do we also manage to erase temporarily, and so easily and quickly, the horrendous examples laid out before us in the news media and in nearly all forms of literature and diverse performances? Is it merely that the expansion of deadly weapons has magnified the likelihood of besting individuals and groups in particular situations to motivate use of every instrument and method to outdo other competitive individuals and groups?

I suggest that this situation can be partly understood by considering together several major aspects of human performance linked to events or circumstances that, evidently across most or all of human history, have generated combinations of human traits and tendencies that bring together diverse components of human nature, and produce the current and now essentially universal and deplorable human situations. Continual expansion of increasingly devastating weapons provides reasons for extending methods and plans for fighting individuals and groups, today including huge and sometimes multiple nations coming together to create incredible havoc in war. Differences in power and motivation from changing weaponry and strategies can continually reinforce tendencies to bully, attack, batter, kidnap, rape, eject, banish, and kill potential or actual competitors. Is it really an exaggeration to say that all we have to do in group-against group confrontations is to change to our rules and proceed?

Aside from bonding of male-female interactions, parent-offspring and other kin relations, and the clustering together of non-relatives in the interest of saving themselves via the strength of their group against other hostile groups, the members of all species seem to have evolved to do their best to outcompete all others in one way or another within their vicinity. It would appear that the members of no species anywhere can even remotely match the "grandly" intelligent members of the human species with respect to continually revising and rendering weapons and strategies increasingly more devastating, and magnifying temptations to kill, bully, or drive away other humans.

Team Competitions in Sports

If humans have become their own principal hostile force of nature, we might easily understand how humans have also become the only species known to play competitively, group against group, in team sports—in groups that win or lose as groups (Alexander 1967–2011). Team sports are forms of play. Most biologists regard play as having evolved as practice—practice for something other than play—instead, for the "real thing," which in this case can only be competition and conflict, perhaps—ultimately—direct or indirect contributions to warfare between human groups.

It is not easy to deny that the raging excitement of such games, including, for example, deliberate distractions by fans (enthusiasts) with loud noises and raucous actions, for example, rapidly flapping devices that create a virtual "field" of noise and confusion in front of a player who is attempting "free"(!) throws. Such devices are designed to confuse and thwart players on the other team.

Teams competing in sports are exhorted to think only of the team, not of themselves as individuals. That is precisely what takes place in the military, in which individuals are dramatically, necessarily, and cleverly (especially when war is imminent or current) subordinated to the service of their functional units and trained to perform as members of closely-knit, single-purpose groups. They are individuals prepared to give their lives to save the group, all of it or part of it. In some sport arenas, the intensity of team sports results in more than 100,000 on-site spectators.

We can consider whether it is possible for widespread and intense team sports to remain true to their apparent original purpose. One question is how to generate such considerations in the interest of promoting social harmony. Games and other activities, including both individual and team sports, may have replaced and reduced the incidence of serious, intense, and negative forms of competition. But hateful conflict, too-frequent killings, and other negative encounters—the latter curiously more often by over-zealous fans than by team members, and with equally avid appetites for increasingly horrendous cinema, radio, and literature—have shown little evidence of receding or disappearing from competitive team sports, or from shocking or gruesome theater in all of its variations.

It might be enlightening to rank the frequencies of incidental or accidental injuries to players and fans in the different sports of basketball, football, hockey, and soccer, and compare them with the frequency and severity of injuries from physical altercations to players and fans. A significant example is the recent report of instructions, or tutelage, by coaches, about pre-arranged attractive cash rewards, offered to professional football players who are then expected deliberately to cause disabling injuries to opponents.

Surely, we all hope and expect that the positive excitement and enjoyment of team sports will continue, and thrive, and that serious team and spectator tensions will continue to identify methods and reasons to soften and adjust their actions without eliminating the excitement and enjoyment of competition. There are reasons for team sports to minimize the potential for generating extreme actions and attitudes that expand and direct the shadows of serious conflict and warfare. It seems possible—and of great significance—that, with care and thoughtfulness, large increases in sporting competitions, on international scales, can become extraordinarily valuable contributions to continuations of positive cooperative and competitive sports, and to reduction or elimination of seriously negative consequences.

Humans Have Created Their Own Principal Hostile Forces of Nature

Unlike members of other species, humans living in groups are extraordinarily— probably in some ways uniquely—competitive within their own species. Wars, genocides, murder, bullying, injustices, dishonesty, deceit, fraud—all such results of hyper-competitive, hyper-patriotic behavior—are "hyper-prevalent" in humans. Again, we are reminded that natural selection tends to maximize reproduction, and that we accomplish this by as nearly as possible out-performing everyone but ourselves.

Following, in some respects, Sir Arthur Keith (1949) and Robert S. Bigelow (1969), I have suggested that modern humans have, for better or for worse, considered other humans, especially in groups, to be their most important hostile forces of nature. Part of my argument began with the possibility that only this feature can explain why the expensive human brain has evolved—or persistently increased in size and complexity—so far beyond the brain sizes of its closest primate relatives (cf. Alexander 1967, 1968, 1979, 1987, 1990, 2008). If this argument is correct, then we can see another way that humans living in groups can be behaving selfishly when they save or help members of their own groups but not those in other groups. When we realize the rates at which people have been killed in wars and genocides, even during the 20th century alone, the emergence of larger human groups and more deadly and catastrophic weapons may become more devastating than we might have anticipated (e.g., Ember and Ember 1990, Scaruffi 2006, Goldhagen 2009, Pinker 2011). If we turn away and simply accept the manner in which differential reproduction functions—and if selflessness, cooperativeness, and social reciprocity do not function so as to build social harmony—we will be yielding to the relentless slow changes of natural selection, and failing in the effort to enhance the possibility of roles for effective social investment and return beneficence in our societies.

Balance-of-power races between different groups of the same species may be uniquely important in human evolution (Alexander 1979, 1987, 1990, 1993, 2005). The particular forms and sizes of group-living to which humans have been driven by this unusual situation have caused dramatic changes in entire collections of related traits, suggesting that whatever caused the human kind of group-living were dire and continuing threats. The organ most dramatically affected is likely the human brain, because of the usefulness of social cleverness and expertise in interactions among individuals of the same species.

SCIENCE AND RELIGION

In my view, it is not necessary, or useful, to disparage the concept of faith. All of us conduct our lives with a large complement of faith. There is no effective alternative because we simply do not have the means to learn for ourselves about everything we would like to understand. The difficulty with the concept of faith is that, particularly in religious contexts, most expressions of it seem to require absoluteness. Thus, by claiming a basis in eternal verity deriving from an unchangeable supernatural source, faith sometimes takes forms that call for permanent incontrovertibility. This assumption is closely related to permanent assumptions of value in the moral rules of humanity. It inevitably becomes adversarial to all of the ways—all of the formal and informal procedures—according to which we learn by exploring the nature of the physical and living universe. Science, whether formal or informal, necessarily changes continually as knowledge accumulates, while moral rules tend to be enforced and maintained, therefore tend to be lasting and unchanging, sometimes by declaration, whether right or wrong. This, and the difference between morality and reality, is a source of conflicts between a science of discovery and a faith supporting agents of supposed unchangeable immortality.

Scientific procedures are effective because they eschew attitudes of "blind faith" in favor of testing repeatedly—and continuing testing—until results are obtained that can be verified by repetition. But we do this not merely in formal scientific investigations but also in all of our ordinary activities. In everyday life even the most devout persons employ "scientific method" almost continually. When problems are encountered, we generate guesses, or ideas, about possible answers, then do whatever is necessary to see if the idea is correct. If it is demonstrated to be in error, a new idea has to be generated and tested. If we are preparing a meal, we may taste the results repeatedly so as to try to improve what is cooking, testing and retesting to get the desired results. Exactly the same procedure is used in trying to find out why our automobile has stopped running, or why a child is brought to a physician to analyze and diminish or remove a serious ailment. As soon as an idea passes all the tests, one by one, the person doing the testing can proceed, using the new knowledge. That is precisely the kind of procedure that is used by scientists at work. Typically, this simply means generating the next hypothesis, or guess—as suggested by the Nobel Prize–winning physicist, Richard Feynman (see Sykes 1994)—and continuing, or else starting the whole testing process again. The more complicated the problem, the more complicated the process of generating worthwhile possibilities and testing them. Willingness to seek alternative explanations, and to revise our understanding of every investigatory process are the hallmarks of success in finding the truth—of discovering how to "get things straight." As already suggested, insistence that faith has to be absolute and permanently unchanging because it derives from supernatural forces—and that science is therefore the enemy of faith (and in this sense, the enemy of rigid and unchanging authority—derived morality)—is probably the most serious conflict between religious and non-religious practices (see the extensive and provocative discussions by Dennett 2006, on religion as a natural phenomenon). This set of conflicts is likely to increase as scientific knowledge expands, while religious faith may risk clinging to long-continued but perhaps no longer acceptable claims.

THE MEANING OF LIFE

Acceptance of the arguments made earlier enables us to repeat two related propositions about the effort to characterize God and the mysteries of humanity: (1) the concept of God as an outcome or consequence of human evolution, derived from the flow of human intelligence; and (2) the meaning (or purpose) of life can be to serve, and probably always has served, as what may be termed a universally acceptable metaphorical spirit, by many thought of as God, and generated via the mystery that is the "hoard" of more or less unstated and unrevealed knowledge that persists, each with its own uniqueness, within every human being's special mind and imagination. The first-numbered statement above has been characterized in such a way as to clarify what is meant. The second statement, however, deserves further discussion.

When the concept of meaning is explored, as by returning to the dictionary (repeatedly!), one finds it treated as mainly synonymous with purpose, aim, and intent. In other words, meaning is virtually always treated as anthropomorphic—as requiring motivation of the sort that humans exhibit when they set out consciously to achieve something, projecting alternative behaviors into the imagined future in a testing way. If one asks "What is the meaning of life?" the answer will be quite different if "life" is meant to refer to all of human life, as opposed to the individual life of a human being. We cannot easily answer the question of meaning for non-human forms of life or nonliving parts of the universe unless or until we accept the idea that the same concept that "created" humanity (using either of the common meanings of humanity) also created the physical universe and non-human life. Otherwise nothing could give meaning to non-living objects or non-human life, other than evolved function *per se,* to all indications never conscious or self-perceived except in ourselves. This is so because, even if we can work out evolved function, we cannot easily translate into intent or purpose the ways in which nonhuman forms of life, especially those apparently lacking anything even remotely resembling consciousness, use environmental stimuli to respond appropriately to events as they arise. As a concept associated with conscious thinking in humans, meaning is possibly without any counterpart among non-human living forms. Only human life can confidently be said to have conscious purpose—or at least very much of it.

Purpose could be ascribed to a human life from two backgrounds. First, I might ask myself what is the meaning of my personal life. In some sense I could make it anything I please. But I suggest we would all agree this would be a difficult attitude to assume and maintain. I think we have to assume that this difficulty is a consequence of the tendencies and capacities allotted to us by our evolutionary history. Thus, it is likely that difficulty would arise because we are not evolved to have such a motivation, or to give such an intent or purpose to our lives. We may not be evolved to answer the question of personal meaning at all, at least in any general or long-term sense—even if we seek it plaintively sometimes. On the other hand many of us have been exhorted at one time or another to give our lives to God—to consider our lives as service to God. But it is difficult to do this—to imagine any meaning for life—unless we accept that an anthropomorphic spirit or attitude formed our lives, or governs them—said differently, created a universal phenomenon, embodying intent among ourselves. This requirement can be met, or accepted and used, not only potentially by a traditional, personified, and in our best thinking, universal God; but also by the spirit or thought process generated out of the human moral capacity, historically residing in the collective minds of the kin group or minimal defensive unit, and expressed from that collection of minds.

Second, and following arguments given earlier, the purpose or meaning of human life as a more general proposition becomes a question. Purpose is something that individuals have so that, as said, we can think about our own lives as having whatever purpose we choose to give to ourselves, whatever purpose we can mount under the *a priori* thrust given to our lives by our personal experiences and by the evolutionary process (*a priori* in the sense of pre-dating or using our personal consciousness to inject purpose or intent). We can also consider the purpose of any outcomes of confluences of interest with other individuals, or group interests, to which we subscribe or accede. Nathan Hale must have been an extreme example when he supposedly said: "I regret that I have but one life to give for my country."

So life *per se* (or in general) probably never had anything legitimately termed a *conscious* purpose until we generated something that could give such purpose to it. As already noted, that "something" was probably generated as an anthropomorphic cause for life—a Creator, designated or named as "God." It is a spirit that gives purpose to life—for many or most people now, but potentially, and most effectively, for all people, as a universal concept (See also, Dennett 2006).

We can wonder if portions of the arguments made here were sufficiently available to the psyches of an ample number of humans across the appropriate portions of history for them to realize consciously that they might have to come up with an anthropomorphic God if they wanted to assure themselves (or others—e.g., potential Nathan Hales) that life in general has purpose—in the same sense as human purpose. That is, there had to be purpose from some source other than the "mereness" of ourselves as individuals or the slightly larger mereness of our local groups—the competitive (defensive, aggressive, warring) and cooperative units.

Further testing of the approach presented here lies partly in persistent efforts over long times to apply the proposed concept in every life situation: (1) to ask in every instance in which God is mentioned or invoked how apt in that instance is the concept of a universal spirit of beneficence; and (2) to see if the concept of such a universal spirit ever becomes impossible, or how often it is more awkward or more reasonable than its alternatives: whether it appears to explain each and every situation, excepting those which turn out upon closer examination to have been based on factual error.

Understanding of the group morality concept of God, and its effective use, also requires that we seek a better understanding of the moral capacity—or its expression as conscience or as willingness to invest socially (and even to hone such willingness by practice in solitary or in anonymity)—that surely characterizes all humans everywhere. Assuming that as individuals we start out socially naive, how do we change under different social stimuli so as to mature what is likely to be regarded as a "normal" concept and effective practice of morality? How can we accomplish this increased understanding so as to promote a concept of God (or a set of such concepts) that can be truly universal? How can we move toward world peace rather than, ironically, primarily enlarging the sizes and frightening capabilities of a smaller number of more horrifically willing and aggressive nations? We need to reflect on how difficult it might be to deny that this is the most we wish to understand about what God and democracy-loving folks have been able to accomplish, while continuing to lack significant prospects of universal peace from the directions taken so far.

I have not discussed the deterioration that causes the invariable termination of the human lifetime. It is well known from biological science that this deterioration results from complex changes in the survivability of the genes and cells of organisms under negative environmental influences across the later parts of life, and that reproductive potential loses its effectiveness as the organism ages (cf. Williams and Williams 1957; Williams 1966; Alexander 1987; and many others). It would seem that being aware of the sad certainty of the end of life may have accounted for the ancient religious belief that death could be turned into a mere transition to a supernatural place called heaven. As biological studies continue on an ever-broadening scale, and are increasingly well understood among the complexities of the evolutionary process, the connections and relationships between the organization of life's *recently analyzable processes* and the assumptions of religion, *generated centuries ago*, will certainly continue to modify our views of life.

MORALITY AND THE EFFORT TO PROMOTE UNIVERSALITY OF GOD

The evolutionary approach to human behavior continues to have serious problems. As suggested earlier, the most obvious one is that evolution is accepted by only a minuscule proportion of the world's population. Even within that minuscule proportion, too many academicians and intelligentsia tend to wall evolutionists off as if they were malignant tumors. This happens partly because only a minority of thoughtful people are extensively educated in biology, and this in turn is partly because the interaction of heredity and development (the generation of the individual) is still poorly understood. Also contributory is the convoluted nature of human efforts to self-understand, not only because we must use the properties we wish to understand to carry out the analysis, but as well the nature of the biological history that few wish to contemplate. It is not easy for anyone to believe, from his or her thoughts about personal motivation and that of other humans, that humans are designed by natural selection to seek their own interests, let alone to maximize their own genetic reproduction. Natural selection appears to have designed human motivation in social matters in such fashion as to cause its understanding to be resisted powerfully.

We lose in analyzing such problems, if we restrict ourselves to discussing only the brighter side of human nature or pretend that the topic is cooperation, and not competition as well. Some moral philosophers and other academicians seem to travel mainly in pleasant worlds, as if with little opaque clouds that tend to admit only the delightful aspects of human intentionality (or their shadows) floating above their heads as they move along the sidewalks of Urbania between their offices and their homes. But the misery in the world is not all there because of pathologies easy to understand or proximate causes easy to remedy. Nor is it all owing to those "other" kinds of people whose motivations (unlike our own, of course!) are sometimes pernicious and self-serving. Moreover, technology and civilization (weapons and war) have created circumstances in which virtually all human striving, designed as it is to better the current quality of life, nevertheless continues to threaten increasingly the future of humans, even that of our planet and life itself.

Analysts of morality must retreat from their subject far enough to examine the reasons for its convolution. We will gain from understanding that the kindness, beneficence, and good fellowship that occurs or remains only locally tends to be selfish, and we must also understand why the idea is repugnant and what to do about that. To solve the problems that human evolutionists have glimpsed so far, we will gain from enlisting a far greater proportion of the world's thinkers. If, as knowledgeable people increasingly suggest, massive beneficence by our generations will be required to ensure the survival of later generations, then unless we don't care we have to know how to reverse the relevant aspects of the striving we have evolved to accomplish. We have to know how to use the fact that no part of biological theory has ever legitimately implied that humans cannot employ their evolution-given traits to set and accomplish goals that are entirely incidental—even contrary—to negative aspects of their history of natural selection. I would suggest that these things will happen only when evolution-minded people have overcome resistance to evolutionary analysis of behavior by explaining, much better than has been accomplished so far, the nature of human motivation and the reasons for its partial concealment and seeming withdrawal.

The capacity to generate an effective moral sense is surely a criterion for all-privileges membership in our species. Suppose that the concept of God does indeed arise out of cooperation and good will, and our reciprocating confidence in the existence of the people involved; and suppose that in the mind of every individual the concept actually stands for that collective of cooperation and good will (whether entirely consciously or not). This would mean that beneficence and charity derive from our sensing of the social power, will, and value of, not an extrinsic human-like being, but the mind (and minds) of the collective itself.

Surely, then, it is a reasonable speculation that the existence and presence of the concept of God—or the overall effect of the existence of a universal moral capacity in the human psyche—created humanity at the "moment" across history when that universal moral capacity, and the appropriate responses to it, became a reality within some of our ancestors, and was on its way to becoming a functional potential within all of us. Perhaps the concept of God was cemented into kin groups all across the human species as a result of feedback from a spreading grasp of the universality of the human moral capacity, even if that understanding was never fully conscious. Perhaps this is the sense in which God (the concept of God) can result in an unusually broad understanding of universality in humanity.

Anyone taking this view can cheerfully—even enthusiastically—accept virtually all phrases in which people currently include the term "God" in social circumstances. Examples are "Thank God," "God bless," "In God's eyes," "With God's help," or "One nation, under God" (more appropriately, from the reasoning being promoted here, "One *World,* under God"—see also terms earlier in this essay). For example, using the above concepts, it would matter little how explicitly thankfulness seems to be attributed to God by an individual for his or her own personal success in any socially significant activity or competition. Whenever, as individuals or subgroups, we invoke or thank a universal God, we would not be acknowledging the will of an entity that for some arcane reason enabled us to win by favoring us personally over all others. We would be hoping to be acknowledging or thanking the collective morality and good will of "All the People." We would be recognizing the value of the fellowship that characterizes our entire social group, including, again, "All the People." We would in a sense be thanking everyone, or everyone in the group producing or overseeing the particular activity or competition under consideration. It is surely difficult to fault anyone for thanking a large number of people collectively for all of their different contributions that happen to bring about a particular successful outcome. The effect is instead likely to be heart-warming and reassuring, and I daresay that this is what people who view themselves as cooperative believers, or who are for other reasons unusually tolerant, tend (and wish) to experience in such situations. Thanking God as the spirit of collective morality would become a unifying and reassuring statement, as it demonstrates itself, rather than a potentially divisive one implying competition for favorable attention from a God that in any way bestows special favors on selected individuals.

The situation just described, though it could include bitter competition between individuals, teams, or cliques, would not meet the requirements of the hypothesis being developed here unless it were part of a larger group interaction that—unlike interactions between individual competitors—could be characterized by a potentially over-riding feeling of unity and cooperativeness among its members. Such a local group would in turn have to be one of several or many competitive groups. It becomes immediately interesting that this problem of how social collectives are constructed and maintained, and how such collectives interact with each other, is the most important and—for some—potentially dismaying consequence of the characterization of God here being described. Yet this problem is consistent with the failure so far of humanity to succeed in promoting the concept of a unified and universal God theme for all humans—or more precisely, since all humans live in social groups, for all of the various internally unified groups of humans on earth.

The apparent match of a large array of terms used traditionally to characterize God, the nature of which might be called a universal concept of moral capacity unique to humans, and the additional match between this hypothesis and the desperate difficulty in using the concept of God to work steadily toward world-wide harmony, may alone demonstrate and justify proceeding further with hypotheses such as those presented here.

I hope it is abundantly clear that I do not seek to disparage the concept of God, at least when it can be understood in the ways described earlier. I began this pursuit with a strong bias that God has to be real in some important (if unusual, unique, and, with hope, universal) sense, that the most important thing to learn is precisely what that sense is, and how the concept of God can apply universally, and merge with the rest of humanity. I also began with a bias that the concept of God, and the views and approaches that underlie it, are central to human existence and human endeavors, have been evolutionarily adaptive to the humans that have appropriated them, and must be clarified if we are to understand ourselves. Obviously, it is my view that the most useful and acceptable version of God is universal, and it ought not to be limited to narrow and proscriptive claims. Nevertheless, considering God as universal raises questions about the likelihood and effectiveness of any version of our efforts to build and support universal beneficence and cooperativeness.

ARE GOD AND EVOLUTION IN CONFLICT?
A Response to The Reverend Gordon Hyslop

I share The Reverend Gordon Hyslop's feeling (*Ann Arbor News,* February 14, 2007) that viewing a new baby grandchild is an awesome experience, and that the human body is indeed a marvel. But one does not have to set the magic and intricacy of human beings against anything at all, as in his statement: "A human life is a miracle from God, not a process of evolution." I would like to convince Reverend Hyslop, and everyone else, that it is not useful to think of God and science as adversarial to one another.

Our evolutionary background and our religion are together anciently responsible for both social cooperation and the extremes of social competitiveness, including internecine battles that have caused immeasurable pain, misery, and suffering all over the world. Perhaps, to reduce serious conflicts and expand harmony, we should join rather than separate our knowledge of evolution and our religious attitudes and beliefs.

As an evolutionary biologist, I have sought for more than 50 years to understand better how evolution has primed our behavior for particular situations. It is a difficult problem, partly because much of our knowledge is not automatically conscious, or has only recently been made conscious. For example, much of what Reverend Hyslop cites—or can cite—as aspects of human complexity could have been discovered only within the last few decades by social, biological, and medical scientists. We became aware of the evolutionary process a mere century and a half ago, and of genes slightly longer than a century ago. So, unfortunately, it is not surprising that we like to think that all of our social attitudes and actions are borne from consciousness, and that genes—as invisible, manipulative, and seemingly alien forces recently brought into our consciousness—do not in any way influence how we behave.

Our uses of the concept of God also involve aspects of our makeup that we do not fully understand. If the concept of God has a real basis—as I believe it must, because of its virtual ubiquity and the passion we associate with it—then surely it behooves us to explore everything about the concept and its influences on our lives, including effects not easily made conscious. Perhaps, rather than continuing to accept that the concept of God can only refer to a supernatural, anthropomorphic, fatherly being, we should ponder straightforward examinations of alternative ideas about the origin and nature of the concept, its role in the mysteries of the human mind, and its significance.

For example, what if the power, guidance, and permanence of God derives, not from a supernatural anthropomorphic being, but from a human-generated and highly effective use of metaphors referring ultimately to the ancient and ubiquitous human kin group, and its replacements and diverse forms in modern society? Thinking of God in our everyday lives evokes primarily emotions such as love and cooperation. So does kinship, and as well the exchanges of reciprocal altruism, the risk of net-cost altruism as potential social investment and return beneficence. God is treated as a concept or spirit, regarded as a source of strength, authority, morality, and protectiveness. So is the kin circle and the sociality of the local community. We expect and wish our families, or kin groups, to continue indefinitely. So is there an assumption that God is eternal. To my knowledge none of the approaches to God rejects that, even if we must attach our affection and our passions through our kin and our collections of cooperative support. The members of every group, or religion, would like to think that their particular view of God will eventually prevail, but this may be true only as long as humanity survives. Members of different groups, on the other hand, sometimes view God so distinctively that intense competition and we-they confrontations are prevalent. Both of these attitudes apply also to our view of kin and benevolent social groups. The desire for wise and infallible leadership, and the roles of older and more influential individuals in kin groups—given reluctance to accept death and the value of believing in communication with deceased leaders or family heads—are possibly long-ago facilitators of the envisioning of God as a supreme individual being with powers beyond our experience in the natural world, including that of negotiating the prospect of eternal life.

The old "local hubrises" of kin groups and religious groups are indeed enemies of broader-scale social accord, and they are sufficiently parallel to suggest a common if not equally ancient background. Local hubris surely has its usefulness, and not merely in the past. But today's environment of calamitous destructive powers—and increasingly rapid and effective world-wide communication and travel— denies us justification for the destructiveness of continuing regional separatism and chauvinism—of extreme-we-they confrontations on pride, stubbornness, economics, politics, religion, kin relationships, unfamiliarity, or even resource distribution.

Returning to Reverend Hyslop's microcosm of grandchild effects, the several adopted grandchildren of my wife Lorrie and myself, whose diverse genetic ancestors derive from the far corners of several different continents, have given us confidence that appropriate social learning—assisted by evolutionary understanding of early imprinting and bonding, and continuing positive association—can yield the pleasures and strivings associated with kinship as surely and as completely as do the usual life situations of formally genetic kin. We are fortunate indeed that the directing of human cooperativeness and competitiveness is largely socially learned, for such learning, as we know, is subject to manipulation by deliberate changes in the circumstances of life. No matter how social learning has worked in the past, knowledgeable application of it in the modern world thus has the possibility of bypassing or minimizing the sometime foibles of our histories with regard to social and religious variations.

Evolution is not the enemy, even if it has not made us perfect. The sciences we use to explain evolution—or anything else—are no one's enemies. Nor is religion, although it may seem surprising that interpretations of God have seldom led religious assemblages to bless everyone, everywhere, equally (as noted later, one minister in my childhood church puzzled me, while I was still a child, by consistently limiting his prayer to, "God bless everyone in this congregation and all those too ill to attend!").

Although I am not a fan of supernaturalism, I am also not surprised that others find it useful or reassuring. Members of my own kin circle rely on a supernatural concept of God, many of them worshipping in the small country church I attended as a child, and where Lorrie and I still visit at every opportunity. Our mutual affection there remains. But for me it is difficult to witness and condone extraordinary and extreme ceremonies, elaborate structures, and assertions that sometimes channel morality and unsupported impressions narrowly, and even pompously, in the effort to accept as factual beliefs that supernatural declarations should override common sense.

I appreciate Doris Lessing's (1992) insight in *African Laughter* that, "Myth does not mean something untrue, but a concentration of truths." But there is cause for concern when myth is used as a weapon against other concentrations of truths. Art, poetry, music, and fiction all utilize myth, and they rarely promote their business by attacking science. Is it too much to expect that novel and ingenious cooperative approaches between science and religion can help lessen the world-wide scourges of unnecessary pain, misery, and suffering, and the continuing sad parade of deliberately premature deaths from human conflict?

SEEKING GLOBAL HARMONY

Considering Efforts to Alter the World's Greatest Problem

Try to imagine the complexity of more than seven billion individual humans, divided among nations and uncountable forms and sizes of groups, all assiduously pursuing an endless variety and complexity of their separate and collective interests, and simultaneously trying to think about achieving the goal of global harmony. Try to imagine the problem of seeking to cause all of those more than seven billion individuals to remove any significant conflicts of interest among their various collectives and groupings, or even merely striving to lessen the worst consequences of their conflicts.

To a large extent, partial solutions to the problem of global harmony depend today on the governed units termed nations. Organized religions, in different countries and situations, can also exercise potent influences in the machineries of nations. Something similar can be said of scientific progress, because scientists and engineers generate and perfect the paraphernalia of medicine, business, and much of everyday life, and as well the instruments and practices of war. As loyal citizens of their nations, and sometimes as staunch members of religious or other authoritarian groups, scientists can also be influential seekers of rewards for the creation and use of increasingly horrific weapons of war.

How can we increase the informing of our populations with regard to willingly competitive and potentially destructive groups in ways that will diminish or terminate devastating conflicts? How do we escape the hyper-competitiveness that we all too often praise, beginning with strong advice, praising the most extreme competitive behaviors to even our young children, as the only effective route to lifetimes of accomplishment? How do we disentangle ourselves from the pernicious influence of diversely and competitively sacrosanct hyper-patriotism, or the divergent and unyielding forms of religions? How do we free ourselves from the view that our readiness for deadly confrontations outweighs the priceless value of our military men and women as we hurry to proclaim and cling at all cost to the sacredness of our essential motherlands? How can we negotiate and modify governments (and ourselves) to seek successfully the means to settle conflicts without acrimony, and with an absence, or at least a minimum, of force or dissension? How, indeed, can we escape from what Abraham Lincoln, in referencing actions during the horrendous U.S. Civil War, referred to "the attractive rainbow that rises in showers of blood"? (Alexander 2011: 286)

Several decades ago I found myself thinking that, in spite of the unique complexity of our brains and our behavior, we humans don't really know who we are or how we came to be as we are. The reason, perhaps, involves evolved tendencies to forget or suppress unpleasantness, and as well be aware of the effects of habit and non-conscious but well-understood practices and actions (Alexander 2011, Trivers 2011). As suggested earlier during comments about mysteries, many such tendencies may be evolved mutings of consciousness, carried out by natural selection, and scarcely available to the contemplation of most persons or situations. It seems clear that we have not evolved to wield all of our prodigious mental capabilities freely and effectively. Nor do we seem continually tuned to understand and contemplate our willingness to engage in serious conflicts that destroy large numbers of our own people. If we don't know who we really are, or how to deal with sometimes galling human extremes, efforts to approach global harmony are likely to fail. We cannot continue to allow the questions that damp or conceal our conscious knowledge to pass fleetingly and unmanaged across our minds. This is why I believe that the most important change that can contribute to global harmony is for humanity in general to learn to know itself better, individually and collectively, as products of our personal backgrounds, including what is derived from the unambiguous, never-ending process of organic evolution.

Natural selection, the principal force that changes us, tends to move slowly—so slowly, and sometimes so imperceptibly—that we cannot easily perceive what is happening or has happened. This has to be one of the reasons for the evolutionary process and its results being more or less overlooked on an everyday basis. But we cannot fully understand ourselves unless we are willing and able to examine and learn how our current life circumstances have come about, and, from that knowledge, how to adjust our lives effectively and profitably. We need to investigate and lay open our capacity for understanding the ways natural selection has manipulated the patterns of our consciousness, canceled our wayward memories in directions that favor reproductive success, and prevented, modified, and all too often warped our potential for the warmth of truly widespread friendliness, empathy, and cooperativeness. We need to support, to whatever possibilities are reasonable and available, all of the members of our world and our species.

> *The manipulations of consciousness that virtually block human self-understanding, presumably consequences of natural selection, leave us with an astonishing prevalence of horrific attitudes and behaviors that, sadly, to many or most people seem either not to exist, or are only moderately and temporarily noticeable. Team competitions, warfare, genocides, and Beilby Porteus's one-murder at a time villains (see earlier) are among the many human conditions that natural selection has apparently modified by manipulating consciousness, causing us to respond almost carelessly and forget quickly when confronted by dramatic and sometimes radical or even monstrous actions of*

> *members of our own species. We seem to live in bubbles of awareness alongside*
> *non-conscious or distortedly conscious strivings, and quick forgetting, the latter*
> *effects products of what* Williams (1993) *called "The Wicked Witch" of "Mother*
> *Nature", demonstrating the selfishness of evolution's differential reproduction*
> (Alexander 2011, *p. 279–80*).

It is not an accident that several of the best-known scientists and thinkers in the history of the world have understood that evolution has been responsible for the worst of humanity's activities. Thomas H. and Aldous Huxley, George Williams, Richard Dawkins, and numerous others have understood that the process of evolution is based on "selfishness," the "selfish" clustering of the tens of thousands of genes that build the organisms we become, and as well can follow a course that mainly generates power and increases access to resources. It would seem that if humanity is to move toward global harmony, it must modify its fate by understanding our human selves deeply, and by building strong desires and capabilities to focus on the positive aspects of humanity, reducing the extreme negatives deriving from hyper-competitiveness and hyper-patriotism, and turning the future of humanity in new directions concentrated on extensive webs of social investment and return beneficence—in other words, on the workings of direct and indirect reciprocity made available by the workings of organic evolution (Hamilton 1964, Trivers 1971, 2011, Alexander 1978, 1987, 1990, and many others; see note on an earlier page, concerning "altruism"). To accomplish this, we must find ways to overcome some of the effects of natural selection, and thereby understand ourselves much more completely, such that we can use all of the knowledge available to us to serve our own interests.

What else can we do to change ourselves—on a global scale—to reduce our ever-ready tendencies to compete at nasty levels, or to commit murder, even in the face of lifetimes in prison or a death sentence; or to wage wars? We surely cannot lose by striving collectively and mightily toward congeniality and negotiation rather than hyper-competitiveness and conflict, and by seeking reduction of aggression, using all reasonable means.

These and other suggestions may have positive possibilities. But it does not seem likely that efforts at such changes will quickly capture our imagination or yield compelling or worldwide outcomes. It is as if everyone believes that she and he are already working as hard toward such outcomes as is reasonable or possible. If that is so, it will not do to expect that the efforts that have been tried so far can solve the problems. Thus, large numbers of small, local, non-overlapping groups, socially close-knit and catering to authoritative moral pronouncements, are unlikely to foster global changes. Such efforts are not likely to solve the problems because humans have been doing these sorts of things more or less in vain for thousands of years. Perhaps we can gain by more effectively identifying broad questions, or knowledge, that can influence a higher proportion of the global population.

The greatest difficulty in seeking global harmony may derive from human groups targeting one another. Humans alone—among all the world's species—plot, plan, and organize massive conflicts to defeat or displace similarly organized and cooperative members of their own species. Can we learn to use the current consciousness of our human background to adjust team efforts of all kinds so that honesty, fairness, and negotiation can increase and lead us toward global harmony? Can we work profitably against the existing minimizing, reversing, and distorting of conscious knowledge generated by natural selection? Surely such efforts would contribute positively toward global harmony.

The curious prevalence of wars between conspecific human groups may have been encouraged by the isolation of human populations that, through extensive migrations during past millenia, became separated geographically but (thank goodness!) did not become so different genetically as to prevent increasingly extensive hybridization after establishment of population mixtures. At least among the distinctive populations forming recently, isolation did not persist long enough to give rise to different species. But early human populations may not have persisted in separations long enough to accumulate differences between diverging populations that increased serious intergroup strife—for example belittling populations living next to groups different in appearance, language, or cultural patterns (for a somewhat similar example in much simpler and different hybridizing [insect] species, see Alexander 2011: pp. 200–201, 205–206). At least during early amalgamations of distinctive populations, such differences almost certainly caused humanity to generate less opportunity and motivation to combine the diversities of our single species peacefully into mixed populations cooperative against the array of non-human enemies. Presumably, in some earlier stages of evolution, humans were still focused almost entirely on non-human enemies. At some point, humans were surely also less likely, or less well equipped, to treat other humans as primary enemies. However, as human population sizes increased, and ecological dominance became a more promising possibility, competition among humans for resources would have become more concentrated and begun to generate small closely-knit kin and social groups developing their own rules and desires to contest against one another.

Whatever the detailed reasons and timing for the incredible tendencies and devices that spawn war in modern humans, our current condition, as already noted, has obviously generated, elaborated, and persisted in supporting massive and horrific within-species, large-group conflicts, along with serial and copycat murders, bullying, and other destruction of humans at many levels and in different numbers. It is unlikely that these socially negative happenings can be easily or quickly reversed.

Any route to world peace, or global harmony, surely depends on a relaxed, tolerant, and unified approach to different attitudes and practices with regard to social life, religion, and the concept of God and, simultaneously, more critical attitudes (a) minimizing tendencies to be rigid or authoritarian about local or group-restricted concepts of right and wrong, and (b) treating more temperately other groups having different views from our own. I think this because it seems evident that people everywhere are inclined to invoke some version of God as a less than universal rallying point, too often in inter-group hostilities, giving us reason to suppose that in some form religion and God may have been involved in inter-group adversity for a very long time.

Most specifically, I hope that self-understanding will reveal to us pathways leading away from our current and evidently historically continuous state of being a world made up of destructively adverse groups and nations that consistently invoke religion and the concept of God as inspirations to social unity, apparently as both conscious and unconscious contributions to efforts to prevail over other similar groups. Part of uncovering such routes includes recognizing the function and sometimes dire correlates of intense patriotism, and the warmth and good feeling that go with the beneficence and cooperativeness that are too often restricted to within-group interactions.

We are not gentle people. But we can be several kinds of people, depending on circumstances. In congenial cooperating local groups, we are mostly kin-helpers, cooperators, and positive reciprocators. When engaging in wars, we frequently, even if only temporarily, become determined killers.

It will surely take all of the capabilities that humanity can muster to accept and complement the unpleasant parts of our collective nature, and to minimize or reverse the unfortunate effects of human history that primed us and set us up indefinitely to continue threatening extremes of human existence. The world's options call for peaceful, casual, and deliberate amalgamations of historically tiny, introverted, and tightly-knit social groups, and, on the other hand, perhaps questioning huge nations unable to refrain from becoming armed beyond sensibility. The opposite outcome—the whole world a unified police state—may or may not be tolerable. Peace may be the goal, but the means and maintenance of peace will require novel levels of statesmanship and, somehow, continuing floods of good will.

It would be wonderful if all of humanity could become sufficiently knowledgeable about its self—positive and negative—to begin to absorb the activities and attitudes of people living in tightly-knit social groups (whether religious or not), discovering ways to transform broadly and definitely the cooperative behavior of the individuals familiar within such groups, along with acceptance of social investment and the responses of return beneficence on a world-wide basis. To the extent that these changes can take place, we might find ourselves comprehending how the building of real cooperativeness and socially positive behavior can turn us toward a sociality reflecting global peace and harmony.

Many different levels and expressions of consciousness can be involved in human selflessness, and in compensating both directly and indirectly reciprocal interactions (Alexander 2011). Promoting the continuance of complex interplay of social investment and return beneficence in religious and other tightly-knit social groups is surely a positive approach to understanding how humans can work toward global harmony.

> *Why should we not encourage the diverse people around the earth to be freely inter-mixed and ready to strive to make all nations democratic—the latter meaning to attempt reasonable correlates, to call attention to the values of elections at suitable intervals, personal and confidential voting, multiple voting political parties, and parliamentary rules and courts that consistently make democratic institutions work for all people? How can we remove the indefinitely continuing and repeating dictatorships that begin to treat resources as the property of the government and as a result not only take up war with their own people but threaten all the rest?*

CONCLUSIONS

The several topics undertaken in this essay—religion, group-living, human minds and mystery, muting of consciousness, science and evolution, concept of God, meaning of life, explorations of natural and supernatural possibilities, universal or restricted moral sense, kinship, social reciprocity, competition and cooperation, and prospects of global harmony—are more than merely difficult. I do not imagine that I have created any broadly credible solutions to the problems I have sought to disentangle. Nevertheless, entering into searches for the purpose of considering these difficult topics—or at least calling attention to them—can potentially be among the worthiest of investigative enterprises.

My effort in this essay has implied that the concept of God arose or became a dependable spirit of cooperativeness, morality, and beneficence, one that is perceived and acted on within groups to aid and protect the group, either directly or indirectly against other such human groups, and whether regarded as supernaturalism, or as a natural but jubilant outgrowth of the invisible clouds of effects and efforts that have generated and persisted out of the stored intellectual mysteries of the human spirit. Considering the concept of God as a universal spirit of cooperativeness within and between social groups should be a repeated and elaborated effort because it facilitates the evolved function of the lifetimes of individual members of the human species, through maximizing the likelihood of persistence of genes in one's own genome, ultimately through assistance to those genomes carrying genes in fractions reliably predictable among the relatives making up our kin groups. Such assistance also involves social investment in unrelated individuals, including spouses, and partners in social reciprocity, when such assistance eventually assists own relatives, and may assist all group members whenever individual and group interests are frequently or permanently similar or identical. The concept of a universal God, whether it be accepted as metaphorical or otherwise, would have become possible only with the evolution of a moral capacity, so, curiously, we might say that God (as the collective expression and coincident awareness of a universal moral capacity) created humanity, or that humanity created God—or both— when moral capacity was engendered, directly or indirectly, consciously or not consciously, by and during the process of organic evolution.

In the sense I have just described them, religion, morality, and attention to the concept of God are parts of what has been identified as kin selection and direct and indirect social reciprocity, in which the returns from acts of benevolence can emerge freely from individuals or groups other than the particular individual(s) being served. Expanding patterns of social reciprocity and kinship behavior can potentially continue moving toward global social harmony because they encourage universal cooperation and fairness (Trivers 1971; Alexander 1974, 1987; Alexander and Borgia 1978; Frank 1995; Irons 1996a, 1996b; Queller and Strassmann 2009; Strassmann and Queller 2010).

At the risk of being judged a hopeless megalomaniac, I repeat here that it is my greatest regret, late in my lifetime of thought and research, that I have continued to be inadequate in my attempts to discover and explain how people everywhere might work to understand themselves sufficiently better from knowledge of evolution, so as to influence the sociality of global humanity in a positive way. Regardless of the pace of technological and other scientific advances, understanding of ourselves in evolutionary terms—understanding sufficiently profound that it requires at least a temporary ability to withdraw slightly and judge ourselves as if we were aliens, or members of a different species—may always be necessary if we are to recognize and accept the most important sources and reasons for change in the social lives of humans. I regret my inability to identify confidently even the first steps of a solution to the long-standing central problem of humanity that derives from the prevalence, throughout our history, of uniquely ferocious and frequent inter-group competitions, and continuing elaboration of ever more deadly and dangerous weaponry within our species (Alexander 2011, slightly altered).

Would that we could shift our attention away from the military, and toward other countries in the example that Paul Krugman has recently illustrated: *"What ails the Arab world is a deficit of freedom, a deficit of modern education and a deficit of women's empowerment. So helping to overcome these deficits should be what U.S. policy is about, yet we have been unable to sustain that. Look at Egypt (cf. the 'U.N. Arab Human Development Report published in 2002 by some brave Arab social scientists...'): More than half of its women and a quarter of its men can't read. The young Egyptians who drove the revolution are desperate for the educational tools and freedom to succeed in the modern world. Our response should have been to shift our aid money from military equipment to building science-and-technology high schools and community colleges across Egypt."* (NY Times, and Ann Arbor.com, March 29, 2012)

Why not more such positive efforts that derive from social investment and the encouragement of return beneficence with the hope of fostering cooperativeness and peace, and the potential to establish a gentler and more peaceful world? Why not?!

It is surely time for the adaptive structures of religion and science
to begin adjusting, finally, into the long-needed partnerships
that can hone their respective capabilities
with the joined skills and emotionalism
of the tangled and still impotent searches
that someday will nurse the gathering fragments
of the potential for global social harmony

ALEXANDER, 2011B:312

Acknowledgements

As with all people interested in understanding life, I am massively indebted to George C. Williams, William D. Hamilton, and Robert L. Trivers, not only for their fundamental and outstanding publications about how evolutionary selection works, but for their friendship and their many discussions with me personally, or in my presence, across several decades. These three biologists are my version of the most inspiringly original contributors to understanding of the evolutionary process during the second half of the twentieth century. Each has summarized his contributions in multiple papers and two or more books. Without their insights I could not have pursued the meager lifetime agenda I set for myself in 1954, three years before George C. Williams, and George C. and Doris C. Williams, published the first two papers developing what I will call the modern view of how adaptation and the spans of lifetimes work.

With delight and special appreciation, I thank Professor Billy E. Frye, former Provost of the University of Michigan and subsequently Chancellor of Emory University, who across several years has been my most important and almost continual discussant and critic through a range of issues in human evolution, religion, biology, and many other topics reflected in this current essay. He has taken me through far too many arguments and corrections for me to remember, including every major topic involving the discussions in this essay, and as well such unusual themes as the ins and outs of reforestation, the special lifetime requirements of orchids and roses, and his unforgettable night-time fox hunts with his father and other men in the Appalachian Mountains, where Billy grew up and where he now lives once again, after serving in many different levels of university administration.

Andrew F. Richards and David Lahti have explored with me the topics of this essay and related others in frequent lunch time free-for-alls across a broad span of years. Across more than a half century, a succession of approximately 50 doctoral students and postdoctoral associates have dealt with all the topics that preceded this current one during as many years of lunches and graduate seminars. Numerous other people, including Lorraine Kearnes Alexander, Holly and Paul Ewald, Michael Martin, Wendy Orent, Naomi Salus, Beverly Strassmann, and Wesley Upton, are among those who have also reviewed and criticized my most recent ideas and manuscripts. I am especially thankful to my friends and associates in the study of human behavior and evolution, William Irons and Napoleon Chagnon, for aiding me in numerous ways, and across several decades.

Beginning in 2003, members of a discussion group of retired University of Michigan professors and other personnel (*Learning In Retirement*), superbly led by Marlin Ristenbatt, contributed extensively by inviting and re-inviting me to discuss evolution and religion with *LIR*, in what turned out to be numerous enthusiastic and probing one- and two-hour sessions.

Finally, I am humbled beyond description by the task undertaken by Kyle Summers and Bernie Crespi, and extremely grateful for the authors who accepted the tasks of selecting, introducing, and criticizing the chapters of the resulting volume.

References I

Some of the labels for the concept of God used in this essay were taken from *Roget's Super Thesaurus*, 2nd edition (Cincinnati, OH: Writer's Digest Books). Otherwise, some concepts and information used in this essay are widely familiar in biology, or for other reasons have not included specific references.

Alexander, R.D. 1967. Comparative animal behavior and systematics. In: *Systematic Biology*. National Academy of Science Publication 1692:494–3517.

Alexander, R.D. 1971. The search for an evolutionary philosophy of man. *Proc. R. Soc. Victoria* 84:99–120.

Alexander, R.D. 1974. The evolution of social behavior. *Annu. Rev. Ecol. System.* 5:325–383.

Alexander, R.D. 1978. Evolution, Creation, and Biology Teaching. *Amer. Biol. Teacher* 40:91–107.

Alexander, R.D. 1979. *Darwinism and Human Affairs*. Seattle: University of Washington Press.

Alexander, R.D. 1987. *The Biology of Moral Systems*. Hawthorne, NY: Aldine De Gruyter.

Alexander, R.D. 1990. How did Humans Evolve? Reflections on the Uniquely Unique Species. *Univ. Michigan Museum of Zool. Spec. Publ.* 1: iii +38 pp.

Alexander, R.D. 1993. Evolution of the human psyche. In: N. P. Mellars and C. Stringer (eds.), *The Human Revolution*. Edinburgh: Edinburgh Press.

Alexander, R.D. 2005. Evolutionary selection and the nature of humanity. In: V. Hosle and C. Illies (eds.), *Darwinism and Philosophy*. South Bend, IN: University of Notre Dame Press, pp. 301–348.

Alexander, R.D. 2008. Evolution and Human Society. *Newsl. Hum. Beh. Evol. Soc.* Summer Issue, pp. 1–20.

Alexander, R.D. 2010. Understanding Ourselves. In: Drickhamer, L. and Donald Dewsbury (eds), *Leaders in Animal Behavior: The Second Generation.* Cambridge: Cambridge University Press.

Alexander, R.D. 2011. *The Mockingbird's River Song: Poems, Essays, Songs, and Stories: 1946–2011.* Manchester, MI: Woodlane Farm Books.

Alexander, R.D. and Gerald Borgia. 1978. Group selection, altruism, and the levels of organization of life. *Ann. Rev. Ecol. System.* 9:449–474.

Bigelow, R.S. 1969. *The Dawn Warriors: Man's Evolution toward Peace.* Boston: Little, Brown.

Boehm, C. 2003. Global Conflict Resolution: An Anthropological Diagnosis of Problems with World Governance. In: R. W. Bloom and N. Dess (eds.), *Evolutionary Psychology and Violence: A Primer for Policymakers and Public Policy Advocates.* London: Praeger.

Burt, A. and R. Trivers. 2006. *Genes in Conflict: The Biology of Selfish Gene Elements.* Cambridge, MA: Belknap Press of Harvard University Press.

Buss, D. M. 2005. *The Murderer Next Door: Why the Mind is Designed to Kill.* New York: Penguin.

Darwin, C. 1859. *On the Origin of Species by Means of Natural Selection.* London: Murray.

Darwin, C. 1871. *The Descent of Man, and Selection in Relation to Sex, 2nd ed.* London: Murray.

Darwin, C. 1965 [1872]. *The Expression of the Emotions in Man and Animals.* Chicago: University of Chicago Press.

Dawkins, R. 1976 (revised 1989). *The Selfish Gene.* New York: Oxford University Press.

Dawkins, R. 1982. *The Extended Phenotype: The Gene as the Unit of Selection.* New York: Oxford University Press.

Dobzhansky, T. 1961. In: J. S. Kennedy (ed.), *Insect Polymorphism.* London: Royal Entomological Society, p. 111.

Elkin, S. 1993. Out of One's Tree. *Atlantic Monthly* (January), pp. 69–77.

Ember, C.R. and Melvin Ember. 1990. *Anthropology.* Englewood Cliffs, NJ: Prentice Hal.

Fisher, R.A. 1930 (revised 1958). *The Genetical Theory of Natural Selection.* New York: Dover.

Flannery, K. 1972. The Cultural Evolution of Civilizations. *Ann. Rev. Ecol. Syst.* 3:399–426.

Frank, S.A. 1995. Mutual policing and repression of competition in the evolution of cooperative groups. *Nature* 377:520–522.

Goldhagen, D.J. 2009. *Worse Than War: Genocide, Eliminationism, and the Ongoing Assault on Humanity.* New York: Public Affairs.

Hamilton, W.D. 1963. The evolution of altruistic behaviour. *Amer. Nat.* 97:354–356.

Hamilton, W.D. 1964. The genetical theory of social behaviour I, II. *J. Theoret. Biol.* 7:1–52.

Hamilton, W.D. 1966. The moulding of senescence by natural selection. *J. Theoret. Biol.* 12:12–45.

Hamilton, W.D. 1971. Geometry for the selfish herd. *J. Theoret. Biol.* 31:295–311.

Irons W. 1996a. Morality, Religion, and Human Nature. In: W. M. Richardson and Wildman (eds.), *Religion and Science.* New York: Routledge.

Irons, W. 1996b. In Our Own Self Image: The Evolution of Morality, Deception, and Religion. *Skeptic* 4: 50–61.

Keith, Sir A. 1949. *A New Theory of Human Evolution.* New York: Philosophy Library.

Kelly, R.C. 2005. The Evolution of Lethal Intergroup Violence. *Proc. Nat. Acad. Sci.* 102: 15294–15298.

Lessing, D. 1992. *African Laughter: Four Visits to Zimbabwe.* New York: Harper.

Queller, D.C. and J.E. Strassmann. 2009. Beyond society: The evolution of organismality. *Phil. Trans. R. Soc., Series B,* 364–3155.

Scaruffi, P. 2006. Wars and Genocide of the 20th Century. (http:/www.scaruffi.com/politics/ massacre html)

Strassmann, J.E., and D. Queller. 2010. The Social Organism: Congresses, Parties, and Committees. *Evolution* 64-3: 605–616.

Sykes, C. (ed.). 1994. *No Ordinary Genius: The Illustrated Richard Feynman.* New York: W. W. Norton.

Trivers, R.L. 1971. The Evolution of Reciprocal Altruism. *Quart. Rev. Biol.* 46:35–57.

Trivers, R.L. 2011. *Deceit: Fooling Yourself the Better to Fool Others.* New York: Penguin Books.

Williams, G.C. 1957. Pleiotropy, natural selection, and the evolution of senescence. *Evolution* 11:398–411.

Williams G.C. and Doris C. Williams. 1957. Natural Selection of Individually Harmful Social Adaptations among sibs with special reference to social insects. *Evolution* 11:32–39.

Williams, G.C. 1966. *Adaptation and Natural Selection.* Princeton, NJ: Princeton University Press.

Williams, G.C. 1993. Mother Nature Is a Wicked Old Witch. In: M. H. and D. V. Nitecki (eds), *Evolutionary Ethics.* Albany: SUNY, pp. 217–231.

References II: Additional Relevant Books

Armstrong, K. 2009. *The Case for God.* New York: Alfred A. Knopf.

Atran, S. 2002. *In Gods We Trust: The Evolutionary Landscape of Religion.* Oxford: Oxford University Press.

Berger, P.L. 1967 (1990). *The Sacred Canopy: Elements of a Sociological Theory of Religion.* New York: Anchor Books.

Boehm, C. 2012. *Moral Origins: The Evolution of Virtue, Altruism, and Shame.* New York: Basic Books.

Boyer, P. *Religion Explained: The Evolutionary Origins of Religious Thought.* New York: Basic Books.

Chagnon, N.A. 1968. *Yanomamo: The Fierce People.* New York: Holt, Rinehart, and Winston.

Dawkins, R. 2003. *A Devil's Chaplain: Reflections on Hope, Lies, Science, and Love.* New York: Houghton Mifflin Company.

Dawkins, R. 2008. *The God Delusion.* New York: Houghton Mifflin Company.

Dennett, D. C. 1995. *Darwin's Dangerous Idea: Evolution and the Meanings of Life.* New York: Simon and Schuster.

Dennett, D.C. 2006. *Breaking the Spell: Religion as a Natural Phenomenon.* New York: Penguin Books.

Dennett, D.C. 2011. *Science and Religion: Are They Compatible?* New York: Oxford University Press.

Diamond, J. 1997. *Guns, Germs, and Steel: The Fates of Human Societies.* New York: Norton.

Dunbar, R. 2004. *The Human Story: A New History of Mankind's Evolution.* London: Faber and Faber.

Gangestad, S.W. and J.A. Simpson (eds). 2007. *The Evolution of Mind: Fundamental Questions and Controversies*. New York: The Guilford Press.

Hamer, D. 2004. *The God Gene: How Faith is Hard-Wired into Our Genes*. New York: Doubleday.

Harris, S. 2004. *The End of Faith: Religion, Terrorism, and the Future of Reason*. New York: Norton.

Hauser, M. 2006. *Moral Minds: The Nature of Right and Wrong*. New York: Harper Perennial.

Hinde, R.A. 1999. *Why Gods Persist: A Scientific Approach to Religion*. London: Routledge.

Hinde, R.A. 2002. *Why Good is Good: The Sources of Morality*. London: Routledge.

Kelly, R.C. 2000. *Warless Societies and the Evolution of War*. Ann Arbor: University of Michigan Press.

King, B.J. 2007. *Evolving God: A Provocative View on the Origins of Religion*. New York: Doubleday.

Kurtz, P. (ed). 2003. *Science and Religion: Are They Compatible?* Amherst, NY: Prometheus.

Meisenberg, G. 2007. *In God's Image: The Natural History of Intelligence and Ethics*. Gateshead, England: Athenaeum Press Ltd.

Pinker, S. 2011. *The Better Angels of Our Nature: Why Violence Has Declined*. New York: Viking.

Putnam, R.D. and D.E. Campbell. 2010. *American Grace: How Religion Divides and Unites Us*. New York: Simon and Schuster.

Rue, L. 2005. *Religion Is Not About God*. New Brunswick, NJ: Rutgers University Press.

Shank, N. 2004. *God, the Devil, and Darwin: A Critique of Intelligent Design Theory*. Oxford: Oxford University Press.

Shermer, M. 2011. *The Believing Brain*. New York: Henry Holt and Company.

Taverne, R. 2005. *The March of Unreason: Science, Democracy and the New Fundamentalism*. Oxford: Oxford University Press.

Trivers, R. 2011. *Deceit: Fooling Yourself the Better to Fool Others*. London: Penguin.

Wade, N. 2009. *The Faith Instinct: How Religion Evolved and Why It Endures*. New York: Penguin.

Wilson, D. S. 2002. *Darwin's Cathedral: Evolution, Religion and the Nature of Society*. Chicago: University of Chicago Press.

Wright, R. 2009. *The Evolution of God*. New York: Little, Brown and Company.

Evolution and the Arts

Creating the Treasures of Your Truths and Mine

Science begins with ideas from observations:
builds scenarios that thrive on journalistic directness.
Its novelties are newly known theories and facts,
the best of them robust, unalterable,
crammed with promises of broad significance
for everyone, said to be valuable
because of objectivity.

Art begins with ideas from observations:
builds scenarios that thrive on metaphor and mystery.
Its novelties are sublime and outrageous:
take-it-or-leave-it, the best of them
boundlessly interpretable, accepted as subjective,
crammed with promises of personal meanings
for everyone, said to be valuable
because of subjectivity.

Science—formal and informal—establishes
the successions and progressions of realities
on which the continued elaboration
of both science and the arts depend.

The mental capacities and tendencies that
facilitate the practice of art,
and render it joyous,
facilitate science as well,
and render it also joyous.

Realities deriving from use of the imagination
in either science or the arts—or merely life—
may seem personal to the scientist or the artist
at every stage.

Realities derived from use of the imagination
in the arts are saved or accepted by personal choice
from the beginning,
can be rejected at will,
and also personally.

Realities deriving from use of the imagination
in science sometimes seem alien,
especially to the populace on which
they are necessarily imposed,
rather than merely invited or sought.

Science and the arts together
enrich the imagination,
imagination drives discovery,

discovery builds knowledge,
knowledge enriches lives.
Art and science each offer novel truths,
foster enthusiastic or fearful incredulity
over the richest of their contents:
one discovered via repeatable hence verifiable

procedures, a tearing down to rebuild further,
on more solid ground, the other on faith
via trust in unrevealed procedures, trust
imposed or accepted as a result of differentials
in the strength of human authority;

sets of truths valued either similarly by all
because they apply to interests common to all,
or offering diverse usefulnesses,
enhancing separately
the unique interests of individuals.
Science progresses with great difficulty
toward explanations of such as meanings,
emotions, likes and dislikes—
so far has tended to resolve problems

largely at physical or physiological levels,
through procedures thriving on reduction and
dissection, to levels conducive to analysis,
followed by slow reconstruction
toward admired and desired wholes.

Art exposes and creates novelty
in personal and emotional propositions,
often at their own level, seeking insights
more through sense-expanding meanings

than through sense-extending technologies.
The novelties of science and the arts
contribute to all our prospects,
adjusting physical and social possibilities,
extending personal and social meanings.

Alexander, 2011, p. 257

INTRODUCTION

Cornerstone to Capstone: Richard Alexander on Social Selection and the Arts

Kyle Summers and Bernie Crespi

(An introduction to an excerpt (Pp. 333–345) from Alexander, R.D., 2008. Evolutionary Selection and the Nature of Humanity, Pp. 301–348 In. (V. Hosle and C. Illies, eds.): *Darwinism and Philosophy*, University of Notre Dame Press, Notre Dame, IN.)

Evolutionary understanding of human behavior was grounded in scientific inquiry by Darwin's (1859) theory applied to the "necessary acquisition of each mental power and capacity by gradation" in his novel process of natural selection. Among all human capacities and achievements, the arts stand alone as furthest from our primate roots, and apparently least amenable to explanation by step by step natural selection and gradation. As such, the arts, as pinnacles of human achievement with no apparent basis in adaptive function, represent an exceptional challenge to theories of how humans have evolved. For these reasons, we consider Richard Alexander's contributions to penetrating the evolutionary underpinnings of art as one of his crowning achievements—an achievement that makes clear his entire framework of thought, from foundations in selection of gene and individuals to capstones in mental powers and capacities that are uniquely human.

Human striving in the context of the arts is one of the most impressive yet puzzling examples of human uniqueness. While tool-making—and hence primitive technology—can be found in many animals, artistic endeavor seems considerably more restricted in its taxonomic distribution. Some animals have created what humans call "art" under artificial circumstances (Morriss-Kay 2010), but there is no scientific evidence that these creations are in any way equivalent to human artistic creations (Zaidel 2010). How has capacity for art evolved—or more specifically, what processes of variation, heritability and, most importantly, selection, could have led to the gradual evolution and maintenance of a human trait at least as unique as language, but apparently much less directly useful?

The evolution of human artistic capacities and skill has remained, until very recently, a surprisingly neglected topic (Dissanayake 1992 Aiken 1998; Boyd 2009; Dutton 2009). Two recent overviews provide insights into current opinion on the evolution of art. First, Moriss-Kay (2010) gives a historical perspective on the evolution of art in the hominid lineage, tracing its development back hundreds of thousands of years. Although art has commonly been held to have appeared less than 45,000 years ago with the migration of modern *Homo sapiens* from Africa to Europe (e.g., Bar-Yosef 2002), Moriss-Kay (2010) argues that the earliest evidence for art far predates this event. She takes a broad view of what constitutes art, including body decoration using pigments. Evidence for human artistic creation has been found from as long as 164,000 years ago, from recent excavations of caves in southern Africa (e.g., Jacobs et al. 2006). These excavations have revealed a variety of traces of artistic endeavors, including color pigments (likely used for self-decoration), engraved bones and decorative beads. Hence, while the flowering of artistic creativity seen in the abundant art created by early European *Homo sapiens* may indicate important changes in human artistic abilities, the evolution of art is likely to have much deeper roots in the hominid lineage. These studies provide important descriptive context for the origins of art, but tell us nothing of why individuals creating or wearing adornment presumably outreproduced those who did not.

Second, Morriss-Kay (2010) points out that early European cave painting, which many researchers find the most striking example of prehistoric art, focused on animals that are either hunted (e.g., bison) or hunters (e.g., lions). These animate artistic subjects suggest functions associated with the observation of, and reverence for, prey species, and admiration and emulation of effective hunters (particularly cooperative group hunters such as lions). She argues that early European art emphasized key components of sociality such as group cohesion, kindness and parental care on the one hand, and hunting skills on the other. With respect to the latter, she suggests that a key function of art—leading to its evolution—was the importance of the development of the "mind's eye," allowing early humans to visualize the final product of tool construction when beginning with an block of unworked material (such as a rock), or allowing hunters to visualize the location of prey even when they have disappeared from current view. Hence, she follows a long tradition of viewing art as a functional extension of human technological capacity, particularly in the context of hunting (e.g., Heinrich 2001). But if art indeed originated in functionally based visualization, what drove the initial transition, and later transformations to arbitrary form?

SEXUAL SELECTION OF THE ARTS

One ultimate explanation for the evolution of art focuses on the idea that art evolved as a display, closely connected to courtship and mate selection in the human lineage (Zaidel 2010). A number of authors (e.g., Zahavi & Zahavi 1997) have likewise proposed that a key function of art, and other manifestations of cognitive complexity, manifests in its contributions to success in sexual selection. For example, some researchers have focused on the potential relevance of Zahavi's Handicap

Principle in explaining artistic accomplishment as a—signal of the underlying "cognitive quality" of the artist (including Dr. Zahavi himself—Zahavi & Zahavi 1997). Perhaps the most well known advocate of the importance of sexual selection in the evolution of artistic ability (and many other aspects of human intelligence, including language) is Geoffrey Miller. Specifically, Miller (2000) argues that artistic creativity evolved to create displays that were evaluated by members of the opposite sex in the context of mate choice.

A key limitation of this approach is that it focuses on male-male competition. Presumably, under this argument, the evolution of artistic abilities in females is an epiphenomenon of the operation of selection on males, rather like the occurrence of horns in females of some species of ungulates (Lande 1980). Strong sexual selection is typically associated with extreme sexual dimorphism (e.g., Andersson 1994), and hence one might predict extreme sexual dimorphism in intellectual and artistic abilities, which does not accord with a substantial body of data (e.g., Charyton et al. 2008). However, Miller (2000) emphasizes that mutual mate choice was likely to be the rule rather than the exception over the course of human evolution, and hence sexual selection could favor similar traits and levels of expression in both sexes simultaneously. He argues that three different forms of sexual selection, including Fisherian Runaway, sensory biases and good genes indicator mechanisms, are all likely to have contributed to the evolution of artistic creativity and abilities (and other human mental attributes), but that indicator mechanisms are likely to have comprised the dominant mechanism over the course of human evolutionary history. For art in particular, Miller (2000) argues that the good genes hypothesis is particularly appropriate to explain the apparent luxuriant extravagance that seems to characterize many forms of artistic creativity and expression.

Miller's arguments have attracted a substantial amount of attention, both positive and negative. On the positive side, several authors have taken his ideas to heart and expanded on them. For example, Dutton (2009) presents Miller's arguments as the most likely explanation for the evolution of artistic ability in his recent book on the evolution of art. By contrast, others (e.g., Driscoll 2006) have criticized Miller's conclusions on several grounds. For example, the claim that art always or even usually functions as a courtship display does not stand up well to scrutiny. Also, art is not always aesthetically pleasing in the way that Miller predicts, and indeed great art sometimes deliberately flaunts notions of aesthetic pleasure. Further, Miller's "sexual equivalence" argument posits that men have preferred artistically talented women as mates throughout human history, but he does not present convincing evidence for this criterion of mate choice, in either historical or current human populations. Miller also makes a secondary argument for a "producer-consumer" relationship between men and women in the context of artistic ability, but this argument simply throws us back to the prediction that there should be sexual dimorphism in artistic ability between human males and females.

If sexual selection cannot substantially or completely explain the origin and evolution of art, might we default to considering it a Gouldian spandrel of selection in some other context, as Dawkins conceives religion? For example, Steven Pinker (1997) has argued for art as a kind of epiphenomenon that arises from having a brain highly adapted in the context of technical problem-solving. Artistic creativity and skill would hence have no ultimate adaptive value. But why, then, would they remain universal across human cultures, costly in terms of energy, time and resources that could be devoted to more fitness-increasing functions, and notably pleasurable, like so may other adaptive phenotypes (Dissanayake 1992).

SOCIAL SELECTION OF THE ARTS

The work of Richard Alexander on the evolution of artistic creativity provides an original, rich, complex and compelling alternative to the hypotheses reviewed above. Alexander first wrote about the evolution of art in several classic works on human behavioral evolution (e.g., Alexander 1979, 1987), and more recently has expanded on the importance of social selection—selection caused by mutualistic, competitive and exploitative social interactions within a species, whereby the same individual can play multiple roles, such as recipient, donor, or observer of social actions—in understanding the evolution of human artistic capabilities (Alexander 2008). These works appear to have gone largely unnoticed by most researchers in this field. Indeed, as Alexander (2008) points out, most of those who write about the arts do not consider its evolutionary foundations at all (e.g., Barzun 2000).

Alexander (2008) begins by warning that the ultimate explanation of the arts that he is proposing is likely to be highly objectionable to many people, because it focuses on the effect of artistic endeavor and accomplishment on the inclusive fitness of the artist and the observers of artistic creations. In the context of human societies, in which cooperation is highly prized and selfishness is considered a fundamental character flaw, the notion that we are, in large part, motivated by a desire to contribute to the general welfare of society is very strong (in and out of academia—see for example Gintis et al. 2003).

Alexander's work on the arts is grounded in the concept that the ultimate goals of human striving—in maximization of inclusive fitness—are normally hidden from us. He emphasizes that it is specifically because our motivations are more selfish than we would like to admit—lest our selfishness be used against us in the continual struggle to form and maintain cooperative alliances—that it is so difficult for us to perceive or accept them. Indeed, Alexander (1989) has argued that the elaborate structure of consciousness, with multiple layers, many of them hidden, evolved in part to keep our true motivations obscure, allowing us to proceed in our daily lives fully convinced of the underlying selflessness of our true motivations. This kind of "self" deception—or lack of conscious awareness of underlying motivations—can be useful in enabling success in the arena of social competition, as also emphasized by both Trivers (1971) and Alexander (1979). Hence, we have evolved a multi-layered and modular consciousness in which our own ultimate motivations are well hidden, even from ourselves. As Alexander argues, our perception that artists are motivated by "pure" striving for perfection should not deter us from seeking an ultimate explanation. Such considerations, of course, apply much more generally to the evolution of human mental powers and capacities, and may motivate non-evolutionary spandrel-based "explanations" of art, religion, and other difficult to explain arenas of human behavior and cognition.

Human motivations and striving are fundamentally social, yet human societies are unusual among animals in the high degree of autonomy of their members (Alexander 2008). Unlike genes in a genome, our reproductive success is often not closely bound to that of other members of the group. Unlike the members of social insect colonies, or groups of cooperatively breeding vertebrates, we are genetically unrelated to the vast majority of the other members of our societies. Recent research emphasizes the point that this social-genetic structure has persisted for much of our evolutionary history (Hill et al. 2011). Such autonomy makes the optimal strategies to pursue in negotiating the complex balance of cooperation and conflict, within and between the multiple levels of hierarchy within and between societies, more difficult to perceive and accomplish. Overlaying the fact that our true motivations are effectively hidden from us complicates the task even further. Nevertheless, humans are highly accomplished at evaluating the costs and benefits of pursuing a huge number of possible courses of action, particularly as they relate to social outcomes. Such evaluations involve forms of more or less conscious "scenario-building," whereby behavioral options can be explored safely within the mind, as a core adaptive function of the human psyche (Alexander 1989).

How do human cognition and sociality relate to the arts? As Alexander (2008) emphasizes, the arts embody elaborate forms of scenario-building, and he places the importance of such scenario-building firmly in the context of the unique nature of human societies, in which close cooperation and mutual dependence are coupled with an over-arching lack of close genetic relatedness between cooperating individuals. Competition and cooperation, under varying degrees of mutual dependence, are thus inextricably mixed. Scenario-building—in the minds of self and others—has led the arts to achieve a level of importance in human societies that makes it possible for successful artists to pursue their art exclusively and be well compensated in turn by those who would like to enhance their own scenario-building and executing skills through observation of the works of a master. Hence the arts, far from representing some sort of trivial pursuit, embody one of the highest levels of intellectual achievement available to our species. Ironically, the extreme importance granted to the arts in many cases transforms them from mere practice to the main event itself—as any parent of a child in a play can attest.

Alexander (2008) develops the concept of artistic endeavor as a key component of social and intellectual play in the context of striving for inclusive fitness returns via self-improvement. This conception of art focuses on the paramount importance of scenario building as part of skill development and task completion, and parallels the focus on the development of the "mind's eye" discussed above. Unlike other researchers, however, Alexander emphasizes the supreme importance of visualizing social relationships, of visualizing the effect of one's own strategies in the context of the strategizing of the multiple others that influence one's fitness in human society. For an evolutionarily significant period of time, the unique ecological dominance of the human species makes this social milieu the key arena in which the inclusive fitness of individuals has been determined in human societies. Given the complexity of evaluating the possibilities and choosing the best course in an intricate array of possible courses of action, the importance of scenario-building cannot be over-emphasized as a key component of social success. As Alexander (2008) argues, this is particularly true given the relative nature of success in human societies, and the fact that all other members of society are continuously updating their scenarios on the basis of those of others, and attempting to take advantage of that information. In this sense, human scenario-building generates positive feedback loops that drive accelerating increases in both social and cognitive complexity (Crespi 2004; Flinn and Alexander 2007).

Alexander (2008) recognizes two aspects of the arts. First, art is used as a means to promote an individual's striving for success within society by enhancing their own physical and mental talents in the context of the elaborate creativity, planning, scenario-building and performance execution required of artists. Second, arts generate potential value to the observers. Observers of artistic creations and performances can obtain a variety of benefits relevant to their own striving in the context of mental scenario building and related activities. Such gains are particularly salient in our species given our unique capacities and proclivities for cultural transmission (Richerson and Boyd 2005). As Alexander (1979) has argued in detail, culture is not a neutral medium, but rather the key setting within which human striving takes place. Cultural adaptations are created and maintained by the considered or unconsidered imitation and anti-imitation of the behavior of members of society (particularly successful members) by other members as they strive to enhance their own status within the social hierarchy (Alexander 1979; Flinn and Alexander 1982). This approach presages similar arguments made by researchers focused on gene-culture coevolution (e.g., Richerson and Boyd 2005). Art, as the embodiment of cultural traditions that take on enduring importance across generations in human societies, constitutes a fundamentally important arena for competition to succeed in the social hierarchy in human social groups. Hence the task of learning the nuances of these traditions—specific to particular societies—from those who have mastered them becomes of paramount importance if one is even to play the game, never mind succeed in it.

Alexander (2008) places the unique importance of artistic endeavor and achievement squarely in the context of his well-developed view of ethical and moral behavior within human societies (e.g., Alexander 1979, 1987). Specifically, he conceptualizes human societies as networks of direct and (largely) indirect reciprocity, in which it is critically important to maintain and enhance one's reputation in the eyes of other members of society. Members of society are constantly observing and evaluating all other members, rewarding those who contribute to society and punishing or shunning those who act to the detriment of the common good. These unique levels of social cooperation take place in (and are driven by) the importance of intergroup competition that make cohesive, cooperative groups the most likely to survive and succeed over the course of human evolutionary history (Alexander and Tinkle 1968; Alexander 1979, 1987). These arguments have (belatedly) been supported via modeling and empirical analyses by a variety of evolutionary anthropologists, economists and psychologists who have recently taken an interest in this area (e.g., Choi and Bowles 2007; Boyd et al. 2009). Scant explicit acknowledgment has been accorded to the pioneering contributions made by Alexander in this domain—partly through specialization and ignorance, but perhaps also because Alexander's perspectives, through his works and those of his students and colleagues, have become so deeply and thoroughly embedded in ways of thinking as to escape conscious notice and citation.

Reputation and status within society thus take central places in Alexander's (2008) explanation of the evolution of artistic creativity and ability in human society. Status is always relative and always in short supply, and high status is accorded to those who can demonstrate exceptional abilities of perception, appreciation, interpretation, translation, and communication via their artistic talents. These traits qualify them, in the eyes of multiple observers, to distill, amplify, and express essential features of the social system through their art. In turn, the ability to appreciate the subtle nuances of artistic performances also requires exceptional abilities of perception and interpretation on the part of observers. The ever-increasing complexity of the interaction between the performer and the observers that characterizes the arts provides an open-ended arena for intellectual play that hones the scenario-building capabilities of both performers and observers, both within the lifetime of particular individuals and across the generations in human societies.

Alexander (2008) takes care to distinguish the process of social selection from sexual selection. Partners in social selection can alternate roles or assume the same roles in their interactions, which center on mutually beneficial forms of direct or indirect reciprocity. These interactions form the main fabric of human society and involve long-term investments in relationships with the potential for both high risk (due to cheating) and high long-term reward. Staying "one step ahead" in such relationships becomes paramount, leading to selection in favor of extraordinary capacities in the realms of social observation, memory, communication (linguistic abilities), empathy, sympathy and other components of scenario-building that form the basis for what we call consciousness. These capacities allow us to be aware of our own thoughts and intentions, to communicate them effectively, to be cognizant of alternative courses of action available to us at different times and places in the future, and to act in our best interests in response to those contingencies.

The term "social selection" implies that the expression and appreciation of artistic creativity ultimately affect the inclusive fitness of the individuals involved, and Alexander (2008) carefully explains how artistic endeavors may connect to heritable traits associated with reproductive consequences. He first reiterates previous arguments (e.g., Alexander 1979) that the learned behaviors comprising culture are not independent of the impact of selection during our evolutionary history, but rather that we have inherited myriad biases, tendencies and capacities that regulate how experience affects learning, which behaviors are adopted (and from whom), and how those behaviors are expressed. Hence culture and cultural transmission are profoundly affected by inherited biases and capacities, which are in turn subject to selection. Rather than being independent, genetic and cultural evolution are linked and influence one another in a reciprocal manner (Richerson and Boyd 2005). The extreme phenotypic plasticity associated with learning makes it appear that activities that require long periods of training and practice (such as art) are independent of any innate variation in ability, but this is not always the case (Ebstein et al. 2010). Heritable traits influencing artistic proclivities and abilities may be specific, or may reflect general overall attributes such as health, immunity, processing efficiency, etc. that can provide a substrate for excellence in the arts as well as other endeavors. Indeed, Alexander (2008) also makes use of Zahavi's Handicap Principle in this context, such that general capacities may underlie the ability of performers to succeed in artistic endeavors. Artistic creativity and capability may thus serve as markers for underlying characteristics that are desirable in a social or sexual partner, such as intelligence and empathy.

RUNAWAY SELECTION AND THE ARTS

Like Miller (2000), Alexander (1987, 1989, 2008) has accentuated the possibility that some form of runaway selection may have contributed to the tendencies toward extremes seen in artistic endeavors and competitions. Unlike Miller, however, Alexander focuses on runaway selection in the context of social competition rather than sexual selection. The possibility of runaway social selection has been discussed for some time (e.g., West-Eberhard 1983; Nesse 2007), but few attempts have been made to model this process formally. As Alexander (2008) points out, systems of direct and indirect reciprocity may be particularly appropriate vehicles for the action of runaway social selection that operates in a manner parallel to that originally suggested by Fisher (1930) for runaway sexual selection.

Two properties may make the comparison of social with sexual runaway selection appropriate: (1) the tendency to choose extremes (to pick the best mate in sexual selection, and the best partner in systems of reciprocity), and (2) feedback between the benefits of choosiness to the chooser and of being chosen to the performer. These features may drive the competition to extremes, whereby the (original) practical nature of the performance is lost as performers seek to exaggerate their art in an attempt to achieve the most extreme performance. In turn, observers may come to appreciate the extremes, not for their practical usefulness, but rather as markers of the extreme capabilities of the performers. This kind of process can occur with or without underlying heritable variation in ability and appreciation, as winning performances may yield direct material benefits for both the performers and the observers. Alexander (2008) discusses fascinating examples of artistic and other types of competitions that have followed bizarre and unique trajectories. For example, in horse-riding competitions, the contrast between "peanut-rollers" (where the horse's head is held low during the ride), and the opposite, where the horse holds both its head and body extremely high, illustrate opposite trends that have both gone to extremes. These competitions have led to extremes that are in no way functional, but allow observers to judge which participant has achieved the most extreme result (and accord status as a result). The proliferation of these kinds of competitions, both in the arts and in sports and other components of human society, has been remarked upon by generations of observers.

Distinct competitive races, in the arts and elsewhere, may arise in a diversity of contexts, at a number of different levels in the social hierarchy, and of course in different societies. These properties vastly increase the number of opportunities for individuals to fruitfully engage in artistic competition, by choosing competitions that are appropriate to their specific type and level of training, skill and opportunity. As Alexander (2008) points out, these considerations, in combination with the tendency to favor extremes in such competitions, may have lead to the diversity of artistic pursuits we see today.

NATURAL SELECTION, THE ARTS, AND UNDERSTANDING HUMANITY

Our brief history and synopsis of Richard Alexander's contributions to understanding the arts serve not just to introduce his contributions in this area, but also to exemplify his incisive and comprehensive manner of thinking about natural selection in human evolution. One cornerstone, inclusive fitness theory, anchors human evolutionary biology from genes to brains, and connects competition with cooperation across all human endeavors, including the arts, its most exceptional. Indeed, understanding Alexander's arguments and hypotheses regarding sexual, social and runaway selection of the arts compels an understanding of his entire perspective on understanding ourselves. In turn, his ideas, which generate testable predictions that interface across scientific disciplines from genomics to neuroscience and anthropology, should motivate and inspire the next generation of studies in evolution and the arts.

References

Aiken, N.E. 1998. *The Biological Origins of Art*. Westport, CT: Praeger Publishers.

Alexander, R.D. 1979. *Darwinism and Human Affairs*. Seattle: University of Washington Press.

Alexander, R.D. 1987. *The Biology of Moral Systems*. New York: Aldine de Gruyter.

Alexander, R.D. 1989. The evolution of the human psyche. In: C. Stringer and P. Mellars (eds.), *The Human Revolution*. Edinburgh: University of Edinburgh Press, pp. 455–513.

Alexander, R.D. 2008. Evolutionary selection and the nature of humanity. In: V. Hosle and C. Illies (eds.), *Darwinism and Philosophy*. South Bend, IN: University of Notre Dame Press.

Alexander, R.D. 2011. *The Mockingbird's River Song: Poems, Essays, Songs and Stories, 1946-2011*. Manchester, MI: Woodlane Farm Books.

Alexander, R.D., and Tinkle, D.W. 1968. Review of *On Aggression* by Konrad Lorenz and *The Territorial Imperative* by Robert Ardrey. *Bioscience* 18:245–248.

Andersson, A. 1994. *Sexual Selection*. Princeton, NJ: Princeton University Press.

Bar-Yosef, O. 2002. The upper paleolithic revolution. *Annu. Rev. Anthropol.* 31:363–393.

Barzun, J. 2000. *From Dawn to Decadence*. New York: Harper-Collins.

Boyd, B. 2009. *On the Origin of Stories: Evolution, Cognition and Fiction*. Cambridge, MA: Belknap Press of Harvard University Press.

Boyd, R., Gintis, H., and Bowles, S. 2009. Coordinated punishment of defectors sustains cooperation and can proliferate when rare. *Science* 328: 617–620.

Charyton, C., Basham, K.M., and Elliott, J.O. 2008. Examining gender with general creativity and preference for creative persons in college students within the sciences and the arts. *J. Creative Behav.* 42:216–222.

Choi, J.K., and Bowles, S. 2007. The coevolution of parochial altruism and war. *Science* 318:636–640.

Crespi, B.J. 2004. Viscious circles: positive feedback in major evolutionary and ecological transitions. *Trends Ecol. Evol.* 19:627–633.

Dissanayake, E. 1992. *Homo aestheticus: Where Art Comes From and Why*. New York: Free Press.

Driscoll, C. 2006. The bowerbirds and the bees: Miller on art, altruism, and sexual selection. *Phil. Psych.* 19:507–526.

Dutton, D. 2009. *The Art Instinct: Beauty, Pleasure and Human Evolution*. New York: Bloomsbury Press.

Ebstein, R.P., Israel, S., Chew, S.H., Zhong, S., and Knafo, A. 2010. Genetics of human social behavior. *Neuron* 65:831–844.

Fisher, R.A. 1930. *The Genetical Theory of Natural Selection*. New York: Dover.

Flinn, M.V., and Alexander, R.D. 1982. Culture theory: the developing synthesis from biology. *Hum. Ecol.* 10:383–400.

Flinn, M.V., and Alexander, R.D. 2007. Runaway social selection in human evolution. In: S.W. Gangestad and J.A. Simpson (eds.), *The Evolution of Mind: Fundamental Questions and Controversies*. New York: The Guilford Press, pp. 249–255.

Gintis, H., Bowles, S., Boyd, R. and Fehr, E. 2003. Explaining altruistic behavior in humans. *Evol. Hum.Behav.* 24:153–172.

Heinrich, B. 2001. *Racing the Antelope: What Animals Can Teach Us about Running a Life*. New York: Harper Collins.

Hill, K.R., Walker, R.S., Bozicevic, M. et al. 2011. Co-residence patterns in hunter-gatherer societies show unique human social structure. *Science* 331:1286–1289.

Jacobs, Z., Duller, G.A., Wintle, A.G., and Henshilwood, C.S. 2006. Extending the chronology of deposits at Blombos Cave, South Africa, back to 140 ka using optical dating of single and multiple grains of quartz. *J. Hum. Evol.* 51:255–273.

Lande, R. 1980. Sexual dimorphism, sexual selection, and adaptation in polygenic characters. *Evolution* 34:292–305.

Miller, G. 2000. *The Mating Mind: How Sexual Choice Shaped the Evolution of Human Nature*. New York: Random House.

Morriss-Kay, G.M. 2010. The evolution of human artistic creativity. *J. Anat.* 216:158–176.

Nesse, R.M. 2007. Runaway social selection for displays of partner value and altruism. *Biol. Theory* 2, 143–155.

Pinker, S. 1997. *How the Mind Works*. New York: Norton.

Richerson, P.J., and Boyd, R. 2005. *Not By Genes Alone: How Culture Transformed Human Evolution*. Chicago: University of Chicago Press.

Trivers, R.L. 1971. The evolution of reciprocal altruism. *Q. Rev. Biol.* 46:35–57.

West-Eberhard, M.J. 1983. Sexual selection, social competition, and evolution. *Q. Rev. Biol.* 58:155–183.

Zahavi, A., and Zahavi, A. 1997. *The Handicap Principle: A Missing Piece of Darwin's Puzzle*. Oxford: Oxford University Press.

Zaidel, D.W. 2010. Art and brain: insights from neuropsychology, biology and evolution. *J. Anat.* 216:177–183.

EVOLUTIONARY SELECTION AND THE NATURE OF HUMANITY

Excerpt from Alexander, R.D. 2008. Evolutionary Selection and the Nature of Humanity. In: V. Hosle and Ch. Illies (eds.), *Darwinism and Philosophy*. South Bend, IN: University of Notre Dame Press, chapter 15.

Connecting the Arts to Evolution

People who are successful in the arts possess a set of basic skills that they use in relation to the social scene: observation, perception, appreciation, interpretation, imagination, prognostication, translation, and communication. They depict the social scene through faithful representations mixed with manipulation, exaggeration, and the parading of incongruity—all of which require extraordinary ability to understand the social scene in the first place. They use visual portrayals in art, dance, and drama. They use language in all forms, from oral narration and music to the literature of poetry and fiction. They use metaphor in every sense of the concept, and the race to catch up with all of the ways they do so will never be finished. They use humor, and to those who are sufficiently attentive they demonstrate its deadly seriousness. They combine most or all of these media, sometimes in a single stunning performance that creates novel comparisons and portrays relationships that enlighten and explain the social world as never before.

Audiences of the arts are accepting particular versions of the social scene secondhand from others whom they perceive to have better basic skills or broader experience. To do this effectively they must recognize the relevant talents and abilities in others. Basic to that task are universal and special features of the human brain: social capacity that includes a fertile imagination and all the talents necessary to use it. These traits are the same as those employed by the artists, leading us to wonder if they derive from a single machinery, even if not always equally functional.

The result of all this is that the arts are a glorious form of interpretive gossip, multifaceted, multilayered, and altogether still complex beyond anyone's comprehension. People in the arts are by definition the best storytellers among us. What they tell us is never superfluous, impractical, or trivial unless we, the audiences, allow it to be. We gain mightily from knowing how and when to listen, to whom to listen, and what to do with the experience afterward. For the arts are theater, and theater in all of its guises represents the richest, most condensed, and most widely understood of all cultural contributions to our patterns of social scenario build-ing through consciousness and foresight. These scenarios, which we build, review, and revise continually every day of our lives, are obligate passports to social suc-cess, and perhaps the central evolved function of the human social brain. We use them to anticipate and manipulate the future—the ever more distant future in ever greater detail.

When Marshall Sahlins (1976) argued that modern hunter-gatherers are mod-els of the original affluent societies because of the surprisingly small amounts of time spent hunting and gathering, he underplayed at least two things. First, time not occupied with securing food and shelter can only be occupied otherwise; and this can be done effectively only by engaging in activities that contribute to repro-duction. Second, sitting around being social is not necessarily trivial or nonrepro-ductive. Sahlins's comments, however, emphasize that the rise of art in its various expressions need not be restricted to recent societies demonstrating leisure time and affluence in some narrow modern sense.

Even if few could say it as well as Doris Lessing (1992, 35), artists are the people who understand that "[m]yth does not mean something untrue, but a concentra-tion of truths."

Status and the Arts

Status is central in human society (cf. Barkow 1989; Dickemann 1979; Henrich and Gil-White 2001; Hill 1984). It is important to every individual in society, hardly ever approaches being "complete," and by its nature is always in short supply. For humans, unlike any other organisms, the numbers and kinds of differences in sta-tus in a complex society can be almost endless and constantly changing; in today's societies, one individual can have different kinds of status in numerous different social circles. These facts create endless and elaborate performances and audi-ences, as well as endless opportunities and competitive races leading to narrow trends and fads about which little more can be said beyond "Second place is the first loser." These things happen because, for various reasons (see below), audi-ences to artistic performance and competitive races can sometimes gain as much as or more than the participants. I would be surprised if any species has anywhere near as many performance-audience interactions as humans.

If status is, as seems evident, a measure of access to the resources of reproduc-tion, we can grasp the importance of understanding it, and how to gain it, in a society like our own. How and why status has become so important in human society is another question. It can be complex and dynamic only when indirect reciprocity has become important. Reciprocity and nepotism are both investments that entail risk. Only a genetic return is required to make nepotism profitable, and unlike social reciprocity, much nepotism can occur regularly with a minimum of cognitive skill. Differential nepotism frequently takes the form of reciprocity, however, when two relatives reciprocally assist one other in times of special need for first one, then the other (see discussion in Alexander 1979a, 53–56). Success in reciprocity, with the simultaneous retention of social acceptance and harmony, requires the highest levels of cognitive skills (see, e.g., Trivers 1971).

From these assumptions it would appear that the underlying reason for direct reciprocity, followed by indirect reciprocity and an importance for status shifts in human society, is actually group living of the sort discussed earlier, which allows and promotes the kind of extensive differential nepotism apparently unique to humans and the most likely precursor of reciprocity.

The existence of status as an important variable in human society sets the stage for interest in any kind of contest or performance that assists in determining status accurately and in changing it to the observer's or participant's benefit. In Alexander (1979a), I hypothesized that much of the arts represents surrogate scenario building for audiences. I have also suggested previously that, for performers, all of the arts can provide opportunities for the elevation of status. Status and the ability to convey items of great importance to others go together. They may represent the key to understanding not only the arts but much of morality as well, because lowered status is effective punishment for behavior deemed immoral (Alexander 1987). Part of the reason this is possible, of course, is that coalitions and alliances capable of meting out punishment, as well as delivering rewards, can form within human groups and are aspects of indirect reciprocity. It is a large part of the complexity of human social life that, as part of the group-against-group aspect of human sociality, these coalitions can also shift and change in virtually every way imaginable.

The Contribution of Social Selection to Human Culture: Is It a Runaway Process?

We need now to engage more specifically the question of the particular flowering of the collective human phenotype usually called "culture." It is obvious that extensive learning abilities in humans, and the long and intimate overlaps of adults and juveniles in social groups, have created a situation in which learned activities are transmitted by learning, thus providing the background for the rapid accumulation of cultural innovations and change, consequently for rapid cultural divergence among variously isolated human groups. But without additional explication these facts seem frustratingly unable to account for the incredible races toward complexity and diversity of human social and cultural activities across the globe—not only the huge specialized human brain but the features that have led to such complex and diverse enterprises as the arts and, indeed, those leading to Barzun's (2000) label of decadence. Something more is needed—another step in our understanding of evolutionary process—to connect everything about the human organism to its complex cultural expressions. I ask now whether this "something more" may be considered appropriately under the label of runaway social selection, already introduced in this context by West-Eberhard (1983) and Alexander (1987, 1989).

Ronald A. Fisher (1930) used the adjective runaway to apply to a hypothetical selective process that he assigned to sexual selection (see also Trivers 1972; Miller 2000). I suggest that sexual selection and reciprocity selection are special forms of the broader concept of social selection. Social selection is a consequence of competition among individuals for rewards arising out of various kinds of social or mutualistic beneficence.

Fisher's runaway sexual selection has two special features. First is the tendency of the choosing parties to begin favoring individuals with traits that are extreme on some axis of desirability, rather than favoring some particular condition (other than extremeness) of the individuals to be chosen. This form of choosing can only occur in an organism that is able to compare an array of individuals and identify desirable extremes (e.g., Alexander, Marshall, and Cooley 1997). It can evolve to become the standard method of choice only if selection continues for a long time in one direction and the best choices continue to be beyond the previous expressions of the trait. Thus, if females that choose extreme males outreproduce those that do not, the choosing of extremes will spread. So will extremeness in males; otherwise the females choosing them would not gain. Spreading or fixing the choosing of extremes will inject a certain degree of inertia into the process. For this reason, once an "extreme-choosing" tendency is in place, extremes in the traits of the chosen individuals can pass beyond the form in which they are adaptive in any other sense and indeed can become disadvantageous in other contexts. Some such conflict actually exists among all the compromises of selection on different traits of the organism because the effect of selection can only be enhancement of the reproductive integrity of the organism as a whole rather than the state of any of its individual traits. When selection is social, however—when it is a matter of individuals choosing other individuals in a mutualistic or reciprocal social interaction rather than, say, competition to detect or capture food, or to escape enemies more effectively—overshoots in adaptiveness in other respects, because of choosing extremeness, will be more prominent.

The second special feature of Fisher's runaway sexual selection is the feedback resulting from the genetic partnership between males and females in jointly produced offspring. Trivers (1972, 166) described it as follows:

> [I]f there is a tendency for females to sample the male distribution and to prefer one extreme (for example, the more brightly colored males), then selection will move the male distribution toward the favoured extreme. After a one generation lag, the distribution of female preferences will also move toward a greater percentage of females with extreme desires, because the granddaughters of females preferring the favoured extreme will be more numerous than the granddaughters of females favoring other male

attributes. Until countervailing selection intervenes, this female preference will, as first pointed out by Fisher (1958), move both male attributes and female preferences with increasing rapidity in the same direction.

We can note, first, that some aspects of human evolution that intrigue us, such as brain functions that result in cleverness in social interactions, including scenario building and testing the social future by weighing alternatives internally, could have evolved partly through sexual selection. In view of the tendency of the human brain to become larger across history, and of human behavior to become more complex—and seemingly ever more rapidly after these features had exceeded their counterparts in other species—we might be concerned to examine the likelihood that runaway sexual selection is involved. A related question is whether some or all features of runaway selection can also occur in nonsexual social interactions such as the high-risk forms of social reciprocity that are prominent only in the human species.

At first one may imagine that there are no parallels in social selection allowing it to become runaway. But the way an individual gains by selecting its social partners parallels the ways an individual can gain from cooperative interactions with a mate and from a mate's parental care. In both cases there is likely to be genetic change in both the ability to choose good partners and the background of the favored phenotypic attributes because a mutually beneficial interaction can be maintained only if, on average, both interactants gain.

Sexual selection is a distinctive kind of runaway selection because joint production of offspring by the interacting pair causes the process to accelerate (see Trivers quote above). The defining feature of runaway selection is not acceleration, however, but the tendency of the process to go significantly beyond adaptiveness in all contexts except within the particular selective race—much further beyond adaptiveness in other contexts than is ordinarily the case in the myriad compromises among the conflicting adaptive traits that create and maintain the unified organism.

This aspect of runaway selection may hold for reciprocity selection, in which, unlike in sexual selection, both parties can carry tendencies not only to choose extremes but also to display extremes. In social selection, again unlike in sexual selection (except in simultaneous hermaphrodites), an individual can play both roles, of chooser and chosen, with respect to the same traits, and alternations of roles can occur during extended interactions between particular partners. To the extent that social success in ecologically dominant humans (see earlier) becomes the central determinant of reproductive success, runaway reciprocity selection may be a more viable possibility than Fisherian runaway sexual selection.

Fisherian runaway sexual selection presumably begins with a likelihood of heritability (i.e., genetic variability) in both variations in choices and variations in chosen traits. It is easy to see that in this circumstance such competitions, or races, will lead to genetic changes, at first changing both the ability and tendency to choose extremes and the nature of the extremes available for choice. Continuing mutations and genetic recombination (outbreeding among temporarily isolated groups) will tend to offer the choosing parties increasingly extreme possible choices, though to a reduced degree as selection continues to remove heritable variations in trait expressions and the trait becomes sufficiently maladaptive in contexts other than sexual selection. From earlier arguments it is not easy to understand the extent to which relevant heritability is likely to disappear, or even become trivial.

Diminution (or disappearance) of heritability in variations will not necessarily change the tendency to choose extremes: in effect, if an ability and tendency to choose extremes in any of a variety of social races (at least in humans) could become genetically fixed in the population (including the ability and tendency to learn from others, or from observation, the values of such choosing), it could still offer advantages to the chooser of extremes (without evolutionary change in the trait per se) because of the usefulness of even nonheritable trait variations chosen in a social partner. Of course, there may be further evolutionary improvements in the ability to identify and use extreme traits even after all choosers are already choosing extremes.

On the other hand, heritable variations in what is chosen will result in a continued march toward greater extremes because this relative quality will not be fixed in the way the ability to identify and favor extremes can be fixed; extremes can be identified only by comparing whatever is available. In Fisher's version of runaway selection, extremes win reproductively at first because they are ecologically superior, meaning functionally superior outside the choice situation itself. Later they continue to win because they are sexually (or, here, socially) superior, even though, as a result of the progress of selection involving choice of extremes, they may have become so extreme as to be otherwise functionally (ecologically) inferior. Here, "sexually or socially superior" means that the choosing parties will have acquired a genetic composition that will cause them to choose the extreme, the resulting choice of the extreme individuals itself causing the chosen individuals to outreproduce.

Extremes, however, will be chosen whether or not, as extremes, they represent heritable variants. If extreme social performances yield special social opportunities to those displaying them (e.g., via their reputations as achievers or winners), then regardless of the basis for the superior performances (i.e., genetic variant or not), winning performances can yield benefits to the kin or other social associates of the individual with the extreme traits. Thus merely joining social competitions or races can pay, though only heritable variations, including the (variant) ability to choose the best from among multiple available races that might be entered, will yield evolutionary change. Heritable variations in the ability to choose appropriate races—including not only those generally likely to be profitable but those in which the choosing individual, because of his or her special traits, has a special likelihood of competing successfully—may be all that is needed to drive runaway reciprocity selection. The more different ways that success can be achieved through reciprocity competition, the more robust will be this type of social selection. This is a kind of selection that will yield at least part of the human type of social intelligence, perceptiveness, and perseverance. As we all know, status (or "reputation") can exist independent of the adoption of a particular extreme in behavior, so the importance of any behavioral extreme in changing some aspect of culture can also depend on whether a prestigious person adopts, favors, or approves of it.

Since Fisherian runaway sexual selection in nonhuman organisms has remained controversial (or theoretical), but social parallels to it may be robust in human society, one may wonder if Fisher actually derived his idea from observing human social situations rather than from thinking about sexual selection in the birds he ultimately used as his examples. If so, it is likely not the only instance in which an evolutionist was inspired by human traits and tendencies to develop a general evolutionary explanation (e.g., Darwin's observations on human selection of variations in domestic animals and the fact that Hamilton's rule applies in its fullest extent as extensive differential nepotism across multiple levels of relatedness only in humans). This suggestion is ironic in another way, in view of the success with which academic biology departments have managed to exclude the human species from their consideration, leaving its analysis to the almost exclusively nonevolutionary approaches of social science and medical departments, and surely delaying acceptable explanations of the human species in evolutionary terms.

When extremes in particular directions are being chosen in social selection, new extremes never before experienced may be chosen above all other expressions of a trait. This behavior was originally referred to in European ethology as a "superoptimal stimulus" effect and is not specifically related to runaway selection. Whenever superoptimal stimulus effects can be identified, however, they suggest a history of choosing extremes, therefore the possibility of some form of runaway selection. If, for example, we perceive that during the past several centuries art—or even human sociality in general—has flourished and become an enormous and diverse enterprise compared to any "ancestral" condition it might have exhibited, we might suspect that this is evidence of a history of choosing extremes continually in short supply and consequently of some kind of runaway process in however artistry affects the securing of the resources of reproduction. Rather than evidence that evolution cannot be used to explain, say, the arts, explanations of recent flowerings of diversity and complexity in human enterprises can be sought by expanding our understanding of the various subprocesses of evolutionary selection. We should not be discouraged because the pathways to explanation become progressively more difficult and call for expansions of our understanding of the workings of the evolutionary process.

The Arts And Competition

Competitive races among humans, which include performances in the arts, can take either of two directions. They can change in the direction that enables individuals to give performances that are unique, not easily compared to those of others, and highly informative to observers about relatively complicated aspects of sociality. These are the races that we would most often term "artistic." Their participants are likely to have carved out special life niches for themselves. In effect, a flourishing of the arts in its most diverse forms reduces direct and confrontational competitive races because it is more difficult to compare creative artists than to compare, say, athletes in the same sports, chess players who contest one on one, or people who try to best one another's record in any uncomplicated or highly patterned and predictable game or contest.

Trends toward individuality in performance and meaning in the arts—as in poetry—can continue, paradoxically, because the value of gaining insight from the brilliance of others is so great to us. The more profound an insight, as well as the more personal or specific, the more difficult it is likely to be to absorb. This is the reason that a poem can be so personal as to be understandable only to its author, yet the author can be accorded high status because of the poem. Because they expand and enhance our imaginations, poetry and art press continually at the boundaries of mystery and incomprehensibility, so that, as audience, we are always faced with the necessity of deciding whether the artist is incomprehensible because of telling us something unusually profound or because of telling us something trivial or wrong in an obscure way. It is a difficult decision because we know in our souls that the scoffer can lose as profoundly as the gull, whether the poetic message is used or rejected directly or attempts are made to use it indirectly during social interactions with the poet in some other context. The same challenge exists in the assessment and use of humor as insight into the social scene, as all those who have laughed too soon on at least one occasion will understand (Alexander 1986, 1989).

Socially competitive races that parallel those involved in the arts can also take a different form, changing instead in directions that progressively narrow the variations available to the performers and the nature of the prizes and diminish the social messages available to observers. Competitors in such races are forced into increasingly or more directly confrontational races.

Modern humans have generated almost endless possibilities of being "world class" in direct competitions, from championships in golf or boxing to first prize in poetry or painting or musical contests to making the largest pizza or longest hot dog in the world, flagpole sitting for the longest time, or establishing the fastest construction times for field latrines in the history of the military. When reputation comes to be based so strongly on relative achievement compared to other humans that the nature of the competition becomes trivial and the relative ranking of the competitor becomes paramount, extremeness in the competition may be represented by activities or "traits" so radical that, in all regards except the specific competition being considered, they are disadvantageous to the individual exhibiting them. Social messages for audiences, moreover, can be restricted to the identity of contestants that finished in particular rankings. Nevertheless, such rankings represent changes in social status or maintenance of social status.

I can give an example from horse competitions. Judged horse competitions are like other judged contests: they can reach maddening levels of faddishness. These endlessly competitive races involve a wide variety of things, including the details of the clothing worn during different competitive classes; precisely how a rider holds hands, feet, and head; minute details of the equipment worn by the horse; and exactly how the horse holds its head. To take only one example here, in certain breeds, in a competition called "Western Pleasure," a tendency arose to favor horses that moved with their heads held somewhat lower than one is likely to see on the random horse galloping across the prairie or walking in someone's pasture. This favoring evidently began because it was undesirable for horses used in ranch work to hold their heads high enough to be in the way of the rancher's work. But lowered heads became a trend that reached such extremes that horses in these Western Pleasure competitions came to be called "peanut rollers" because their heads were so low they were jokingly regarded as able to roll peanuts as they walked or jogged along. Meantime, in breeds originally bred to move elegantly in parades and the like, and in others to show off unusual gaits, precisely opposite trends were set in motion in similar competitive classes. The horses were encouraged or forced to hold their heads higher and higher until the neck was essentially vertical. While some horse people were drugging their horses, starving them, or bleeding them (yes, I do mean those things) to get their horses to keep their heads down and move ever more slowly (and dejectedly!), those working with other breeds were letting their horses' fore hooves grow long so as to lift their front ends and their heads even higher. Eventually some of the latter people began placing plastic and other kinds of extensions on their horses' front hooves so that they appeared to be walking with their forefeet on boxes. And the riders began to sit farther and farther to the rear on the horse's back so that the horse would have a greater tendency to lift its front end in a fancy way while traveling around the show ring. This position became so extreme that riders began to wear capes that flowed over the horse's rump and to place bulky garlands across the saddle in front of them, both items evidently to conceal how embarrassingly far back the rider was sitting on the horse. Finally, some of the people with a certain breed of horses began to "sore" the tender portions of their horses' forefeet with acid or a knife, and to train the horses with heavy bruising chains around the tops of their hooves, in both instances so that they would lift their feet (and their heads) even higher. It seems sometimes that no judged contest of humans can maintain a happy medium in any one trait or action or fail to become narrow and faddish. Extremes appear, are quickly favored, and eventually become so ridiculously exaggerated as to discourage many people from participating. Perhaps only periodic dramatic changes in direction seriously retard such extreme trends.

Extremes of judging in narrow competitions are not restricted to small audiences or small-time events, as anyone knows who watches events such as figure skating or gymnastics in the Olympic Games. But for individuals or teams representing institutions (or nations), the social message has to do with competition between the institutions, meaning that, because of their symbolism, even narrow and confrontational races with little relationship to social questions or broadly important qualities of the contestant can attract huge audiences and yield high status for excellence in performance.

The incredible tendency of humans to engage in competitive races and carry them to extremes tells us that such competitions must be a very old part of our makeup, and a part that somehow furthered our reproduction. A point with which such races are entirely compatible is that organisms do not evolve just to reproduce successfully but to outreproduce everyone like them.

Whenever tendencies exist to recognize and admire "winners," whether they are outstanding artists, musicians, actors, dancers, poets or writers, shamans, humorists, athletes, war heroes, or thinkers, the tendencies may result in prolonged evolution, exaggerating the bases for the skills represented, perhaps producing the extreme skills and capabilities exhibited in the arts today. To the extent that such capacities tend to have a common genetic basis, the selection could be even more effective. Thus a wide variety of artists might achieve their special abilities partly through a common ability to construct and retain mental scenarios. Whether or not the most talented and specialized artists could use their special abilities to succeed in the social scene, their (genetically recombined) offspring might have superior skills in a variety of social situations. The favoring of seemingly different capabilities could have had common ground that led to the exaggeration of rather specific mental capacities, to which we have given labels such as "intelligence," "genius," or "perspicacity." The prediction from this suggestion is that people unusually capable in one intellectual or other endeavor in the arts are likely to be unusually capable in others as well. To the extent that this generalization is accurate, and tends to involve the useful skills of humans very broadly, the concept of "maladaptive in other contexts" because of dramatic overshoot in particular traits is diminished. While reciprocity selection is evidently distinctive and influential in humans, the question of whether it involves a runaway process, at least in the Fisherian sense, remains unclear.

In the 1930s and 1940s, the newsreels that preceded showing of movies in theaters proclaimed that in the twenty-first century people would not have to work more than fifteen or twenty hours a week because technological devices such as robots and calculating or computing machines would take care of most of the mundane tasks humans had to do then. As a specific example, when the labor-saving three-point hitch was developed for small farm tractors (in England by Harry Ferguson), its glory was assumed to be that it would allow the small farmer to do all the necessary work on his 160 or so acres with considerably less time and effort, giving him extra leisure time. In both this special case and in general, the virtual opposite has happened. Today in the United States, in many or most families both parents work even more than one parent worked in the early part of the twentieth century, and children are—seemingly "necessarily"—relegated to day care and preschool. Farmers found that the three-point tractor hitch actually helped set off a new competitive race that required farmers to multiply the number of acres worked in order to make a living wage. Instead of having more leisure time, farmers were able to do more work in the same time, and, like everyone else in the same situation, they seemed to be essentially forced to do that to keep up with or forge ahead of everyone else. More recently, the advent of efficient hydraulic systems on tractors has largely superseded the three-point hitch and set off a new race in which a single family may find itself working one or two thousand acres while selling such products as corn and other grain at prices not too different from those paid when the three-point hitch was invented more than a half-century ago. All of this speaks again to the fact that, like other organisms, humans must be presumed to have evolved not merely to reproduce successfully but to outreproduce their most significant competitors, leading to the proliferation of unending competitive races.

As implied in the above discussion of so-called labor-saving devices, we can perhaps be bold enough to ask whether concatenations of some human socially competitive races, because of their combined diversity, narrowness, and runaway and faddish effects, may even have value in comprehending why Barzun (2000)—and others—have felt it necessary to interpret both long- and short-term trends in human sociality as decadent.

Throughout the history of such narrow and frantic competitive races, something more properly called "art" has persisted, apparently because of a second attitude toward competitive races (both strategies can be witnessed being employed by children in that microcosm of the human social world that consists of competing for the approval of a single set of parents). One expression of this second attitude I have earlier described as avoidance of directness in competitiveness rather than the seeking of ever-more confronting forms or expressions of competition.

Another similar expression may arise, not through avoidance, but through the incidental perception and adoption of new and appealing avenues of creativity. Whatever their origin, unique life niches are created by some humans for themselves, from which they are able to tell the world things that no one has ever known before. Perhaps, as individuals, these artists can indeed be said to have mounted the "single-minded effort to render the highest kind of justice to the visible universe" alluded to by Joseph Conrad (quoted in Barzun 2000, 791). Incidentally, they also magnify the social diversity that we humans euphemistically refer to as "division of labor."

CONCLUSION

In sum, I have tried to consider every relevant aspect of the topic I originally proposed. At least I don't think I left any gaps that might require miracles. If that is true, perhaps this essay helps make it more likely that some day we will know how to connect the entire array of human activities to a base in reproductive effort.

The undertaking is important because it promises dramatic insights into our most precious thoughts and strivings, and those most stubbornly resistant to change. Such human self-understanding may be the most important change imaginable in our universe, not merely to promote happiness and well-being but to lengthen human lifetimes by reducing needless strife and aggression at all levels.

We think of the arts as differing from the sciences in seeking meanings—social facts that are often restricted in their usefulness, or personal rather than general. Part of the reason for the restrictions may be that the uniqueness of individual human genomes has resulted in each of us evolving to behave (in evolutionary terms, appropriately) as if, for each of us, the collection of our personal life interests were unique. We have evolved to use the myriad facts available to us in ways unique to ourselves, which in some sense is the meaning of meaning. Human social races are prevalent because of the importance of the prestige thought to be derived from doing well in almost any such race and the consequences of prestige that bring benefits (power) through the action of direct and indirect reciprocity. Evolutionary social selection, including nepotistic selection and mutualistic and reciprocity selection, may be the last important set of mechanisms required for the connecting of complex aspects of cultural phenomena, such as the arts, to a life basis in evolutionary selection favoring reproductive success. It leads to the exaggeration of all the capabilities that are displayed in artistic works and performances.

I recognize that all of my discussions are imperfect and incomplete. They will be worthwhile, however, if the shortcomings help point the way for those with the next round of insights. I conclude that we seem not to be lacking in mechanisms to approach analysis of our own activities in evolutionary terms, only in knowing how to use them. Perhaps the next generation can make sense of the rudimentary arguments that I and others have generated so far on these difficult topics.

NOTE

I am deeply indebted to four of my former students. Dr. Andrew F. Richards has considered most of these issues with me across the past several years nd has criticized the manuscript at several stages. Beverly Strassmann, David Marshall, and Mary Jane West-Eberhard provided stimulating criticisms of the chapter on short notice. David Lahti has also helped me immensely by consistently putting forth useful and constructive arguments. Because I did not accept all of the efforts to help me, I stress my own responsibility for any remaining errors or awkwardness.

POSTSCRIPT

Epitaph

> Who gains the privilege to pass,
> With mind still undiminished,
> Submits to this sad epitaph,
> Alas! I was not finished.

References

Alexander, R.D. 1979a. Evolution and culture. In: N. A. Chagnon and W. G. Irons (eds). *Evolutionary Biology and Human Social Behavior: An Anthropological Perspective*. North Scituate, Mass.: Duxbury Press, pp. 59–78.

Alexander, R.D 1986. Ostracism and indirect reciprocity: The reproductive significance of humor. *Ethology and Sociobiology* 7: 253–270.

Alexander, R.D 1987. *The Biology of Moral Systems*. Hawthorn, New York, Aldine de Gruyter, xxii + 303 pp., 8 figs.

Alexander, R.D 1989. The evolution of the human psyche. In: C. Stringer and P. Mellars (eds). *The Human Revolution*. University of Edinburgh Press, pp. 455–513.

Alexander, R.D., David Marshall, and John Cooley. 1997. Evolutionary perspectives on insect mating. In *The evolution of mating systems in insects and arachnids*, ed. J.C. Choe and B. J. Crespi, 4–31. Princeton, N.J.: Princeton University Press.

Barkow, J. H. 1989. Darwin, sex, and status: biological approaches to mind and culture. Toronto: University of Toronto Press.

Barzun, J. 2000. *From Dawn to Decadence. 500 Years of Western Cultural Life*. New York: Harper Collins.

Dickemann, M. 1979. The reproductive structure of stratified societies: A preliminary model. In: N. A. Chagnon and W. Irons (eds). *Evolutionary Biology and Human Social Organization: An Anthropological Perspective*. North Scituate, Mass: Duxbury Press, pp. 331–367.

Fisher, R. A. 1930 (1958). The Genetical Theory of Natural Selection. 2nd edition. (1958). New York: Dover Publications.

Henrich, J. and F. J. Gil-White. 2001. The evolution of prestige. Freely conferred deference as a mechanism for enhancing the benefits of cultural transmission. *Evolution and Human Behavior* 22:165–196.

Hill, J. 1984. Prestige and reproductive success in man. *Ethology and Sociobiology* 5:77–95.

Lessing, D. 1992. *African Laughter: Four visits to Zimbabwe*. New York: Harper.

Miller, G. F. 2000. *The Mating Mind: How Sexual Choice has Shaped the Evolution of Human Nature*. New York: Doubleday.

Sahlins, M. 1976. *Culture and Practical Reason*. Chicago: University of Chicago Press.

Trivers, R. L. 1971. The evolution of reciprocal altruism. *Quarterly Review of Biology* 46:35–57.

Trivers, R. L. 1972. Parental investment and sexual selection. In: B. Campbell (ed). *Sexual Selection and the Descent of Man*. Chicago: Aldine, pp. 136–179.

West-Eberhard, M. J. 1983. Sexual selection, social competition, and evolution. *Quarterly Review of Biology* 58:155–83.

Index